C0-AYI-686

SOCIAL SCIENCE RESEARCH AND CLIMATE CHANGE

SOCIAL SCIENCE RESEARCH AND CLIMATE CHANGE

An Interdisciplinary Appraisal

Edited by

ROBERT S. CHEN

Department of Geography, University of North Carolina, Chapel Hill, North Carolina, U.S.A.

ELISE BOULDING

Department of Sociology, Dartmouth College, New Hampshire, U.S.A.

and

STEPHEN H. SCHNEIDER

National Center for Atmospheric Research, Boulder, Colorado, U.S.A.

D. Reidel Publishing Company

A MEMBER OF THE KLUWER ACADEMIC PUBLISHERS GROUP 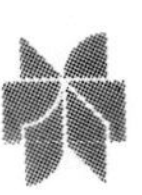

Dordrecht / Boston / Lancaster

Library of Congress Cataloging in Publication Data

Main entry under title:

Social science research and climate change.

Includes index.
1. Climatic changes—Social aspects. 2. Atmospheric carbon dioxide—Social aspects. I. Chen, Robert S. II. Boulding, Elise. III. Schneider, Stephen Henry.
QC981.8C5S63 1983 304.2′5 82–24149
ISBN 90–277–1490–8

Published by D. Reidel Publishing Company,
P.O. Box 17, 3300 AA Dordrecht, Holland.

Sold and distributed in the U.S.A. and Canada
by Kluwer Academic Publishers
190 Old Derby Street, Hingham, MA 02043, U.S.A.

In all other countries, sold and distributed
by Kluwer Academic Publishers Group,
P.O. Box 322, 3300 AH Dordrecht, Holland.

Printed in The Netherlands

TABLE OF CONTENTS

PREFACE AND ACKNOWLEDGEMENTS

This volume is the result of efforts over the past several years by a small group of social and natural scientists. The group's object was to identify and assess general research strategies and specific projects to explore the potential social and institutional responses to the advent or prospect of a climatic warming caused by human-induced (anthropogenic) emissions of carbon dioxide (CO_2). The effort was motivated by the recognition that, in order for society to be able to make informed decisions about energy and land use and other activities that could cause significant changes to the global environment, it would be necessary to understand in advance the possible range of impacts of such changes on human activities and welfare and the ability of society to deal with these impacts. Thus, a research planning project on the general environmental and societal impacts of a CO_2-induced climate warming was organized and managed by the Climate Project of the American Association for the Advancement of Science (AAAS) with the support of the Carbon Dioxide Effects Research and Assessment Program of the U.S. Department of Energy (DOE). A Panel on Social and Institutional Responses was formed under the joint leadership of sociologist Elise Boulding and climatologist Stephen Schneider. Its findings are presented in this book.

The Panel first met at a major interdisciplinary workshop organized by the AAAS at Annapolis, Maryland in April of 1979. Using as a reference a hypothetical scenario of how the climate might change as the result of CO_2 emissions, the panel identified a variety of important issues and research questions pertaining to the nature of possible societal perception of and responses to a climate change. The Panel's report, published in a DOE document, *Workshop on Environmental and Societal Consequences of a Possible CO_2-Induced Climate Change* (Carbon Dioxide Effects Research and Assessment Program, Report 009, U.S. Department of Energy, CONF-7904143, 1980), emphasized the unusual characteristics of the "CO_2 problem", including its long-term, slowly developing, and irreversible aspects, and underscored the importance of viewing the problem in the general context of other societal problems and rapid societal change.

Subsequent to the Annapolis workshop, a small core group of individuals continued to meet under the AAAS's aegis and Boulding's and Schneider's guidance to examine in greater detail specific research opportunities from the perspective of selected social-science disciplines. Six different papers were commissioned, and initial drafts of each were discussed by the group. The group's discussions also led to the realization that a specific treatment of the interdisciplinary aspects of research on the CO_2 problem would be desirable, a task which was undertaken by the group's rapporteur, Robert Chen. Each of the papers was then independently reviewed by several scientists from the relevant social-science fields and modified as appropriate before inclusion in this volume.

Based on the group's discussions and the individually authored papers, the group's leaders also developed an overall research agenda for social-science research on CO_2-related

R. S. Chen, E. Boulding, and S. H. Schneider (eds.), Social Science Research and Climate Change: An Interdisciplinary Appraisal, vii–viii.

climate change in conjunction with the continuing work of other groups. This research agenda includes specific suggestions of activities and projects that the group leaders felt important to ensure that the most critical questions that could place policy making on a firmer scientific basis would be addressed in the most efficacious manner. The research agenda and suggested activities are described in another DOE document, *Environmental and Societal Consequences of a Possible CO_2-Induced Climate Change: A Research Agenda, Volume I* (Carbon Dioxide Effects Research and Assessment Project, Report No. 013, U.S. Department of Energy, DOE/EV/10019–01, December 1980), and are summarized in the introduction to Part II of this volume.

The editors of this volume who, as mentioned, were the leaders and rapporteur of the AAAS Panel, wish to express their appreciation to the many individuals who have contributed to this work over the past several years, including the participants in the Annapolis workshop, the authors of the papers in this volume, the reviewers, and the other members of the AAAS Steering Group. Special thanks are due to Roger Revelle, whose intellectual leadership played a key role in the success of the AAAS efforts, to David Burns, who provided incomparable staff support and critical comments throughout the project, and to David Slade of the DOE, without whose foresight, leadership, and enthusiasm this work would never have been performed. David Sleeper helped considerably with the editing of the research agenda, Barbara Riley of the AAAS and Mary Rickel of the National Center for Atmospheric Research (NCAR) provided prompt and efficient secretarial and editorial support, and Justin Kitsutaka of the NCAR drafting group prepared the figures in the papers by Warrick and Riebsame and by Chen. (NCAR is sponsored by the National Science Foundation and any opinions, findings, conclusions or recommendations expressed in this publication are those of the authors and do not necessarily reflect the views of the National Science Foundation.)

Finally, R. Chen wishes to acknowledge the support of the Climate Board of the U.S. National Academy of Sciences which made his participation in many of the AAAS activities possible.

PART I

RESEARCH IN CLIMATE CHANGE AND SOCIETY: TWO PERSPECTIVES

Introduction

Many questions of importance can be roughly divided into two parts: "what if?" and "so what?". *What if* carbon dioxide pollution doubles? *So what* if it does? Before panicking over a CO_2 increase or related climate changes, we should most certainly know something about the likelihood of such events and their potential consequences for our health and well-being.

The natural sciences have generally focused on the "what if" side of things. What is it that is likely to occur – or could be made to occur? What are the underlying causes and important influences? With respect to the earth's climate – including its atmosphere, oceans, and ice – natural scientists have a long history of speculation and theorizing about potential climate changes, their origins, and their consequences for physical and biological systems. Indeed, some have proposed actions to prevent or counteract changes.

The social sciences have also devoted much effort to the "what if" question. What direct costs and benefits might arise from climatic changes? What societal developments might be triggered by extreme climatic events or episodes? Notably, questions like these do begin to delve into issues of "so what". Will people and their institutions be able to work together to prevent climate changes and, if so, with what consequences? Or, will they be capable of adapting to climatic changes, which are likely to be slowly developing, unevenly distributed, uncertain in nature, irreversible, and so forth? Clearly, these are much more difficult questions to answer, among other things because of the added elements of human perception and choice. Moreover, they require an *interdisciplinary* understanding that bridges not only the gaps between the natural and social sciences, but also, as Boulding points out in the first paper, many often considerable differences among the individual social-science disciplines themselves (and by the same token among the natural-science disciplines).

Both parts of this volume represent attempts to develop interdisciplinary communication and understanding and thus to come to grips with the "so what" question. Part I consists of two perspectives on the climate change problem, one by a social scientist, another by a natural scientist. Part II contains the individual contributions prepared by the AAAS group members, each of which deals with some aspect of social science research on the likely societal response to climate change in the context of increasing CO_2.

In the first paper of Part I, Boulding emphasizes the need for interdisciplinary interaction not just with respect to the climate change issue, but also relative to other pressing problems such as the possibility of nuclear war. Thus, she implicitly applies the "so what" question to social science research on climate change itself – and her answer is that such research should in fact help scientists improve their ability to deal with any global problem.

R. S. Chen, E. Boulding, and S. H. Schneider (eds.), Social Science Research and Climate Change: An Interdisciplinary Appraisal, 1–2.

The second paper by Schneider deals more directly with the "so what" of possible climate change. From the natural science perspective, potentially large impacts on food production, water supply, energy use, and so forth are clearly too important to ignore. Some estimates must be fashioned before a rational response is possible. Indeed, both problems and opportunities are likely to arise from any such impacts, and so one important focus should certainly be to examine how society can respond constructively and effectively to environmental change. Scientific research is part of this response process, but unfortunately the uncertainties are considerable and the accumulation of new insight is slow and painstaking. Innovative efforts are thus needed to ensure that time is not wasted and at least some bounds to the "so what" question are available far enough in advance for reasonable actions to be considered.

A major goal of this volume, then, is simply to collect and document a broad range of issues related to climate and society. In particular, we focus on areas of both certainty and uncertainty uncovered by the AAAS group, and focus our contributions so as to guide future research activities. It is our belief that a categorization of what has already been done, as well as to what still needs to be done, can greatly improve society's ability to learn and respond to change of any kind.

ROBERT S. CHEN

ELISE BOULDING

SETTING NEW RESEARCH AGENDAS: A SOCIAL SCIENTIST'S VIEW

While we are primarily speaking to the social science community in this book, we hope it will be read by physical scientists concerned with climate issues as well. Interaction between the two communities is needed on both sides. For the social sciences, this is an era of emergence from "inside" the social system – from that previously self-enclosed social world which treated natural phenomena as inputs to be taken as given. Social scientists have had a lot to deal with lately. Social systems are not what they used to be now that international political debates on the Global Problematique throw such concepts around as the international economic order, the international information order, and the international cultural order. When the international environmental order is added as well, such intrusive concepts as society-nature interactive systems begin disturbing the social science peace of mind. In earlier days economists handled nature by simply tracking resource flows through the social system. Then resources and environments turned interactive with the societies to which they had been inputs, and geography began to take center stage as the discipline ready to handle interactive analysis, having the tools of both the physical and the social sciences within its own discipline. In the climate-society project which gave rise to this book, geographers have played a key role in bridging the gap between physical and social concepts and giving operational meaning to an interactive nature-society feedback system. Understanding the nature-society interface has become an important new agenda item for the social sciences.

Because the international physical science community has an ancient lineage, its own communications system and its own traditions of interdisciplinarity, this community began identifying long-term state-of-the-earth problems well before researchers, designers and practitioners of societal arrangements were aware of these issues. The particular sequence of events that made the International Council of Scientific Unions aware of the possible problem of CO_2-induced global climate warming will not be told here, but the ICSU initiatives have produced a variety of national and regional research projects, of which the current American Association for the Advancement of Science Climate Change project is one. It was recognition that the climate change problem was related to behaviors and choices within social systems that led to an invitation to social scientists to participate in the project. There was also very little confidence on the part of the inviting scientists, it must be said, that the social sciences had any relevant scientific methodologies to bring to the analysis of the problem.

Part, although by no means all, of the blame for this lack of confidence must lie with the social scientists. In the development of social impact assessment, which became a way of identifying interactions between new physical technologies and social systems, there was little effort on the part of social scientists to understand the physical technologies in their own right. Physical and social scientists concerned with the same problem thus have talked right past each other. Thus social scientists located in Schools of

R. S. Chen, E. Boulding, and S. H. Schneider (eds.), Social Science Research and Climate Change: An Interdisciplinary Appraisal, 3–8.

Agriculture and Home Economics often had a very good understanding of the operation of relevant bio-physical systems, but they tended to be looked down upon as "mere" practitioners of applied social science by their more theoretically oriented colleagues in the Arts and Sciences Divisions of the same schools. Under such conditions they perforce kept their knowledge to themselves.

There has been no lack of research settings where environmental impacts are critical to local communities. Boom towns are an obvious example. Rarely, however, do social scientists doing boom-town research include physical scientists in their study teams, and if they try, funding agencies are unresponsive.

The interdisciplinary teams studying natural disasters such as floods, earthquakes, and volcanic eruptions have, in theory, produced social scientists with skills in multidisciplinary analysis utilizing physical as well as social data, but generally the social data are compiled separately and little interaction takes place.

The time has passed when so-called interdisciplinary projects can be content with parallel but separate data collection projects. While social and physical scientists can't simply master each other's specialties, they can try to learn to think with each other's concepts. The crux of the matter is getting some gut-level understanding of the meaning of interactive systems at the planetary level. Every social scientist in the current climate project has had a mind-stretching experience in considering even at the simplest descriptive level the climate-determining interactions of lithosphere, geosphere, hydrosphere, cryosphere, biosphere, sociosphere, lower atmosphere and stratosphere. The "hard-soft" dichotomy between the physical and the social sciences melts away in the face of the prediction-resistant complexities of planetary climate processes. Predictability belongs in the laboratory, or when one is dealing with large-scale processes taking place under familiar and well-defined conditions. The economist, sociologist and the psychologist can predict as well as the physicist when the variables are under control. In the study of these macrolevel interactive processes on which our survival as a species may depend, we are all learners, from whatever discipline, including the disciplines of the humanities. We all need each other. There is a great difference, however, between talking about the need for interdisciplinary research, and analyzing the conditions that can foster it. One of the most important papers in this book is the concluding paper by Chen on Interdisciplinary Research and Integration, precisely because it spells out what is needed to produce good interdisciplinary work. As a climatologist who has had a good deal of recent exposure to social science thinking, he has put his finger on what can make physical-social science interaction function productively.

But why single out the climate for interdisciplinary study? Given the threat of nuclear extinction, as several authors suggest, it is hard to see why understanding climate processes should be a top priority. CO_2-induced global climate warming is only a hypothesis. The threat of nuclear war is a fact. It is one of the ironies of industrial society that we have not been able to mobilize a massive interdisciplinary research operation to remove the threat of war. The "conditions of choice," as Warrick and Riebsame would put it, have not made that possible. But the skills and understanding of interactive relationships which the international scientific community are developing as it works on the climate problem are needed for all the global problems we will be facing in the coming century, including the problem of war.

It is the scientific community that has set the climate agenda. Governments will set other agendas. The UN also has its own agendas, since the UN as a whole is more than the sum of its constituent member states. The four thousand or so transnational citizens' associations (NGO's) also have agendas. The knowledge and skills needed to respond to these agendas must come from national and international, interdisciplinary scientific communities. The priorities among agendas come from human values, but science can help operationalize the values and clarify the priorities through its analysis of interactive effects among the physical spheres, and between the social and physical spheres. What this project, and innumerable other projects dealing with global problems in an interdisciplinary mode, can do is to begin, however humbly, to develop some new competence in global problem-solving.

How Interdisciplinary?

The Climate Project was organized by panels representing different clusters of disciplines. The clusters were as follows:

(I) The Oceans and Cryosphere.
(II) The Unmanaged Biosphere.
(III) The Managed Biosphere.
(IV) Social and Institutional Response to Climate Change.
(V) Economic and Geopolitical Considerations.

The first three panels consisted of physical and biological scientists, the last two of social scientists. Panel IV, however, benefitted immensely from having two climatologists as key participants, one as co-convenor and one as rapporteur. There was considerable argument about whether the economists should be a separate group. Because they were further along in their own specialized analyses, they remained separate but the social scientists in Panel IV missed them. It will be noted in the papers that follow that each author painstakingly includes some discussion of the economic dimensions. Because the physical scientists had already been working on climate problems, and most of the social scientists were new to the field, the latter group had a lot of learning to do. During the first workshop we spent most of our time listening to, and reading papers by, the other panels. Our own input was relatively minimal. By the second workshop the social scientists had more to say. Even so, the panel was still far from complete in relation to relevant social science expertise. A demographer and an environmental sociologist were in process of being added but their contributions are not reflected in this book because funding ran out. Interdisciplinary interaction in the broad sense took place primarily among the convenors of the five panels, who met regularly to discuss and evaluate ongoing work. In the third stage, indefinitely postponed due to withdrawal of funding, there would have been direct interaction at the problem level between physical, biological and social scientists.

The papers in this book are a product of Stage II, when scholars who had listened, learned and had some interaction with specialists from the other four panels turned back to their own disciplines and addressed the questions: (1) what does my discipline already do that is relevant to this problem; (2) what should it do; and (3) how could interdisciplinary work be designed? Because the working sessions at which the papers were thoroughly discussed and reworked consisted primarily of the social science panel members, actual conceptualization of how to work with the other panels did not enter in at this phase. The papers are therefore interdisciplinary only at the social science level. Even that is no mean achievement: The influence of the problems set for us by our physical science colleagues is pervasive in the papers.

The papers in this book are really an invitation to the rest of our colleagues out there in the social sciences to join us in seeking collaborative settings where we can work together across our own disciplinary boundaries and across the physical-social science boundary, in order to address more adequately the serious problems that beset our societies nationally and globally. Each of us has tools and concepts that can strengthen the work of our colleagues.

The book is also our way of putting on record some important preliminary work that should not have to be reduplicated when work on pressing environmental problems resumes. A series of interdisciplinary research agendas is offered which can be "ready to go" at any time. We consider the bibliographies accompanying the papers as valuable as the papers themselves, as a way of bringing together who is doing what in a certain set of climate-related environmental arenas.

Reflections on Conceptual Problems in This Project: Systems Stability, Instability and Rates of Change

Climate change is a long slow process. Social scientists are used to studying short-term changes, particularly crisis situations, because these are the situations in which their expertise is called upon. How then can social science knowledge be drawn upon for these long slow processes? The answer lies precisely in the characteristics of long-term system change. What we call system stability in both physical and social systems represents a social perception resting on fairly short-term observations. United States agriculture, for example, has benefitted from a twenty year period of reasonably favorable rainfall and growing season length which has been mislabeled by many the "normal" climate condition. Much greater climate stress preceded – and is likely to follow – this "stable" period, requiring major adaptations in agriculture, and in the sectors providing housing, clothing, processed food, energy needs and commerical recreation. This will likely be true whether the long-term trend is a cooling or a warming. However, the long-term trend is never what people adapt to, as Warrick and Riebsame point out. It is the more extreme of the short-term fluctuations within a trend that require adaptation. Thus behavior in drought, floods and other extreme weather conditions which we label crises are the raw materials from which we can draw conclusions about what kinds of adaptive behavior could be mobilized to deal with long-run change. Cushioning short-term fluctuation effects too completely for impacted local communities with a massive infusion of helping services may dampen skills of adaptation needed for the long run.

Local short-term adaptation and long-term societal planning are however very different orders of phenomena. Every author in the papers to follow deals in one way or another with the difficult problem of communicating long-term trend changes to planners accustomed to dealing with assumptions of system stability. The fact that it is human behavior with regard to energy use that may be setting in motion the system change, and that change in human behavior might, but cannot certainly, prevent that change, makes information input to planners all the more difficult. Dealing with these levels of risk and uncertainty creates essentially new questions for social science, questions that make the present research of great importance.

Time as a Resource

The past is an important resource for dealing with change, since past fluctuations and past system states with the adaptations they engendered are in the historical record. The Data Bank on Climatic History proposed by Rabb may be the most important data bank available to the twenty-first century in that turns out to be a century of wildly fluctuating climate. People can deal with almost anything if it is put into intelligible time frames.

The future is also an important resource for dealing with change, since the future is the space-time in which patterns of adaption can be worked out. There has to be a sense that there is *time*. There also has to be a sense of an Other Possibility. The future is the only place where that Other Possibility can reside. Therefore, people's ideas about the future are also an important data bank. Something for decisionmakers to consider is whether the future should not be appreciated instead of depreciated. Time discounting may not only harm future generations, it may harm our own.

The present is when decisions are made. There is no other time in which to make them. Decisions are what everything follows from, yet we know little about how we evaluate data when we make decisions, as Fischhoff and Furby point out. Everyone stacks the cards in decision-making – scientists, planners and parents. We need to know more about the card-stacking process, and more about how to use the past and the future in that present decision-making moment.

Units and Sectors of Adaptation

We don't have to worry equally about all sectors of society in terms of adaptation to change. The socio-technical sector, say Warrick and Riebsame, will make adaptations on their own in a matter of a few decades (i.e., energy, transportation, construction industry). "Hybrid" sectors like agriculture and water supply may need the most help. But will the help help? The ultimate unit of adaptation, as Torry points out, is the household-cum-local-organizational environment. When can local households do better than regional planners in managing environmental change? There is always a knowledge gap between local and regional. Mann sees planning as the only way to get resources where they are needed, but also sees governments as highly resistant to changing needs. How can we optimize both local and governmental resources?

Perception and Conditions of Choice

Every author in this book sees perception of change, perception of crisis, as problematic. What determines what people *see*? Past experience, gender, age, family, group and class interest, national interest, world interest, all play a part. Gender interest is never mentioned in these papers, but women's perceptions should be studied since women are the primary managers, adapters, buffers, in times of shortage and hardship as this impinges on a household. Much of the drama of the Little Ice Age in Europe was played out in households and is reflected in the letters written by women to family members elsewhere. What the poor perceive in times of stress is rarely studied, although there is general agreement that they endure the biggest adaptation costs. How do economic, information and cultural orders create "conditions of choice" (Warrick and Riebsame) that determine what options people see and what choices they feel able to make?

Conflict and Change Related Skills

People don't just adapt to change, they create new realities. Weiss describes the new international regimes that have evolved to deal with unprecedented pollution problems in the twentieth century. When interests conflict in the face of environmental and resource changes, whether between households or nations, what determines the development of skills of conflict resolution rather than a resort to violence? What is the relationship between societal vulnerability and societal creativity?

These are only a few of the many questions raised by the material in this book. As you read, add your own questions – and join the interdisciplinary enterprise.

Darmouth College
Hanover, New Hampshire

STEPHEN H. SCHNEIDER*

CO_2, CLIMATE AND SOCIETY: A BRIEF OVERVIEW

For about a century the academic community – or at least a segment of it – has been aware of the possibility that increasing atmospheric carbon dioxide from fossil fuel burning and other human activities could significantly alter global climate and thus become a major societal problem. Now, however, the CO_2 issue is frequently reported – often in ominous tones – in the popular press, and its implications have even penetrated to the halls of Parliaments and the U.S. Congress. [1] What causes the climate to change – and the potential contribution of carbon dioxide – is self-evidently an important problem for climatologists and other environmental scientists. But why should the "average citizen" – or his or her elected representatives – worry about the CO_2 issue? Food production, which depends significantly on both climate and carbon dioxide per se, is one such reason.

The Corn Belt and Great Plains of the United States, to take one example, comprise a principal food producing and exporting region in the world. We owe much of this to favorable climatic resources of the region. For a number of years, particularly in the mid-1930s and the mid-1950s, and to a lesser extent in the past few years, climatic conditions have not always been so favorable; economic and social hardship followed in the wake of these episodes of bad weather. But some 5000 to 8000 years ago, when considerable proxy evidence suggests that mean global temperatures were perhaps 1–2 °C warmer than today [2], dunes of drifting dust covered portions of the high plains. Prairie-like conditions now found generally west of the Missouri River invaded deep into the present Corn Belt, stretching all the way into Indiana, in what some archaeologists have called the "prairie peninsula". While CO_2-induced climate change is certainly not the only – nor even necessarily a principal – cause of this event, the important message for our purposes is that seemingly small changes in global temperature can be reflected by large, regional changes in temperature and precipitation patterns. While such alterations are not necessarily bad, depending upon the nature of economic activities in each region when the climate changes, the important point is that we will have to learn to adapt – or migrate – should such climatic shifts occur in the future. If they occur rapidly or we are inflexibly dependent on the climatic status quo, then considerable environmental and societal impacts are likely.

In order to help put some perspective on the overall CO_2 problem we can portray each of its many subcomponents as part of an inverted pyramid, as in Figure 1. [3]

Behavioral Assumptions

At the very base of the pyramid of CO_2 issues is neither physics nor chemistry nor biology, but rather social science. That is, the amount of carbon dioxide injected into the atmosphere over the next several decades is not known – nor is it precisely knowable.

R. S. Chen, E. Boulding, and S. H. Schneider (eds), Social Science Research and Climate Change: An Interdisciplinary Appraisal, 9–15.

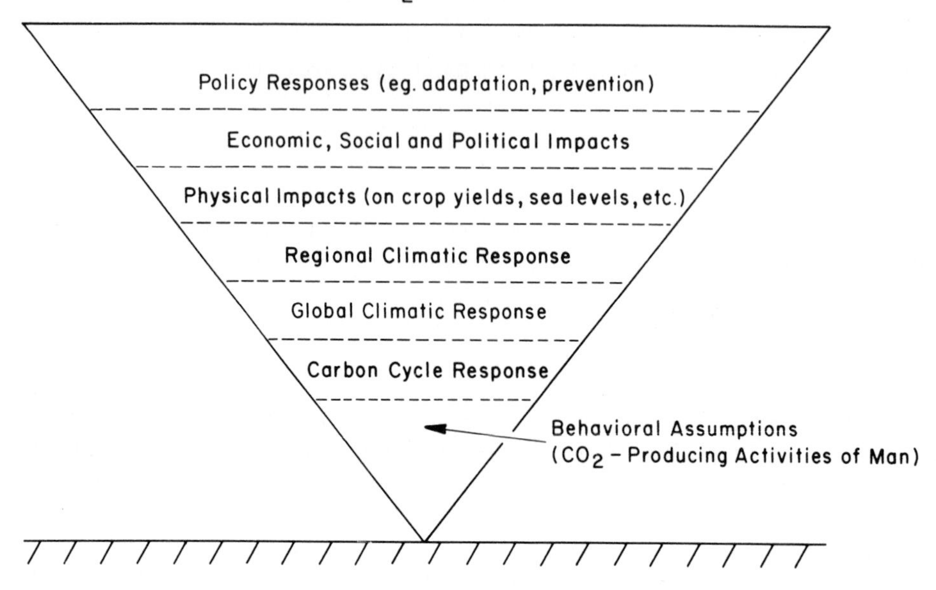

Fig. 1. A schematic representation of the cascading pyramid of uncertainties associated with the CO_2 problem.

It depends upon underlying behavioral assumptions as to what will be the CO_2-producing activities of man. These depend upon projections of (a) human population, (b) the per capita consumption of fossil fuels by that population, (c) the deforestation and reforestation activities of man, and (d) even any engineering countermeasures to remove CO_2 from the air. These, in turn, depend on issues such as the likelihood that alternative energy systems – including more efficient end usage – will be available, their price, and their social acceptability. It also relates to food. For example, before a significant fraction of the world's coal can be used by most Third World countries, it will have to be shipped from the principal coal-endowed nations: the USSR, the US, and China. In order for most countries to purchase coal after oil production is projected to taper down around the turn of the century, it will be necessary to have sufficient foreign exchange to buy it. This could be a serious constraint on global coal usage. [4] Whether such financial resources will be available depends upon the food situations of these countries, particularly the most populous, which in turn depends upon technological, economic, and political assumptions as to how food demands will be met. Another critical assumption is the price of fossil-fueled energy supply systems relative to other alternatives, particularly increased energy end use efficiency. Some believe that the latter will virtually eliminate any likelihood of CO_2 increases beyond some 50% or so of present values. [5] To address these issues – which are at the bottom of the CO_2 question – requires an interlocking set of assumptions behind projections of how people *could act over time*: simply, a set of integrated scenarios.

Carbon Cycle

Assuming we can (at least to an accuracy of 50%) estimate how much CO_2 might be produced in the next few generations, we must then analyze how the global carbon cycle will dispose of the CO_2. This involves questions such as the uptake rate by the oceans, the uptake by photosynthesizing green plants, and the deposition of carbonate chemicals in marine or other sediments. At present there is considerable controversy over the relative roles of the oceans and biosphere in accepting – and producing – CO_2. [6] Furthermore, if CO_2 increases, it will change the rates at which plants grow, and perhaps change the climate in such a way as to alter both biological and oceanic sources and sinks of CO_2. [7]

Global Climate Response

If such CO_2 increase scenarios are derived, they can then be fed into models which predict the resulting global climatic response. Such CO_2 increases as typically projected are unprecedented – at least in the sense that if similar CO_2 concentrations occurred in the past we have insufficient measurements to verify either how much CO_2 was present or what climate changes it induced. As, in this sense, they are unprecedented, we thus have little choice but to build models of the earth's climate in order to simulate the sensitivity of the present climate to projected CO_2 increases. There is considerable uncertainty as to the reliability of such models since there is no *direct* verification possible. Nevertheless, indirect verification tests of such models can and have been done. [8] Indeed, critical assessments by national and international groups of scientists almost always produce the same conclusions: that a fixed doubling of carbon dioxide would increase equilibrium global surface temperature by something between 1.5 and 4 °C.

Regional Climatic Response

Although we may now have achieved some consensus on the global surface temperature rise from a given carbon dioxide increase, what is important for economic and social impact studies is the regional response. Unfortunately, to predict regional responses of temperature, rainfall and so forth requires climatic models of greater complexity – and expense – than is needed to make globally averaged predictions. Nevertheless, those few models which have been applied to this problem tend to suggest certain coherent features. These include generally wetter subtropical and monsoonal rainbelts, longer growing seasons in the high latitudes, wetter springtimes in high and midlatitudes, and – most problematically for future agriculture in developed countries – drier midsummer conditions in mid- and some higher latitudes. [9] Considerable uncertainty still remains, however, particularly since regional climatic effects inferred so far from detailed climatic models are obtained from fixed – usually a doubling or quadrupling – increases in CO_2 for which the model's climate was allowed to reach a new steady state for this enhanced level of CO_2. Unfortunately, CO_2 in the real world does not grow as an instantaneous doubling or quadrupling, then held fixed in perpetuity, but rather as a smooth increase over time – which has been roughly exponential over the past few decades. The effects of the oceans in altering the actual course of the climate change over time relative to the effects predicted with equilibrium model calculations in which CO_2 is doubled and held fixed forever are now a subject of debate in the scientific community. [10] It is important to predict the actual time-evolution of CO_2-induced climatic change because it affects what is needed for human societies to determine how to react to the advent or prospect of a CO_2 problem: namely, scenarios of regional climatic changes. [11].

Physical CO_2 Impacts

Given a time-evolving scenario of regional climatic change, we can next estimate its physical impacts on the environment and society. Most important, of course, would be its effects – direct and indirect – on crop yields. Next is the potential for altering the range or population sizes of some pests which could impact on plant, animal, or human health. Also of interest would be estimates of how yet undeveloped environments, such as certain tropical forests, might be altered. In addition, changes in patterns of heating and air conditioning energy demand would occur with climatic change, as might the potential for using some renewable energy sources which are dependent on climatic conditions. Perhaps the most dramatic, even if far off, potential event, would be the breakup of part of the West Antarctic Ice Sheet, possibly resulting in a five-or-so-meter sea level rise over a century or so. One estimate suggests that several percent of the U.S. land area above could be lost if such an event occurred, along with economic losses in the trillion dollar range. [12]

Economic, Social and Political Impacts

Before one can seriously suggest policy responses to the advent or prospect of CO_2 increases, all of which might entail costs, one first needs to examine potential costs – and benefits – of CO_2-induced environmental or societal impacts so that these can be weighted against the costs of mitigating strategies. This is not a simple question of aggregating total dollars lost or gained, but also involves the important equity question of who wins and who loses, and how the losers can be compensated and the winners taxed. Moreover, even the *perception* that the economic activities of one nation created climatic changes which might be interpreted as detrimental to another has the potential for being divisive to international relations. [13] Nevertheless, given *scenarios* of behavioral assumptions, carbon cycle response, global climatic response, regional climatic response, physical impacts, and economic infrastructure in the future, we can estimate, particularly for the agricultural sector, some potential economic, social and political impacts. These can then help a spectrum of decisionmakers to formulate a set of appropriate responses.

Policy Response

Finally, at the top of our inverted pyramid on Figure 1 sits a category, "policy responses". These include the possibility to adapt, prevent, or even attempt to ameliorate the buildup of CO_2 or its consequences. Schneider and Chen suggested a hierarchy of four possible policy responses [14]:

(1) Do nothing.

(2) Study and monitor.
(3) Build resilience (that is, invest some present funds to increase available future options to deal with the advent or prospect of CO_2 buldup).
(4) Law of the air (that is, set up national emission quotas for CO_2 based on a complex, difficult set of international negotiations).

One thing that is clear from this brief perspective on the CO_2 issue: it is international and interdisciplinary. This point has been emphasized by all of the more than a dozen national and international groups which have seriously looked at the CO_2 question. Moreover, it cannot be addressed from the point of physical science alone, but also must *equally* consider biological and social science components. This was explicitly recognized, for example, in the U.S. National Climate Program Act. [15] Section II of the Act reads:

> The Congress finds and declares the following: . . . an ability to anticipate natural and man-induced changes in climate would contribute to the soundness of policy decisions in the public and private sectors.

In Section V, under "Program Elements", the Act declares:

> The program shall include, but not be limited to, the following elements: . . . basic and applied research to improve the understanding of climate processes, natural and man-induced, and the social, economic, and political implications of climate change . . . measures for increasing international cooperation in climate research, monitoring, analysis and data dissemination.

It is important to note that this Act, and its subsequent several renewals, had been passed by both houses of Congress with virtually no dissent. This recognition on the part of the U.S. Congress that CO_2 and other climate research problems are not partisan, but rather global, future issues is reassuring.

A number of national and international groups have met to discuss and assess the CO_2 issues. Recently, for example, an international workshop on energy and climate held in the Federal Republic of Germany concluded in the preface to its findings that "in the decades ahead, decisions have to be made to reduce or avert the impacts of climatic change before all the answers have been obtained. Although a climatic impact assessment program is faced with many uncertainties, it nevertheless has to be started now, because society cannot afford to wait until all variables are quantified to the satisfaction of all parties involved." [16]

But how do we proceed to do the long-term physical, biological and social research on various aspects of the carbon dioxide/climate problem called for by all these august bodies? Suggestions that deal with CO_2 and its effects are often either too vague – for example, the advice to "build resilience" – or too restrictive – for example, the suggestion to study how one variety of corn plant responds in a greenhouse to doubling of carbon dioxide. While, of course, many such specific studies, when combined with concepts like "build resilience", can eventually improve the scientific basis upon which future policy making will be made, a more efficient way to proceed has been suggested by several different groups: *Scenario analysis*. This general approach was highlighted in the findings of the joint World Meteorological Organization/International Council of Scientific Unions/United Nations Environment Program Group of Experts meeting held in Villach, Austria in November 1980 [17] and also in the conclusions of the

Steering Committee of the American Association for the Advancement of Science/ Department of Energy Study on environmental and societal consequences of CO_2 increases. [18]

It could certainly be argued that any particular climate scenario, as well as the food, water, energy and other societal impacts estimated from such scenario analyses, could well be significantly different from what would happen if CO_2 increased as assumed in any scenario. There are just too many uncertainties to accept any prediction as reliable. Therefore, analysis of *several* differing but plausible scenarios probably provides the best way to begin the process of identifying what activities in society could be most vulnerable to CO_2 increases. Such analyses could also help sort out what options, both now and in the future, might minimize our vulnerability to increasing CO_2 or could even help us to take advantage of new opportunities that an altered climate may create. If such studies are not made, then society will merely "perform the experiment" of unprepared, "post-crisis" adaptation with little lead time to minimize preventable damage or maximize available advantage.

*National Center for Atmospheric Research***
Boulder, Colorado 80307

Notes

* Any opinions, findings, conclusions or recommendations expressed in this publication are those of the author and do not necessarily reflect the views of the National Science Foundation.

** The National Center for Atmospheric Research is sponsored by the National Science Foundation.

Bibliography

[1] For example, see Committee on Governmental Affairs, United States Senate, 1979, *Carbon Dioxide Accumulation in the Atmosphere, Synthetic Fuels and Energy Policy, A Symposium*, July 30, 1979. U.S. Government Printing Office, Washington, D.C.

[2] Kellogg, W. W. and Schware, R.: 1981, *Climate Change and Society: Consequences of Increasing Atmospheric Carbon Dioxide*, Boulder: Westview Press. Appendix C; and Butzer, K. W. 1980. Adaptation to global environmental change. *The Professional Georgrapher* **32**, 269–278; Schneider, S. H.: 1983, 'On the Empirical Verification of Model-Predicted CO_2-Induced Climate Effects', in J. Hansen and T. Takahashi (eds.), *Climate Processes: Sensitivity to Irradiance and CO_2*, American Geophysical Union, Washington, D.C., submitted.

[3] This figure was taken from Schneider, S. H. and Londer, R. S. 1984, *The Co-evolution of Climate and Life*, San Francisco: Sierra Club (in press). Figure also accompanied the prepared testimony of S. H. Schneider. *Carbon Dioxide and Climate: The Greenhouse Effect*. No. 45. House Committee on Science and Technology, 97th Congress, Joint Hearing of Subcommittee on Natural Resources, Agriculture Research and Environment and the Subcommittee on Investigations and Oversight, July 31, 1981. Washington, D.C., pp. 31–59.

[4] Ausubel, J. H.: 1980, *Climatic Change and the Carbon Wealth of Nations*. Working Paper WP-80-75. Laxenburg: International Institute for Applied Systems Analysis.

[5] Lovins, A., Lovins, H., Krause, F., and Bach, W.: 1981. *Least-Cost Energy: Solving the CO_2 Problem*, Andover: Brick House Publishing Company.

[6] Bolin, B. (ed.): 1981, *Carbon Cycle Modeling*, SCOPE 16, Chichester: Willey; see also a number of chapters in Clark, W. C. (ed.), 1982: *Carbon Dioxide Review 1982*, Oxford University Press, New York.

[7] Rosenberg, N. J.: 1981, 'The Increasing CO_2 Concentration in the Atmosphere and Its Implica-

tion on Agricultural Productivity: I. Effects on Photosynthesis, Transpiration and Water Use Efficiency', *Climatic Changes* **3**, 265–279.

[8] National Academy of Sciences: 1982, *Carbon Dioxide and Climate: A Second Assessment*, CO_2/Climate Review Panel, Washington, D.C.

[9] Manabe, S., Wetherald, R., and Stouffer, R. J.: 1981, 'Summer Dryness Due to an Increase in Atmospheric CO_2 Concentration', *Climatic Change* **3**, 347–386.

[10] Bryan, K., Komro, F. G., Manabe, S., and Spelman, M. J.: 1982, 'Transient Climate Response to Increasing Atmospheric Carbon Dioxide'. *Science* **215**, 56–58; Thompson, S. L. and Schneider, S. H.: 1982, 'Carbon Dioxide and Climate: Has a Signal Been Observed Yet?' *Nature* **295**, 645–646; Thompson, S. L. and Schneider, S. H.: 1982, 'CO_2 and Climate: The Importance of Realistic Geography in Estimating the Transient Response', *Science* **217**, 1031–1033.

[11] Schneider, S. H. and Thompson, S. L.: 1981, 'Atmospheric CO_2 and Climate: Importance of the Transient Response', *J. Geophys. Res.* **86**, 3135–3147. See also ref. [2].

[12] Chen, R. S.: 1982, 'Risk Analysis of a Global Sea Level Rise Due to a Carbon Dioxide-Induced Climatic Warming', Master's Thesis, Massachusetts Institute of Technology, Cambridge. MA.

[13] Perry, J. S.: 1981, 'Energy and Climate: Today's Problems, Not Tomorrow's', *Climatic Change* **3**, 223–225; Schneider, S. H. with Mesirow, L. E.: 1976, *The Genesis Strategy: Climate and Global Survival*, Plenum, New York, Chapters 6 and 7.

[14] Schneider, S. H. and Chen, R. S.: 1980, 'Carbon Dioxide and Coastline Flooding: Physical Factors and Climatic Impact', in Hollander, J. M., Simmons, M. K., and Wood, D. O. (eds.), *Ann. Rev. Energy* **5**, 107–140. Please see this article for more thorough discussion and specific examples of these four policy response options.

[15] National Climate Program Act, Public Law 95–367, September 17, 1978.

[16] Bach, W., Pankrath, J., and Williams, J. (eds.) 1980, *Interactions of Energy and Climate*, D. Reidel Publ. Co., Dordrecht, Holland, p. VIII.

[17] *On the Assessment of the Role of* CO_2 *on Climate Variations and Their Impact*: 1981, Joint WMO-ICSU-UNEP Meeting of Experts, Villach, Austria, November 1980. Geneva: World Meteorological Organization.

[18] *Environmental and Societal Consequences of a Possible* CO_2*-Induced Climate Change: A Research Agenda*, December 1980. No. 013 of Carbon Dioxide Effects Research and Assessment Program, DOE/EV/10019–10, Washington, D. C.: United States Department of Energy, pp. IV–1.

PART II

RESEARCH IN CLIMATE CHANGE AND SOCIETY: INDIVIDUAL CONTRIBUTIONS

Introduction

Perhaps the principal motivation for this volume – and indeed for most research on climate change – is the intuitive feeling that at some point some actions by society may well be necessary to deal with the possibility, if not the onset, of climate change. Major natural upheavals in the earth's climate have certainly occurred throughout the geologic past, and even relatively minor climatic fluctuations have greatly influenced societal development on innumerable occasions in recorded history. Should society fail to respond to future changes with sufficient wisdom, effectiveness, or rapidity, the consequences for some could be most serious, perhaps disastrous.

The possibility that human activities such as fossil-fuel burning, deforestation, and chlorofluorocarbon usage might significantly affect the climate adds another dimension to the more general problem of climate change. Society itself, it is now clear, could be a major initiator of global environmental changes. Society can choose to assume responsibility for managing its activities to avoid such changes – or else suffer their consequences. Ethical ramifications aside, an unavoidable implication of this responsibility is that climate change could become a consideration in virtually all facets of everyday life, from individual consumer decisions to national energy policies. Society's perception of or response to the possibility of climate change at a variety of decision-making levels thus emerges as a critical issue, one that is a key focus of this volume.

In the course of its discussions, the AAAS group identified four major underlying research questions:

(1) What stresses on society would result from the advent or prospect of increasing atmospheric carbon dioxide and associated climate or other environmental impacts?
(2) What determines differential vulnerabilities to stress of different social groups?
(3) What stress-response capabilities and patterns are presently available in different sectors and at different levels?
(4) What response capabilities might be developed for the future?

These four questions illustrate an important assumption of the group and of this volume, namely that "it is society that is the subject of research – not climate." This focus on society is particularly critical because it suggests that there is indeed a wealth of careful research and thinking available that may be relevant to specific facets of the general problem of climate change and society. In particular, considerable knowledge and experience exists in a variety of social science disciplines that pertains to the perception and behavior of specific sectors and groups in society and the institutions and processes on hand for coping with change. For example, much research has been conducted in recent years on what might be termed "surrogates" for climate change – that is, situations

R. S. Chen, E. Boulding, and S. M. Schneider (eds.), Social Science Research and Climate Change: An Interdisciplinary Appraisal, 17–18.

such as droughts, soil erosion, pollution, and natural hazards that should provide important insights into climate's potential impacts on society and the likely behavioral responses at a variety of levels. Such research "tie-ins", discussed more extensively in the paper by Chen, supply a crucial starting point for all of the papers in this volume: simply that we, society, can learn from a variety of relevant experiences.

However, it was also clear to the group that each social science discipline cannot make much progress alone. Some innovative, interdisciplinary research plan or program is clearly necessary to ensure that the key research questions are addressed broadly and effectively. Such an effort would need to incorporate continual communication among many different natural and social scientists. Moreover, some form of integrative framework, such as that proposed in this volume by Warrick and Riebsame, would certainly be required. Another approach might be to develop "scenarios", or hypothetical depictions of future conditions and events, to provide a framework for thinking about the complex and uncertain future (see Lave *et al.* in DOE, 1980a).

As a point of departure for a research program, our Panel proposed a set of seven major interdisciplinary research areas to bring together research efforts and people scattered in many different disciplines. The initially proposed areas were:

(1) Data collection.
(2) Impacts on society of past climate changes.
(3) Current climate-stress surrogate situations.
(4) Societal response mechanisms.
(5) Governmental response mechanisms.
(6) International system.
(7) Risk perception, information, and decision making.

These research areas are not intended to be comprehensive, but do encompass the major research issues that the group felt to be important. The interested reader is referred to the complete Social and Institutional Panel report (DOE, 1980b) for further details on these areas.

Let us proceed then to the individual contributions.

R. S. CHEN

Bibliography

DOE 1980a. *Workshop on Environmental and Societal Consequences of a Possible CO_2-Induced Climate Change*. Carbon Dioxide Effects Research Assessment Program, Report 009, U.S. Department of Energy, CONF-7904143, 1980.

DOE 1980b. *Environmental and Societal Consequences of a Possible CO_2-Induced Climate Change: A Research Agenda, Volume I.* Carbon Dioxide Effects Research and Assessment Project, Report No. 013, U.S. Department of Energy, DOE/EV/10019–01, December 1980.

INTRODUCTION TO: SOCIETAL RESPONSE TO CO_2-INDUCED CLIMATE CHANGE: OPPORTUNITIES FOR RESEARCH

Research on climate and society from several different disciplines needs to be integrated before it can realistically be considered interdisciplinary. Perhaps it is most appropriate, then, to begin our disciplinary contributions with the one from geography, itself an "interdisciplinary discipline". Geographers Warrick and Riebsame have assembled into one meaty document an amazing breadth of concepts and proposals to diagnose, understand and, ultimately, deal with climate and society issues in the CO_2 context.

They first offer a conceptual framework from which specific research questions and topics flow. Response strategies to the advent or prospect of CO_2-induced climatic changes would arise, Warrick and Riebsame argue, from a variety of interconnected sectors (agriculture, energy, etc.). Choices eventually are made depending on perceptions of individual decision makers, "subject to a complex set of incentives and constraints imbedded in economic, social and political institutions." The principal issue to study, they contend, "is the process by which society adapts to climatic change." They then offer several examples.

Next, Warrick and Riebsame list eight key questions to address in the CO_2 context. (These questions are even more general than "climate and society" issues, as they would provide a good starting point for other environmental impact assessment areas like soil erosion or acid rain.) Then, each question is examined in the climate and society context in some detail, and research opportunities are identified for each of the eight. The net result is that the Warrick/Riebsame contribution provides both an appropriate beginning for this part of our Volume and an independently valuable introduction to the overall problem of societal response to climatic changes.

S. H. SCHNEIDER

R. S. Chen, E. Boulding, and S. H. Schneider (eds.), Social Science Research and Climate Change: An Interdisciplinary Appraisal, 19.

RICHARD A. WARRICK AND WILLIAM E. RIEBSAME

SOCIETAL RESPONSE TO CO_2-INDUCED CLIMATE CHANGE: OPPORTUNITIES FOR RESEARCH

Abstract. How might a climate change, induced by increased CO_2 in the atmosphere, affect societies? What is the range of existing and potential mechanisms for societal response? And how might research contribute to a reduction of the adverse impacts (or enhancement of the unique opportunities) of a climate change by providing greater understanding of the processes involved in climate and society interaction? This paper reflects an initial effort to shed light on these questions. It offers first a framework for identifying key issues in climate-society interaction; eight major questions are suggested by the framework. A discussion of each major question is then presented with the purpose of reviewing the current state of knowledge, identifying the gaps in understanding, and offering opportunities for research to fill those gaps. In all, twenty-two research needs are outlined and are summarized at the conclusion of the paper. The perspective is interdisciplinary, but the review draws heavily from the geographic literature, reflecting the disciplinary bias of the authors.

0. Climate and Society Relationships

As the need for, and interest in, climate impact assessment has become evident during the last several years, a flurry of research activity has taken place. These efforts, however, often rely upon a rather simple concept of climate and society, a concept we term "neo-environmental determinism", which depicts a one-way, causal relationship between climate (or environment, if you will) and human activity. Although the roots of such a conceptual model can be traced to the classical Greeks, a resurgence of intellectual environmental determinism was strongly evident in the first decades of the 20th century. Ellsworth Huntington, a geographer, was a noteworthy proponent of this view, as reflected in his *Civilization and Climate* (1915). He assumed that climate is the independent variable whose explanatory power is pervasive in understanding social structure, settlement patterns, and human behavior. Such arguments led easily into racial or cultural stereotyping (e.g., tropical climate fosters lazy, unmotivated peoples, and midlatitude regimes promote vigorous, industrious humans), and finally to explanations of the differences in the global patterns of economic development we see today. These views fell out of intellectual favor as being too simplistic and mechanistic, and largely disappeared (at least in their overt form) for many decades.

With the recent concern over the steadily rising levels of atmospheric CO_2 and other climate-related issues, however, similar notions seem to be emerging – some strongly, others subtly. Such "neo-environmental determinism" is hinted at in the works of Carpenter (1968) and Bryson *et al.* (1974), who suggest that drought was the chief cause of decline of Mycenaean Greece during the late Bronze Age, as well as in the works of Biswas (1980a) and Biswas and Biswas (1979), who argue for renewed attention to climate as a determining factor in agricultural underdevelopment and in human health.

R. S. Chen, E. Boulding, and S. H. Schneider (eds.), Social Science Research and Climate Change: An Interdisciplinary Appraisal, 20–60.

Lambert (1975) attempts to demonstrate the linkage between climate change and the growth of civilizations. Climatic determinism in its most extreme form is exhibited in the popular literature; an article in *New Scientist* by Harrison (1979), for example, attributes everything from labor efficiency to agricultural productivity to climate differences (not unlike the Greek theories). Kates (1980a) suggests that most of the work in climate impact assessment to date adheres implicitly to this conceptual model.

In many cases, such explanations are too simplistic, ignoring factors of human choice and societal adaptability. This is not to say that climate does not exert a strong influence on society. Indeed, environment as an explanatory variable was perhaps shunned unnecessarily for many years because of its association with the earlier, discredited theory of environmental determinism. In fact, a simple model of this sort is not necessarily undesirable, since it is easier to put into operation and leads to more tractable research problems. However, caution should be exercised in relying too heavily upon such a formulation, for it tends to influence which research questions are asked and which methodologies are employed. More importantly, it neglects the many interacting variables involved in climate change and human adaptation. Critical questions concerning societal vulnerability, response mechanisms, adjustment choice processes, and feedbacks of societal responses to natural and human systems are not integral to the model. For these kinds of questions, we must formulate our concept of climate and society differently.

0.1. A Framework for Identifying Research Needs

The conceptual model illustrated in Figure 1 portrays climate effects – beneficial as well as detrimental – as products of variation in both the Natural and Social Systems. Thus, the consequences of climate change are just as much a result of societal characteristics as of environmental fluctuation, as indicated in the lefthand side of the diagram. The initial effects on selected sectors (e.g., agriculture, energy, water supply, etc.) diffuse through social systems via pathways and linkages ingrained in social organizations at any given time and place. Eventually, these are translated into culturally defined effects – both positive and negative – on human well-being. As depicted by the "major information components" in Figure 1, response strategies are formulated on the basis of information on climate change, on the level and distribution of societal effects, and on the range of mechanisms which could be adopted. How this information is used depends partly upon the ways in which it is channeled and upon its perception and interpretation by decision-makers (e.g., farmers, energy planners, water supply managers, and politicians). Final choice of strategies is subject to a complex set of incentives and constraints imbedded in economic, social, and political institutions. Those strategies alter the natural systems, the social systems, or both, as the dashed lines in Figure 1 suggest. According to this view, the entire system is interactive and dynamic, changing as societies change and adjust. Generically, Kates (1980a) would call this formulation an "interactive model with feedback"; it is similar to the model proposed by the Scientific Committee on Problems of the Environment (SCOPE) Workshop on the Climate/Society Interface (SCOPE, 1978).

However, our aim is not to define and defend one particular descriptive model of climate and society. Other models exist and can be argued convincingly. Rather, our intention is to provide a framework for systematically organizing our thoughts about

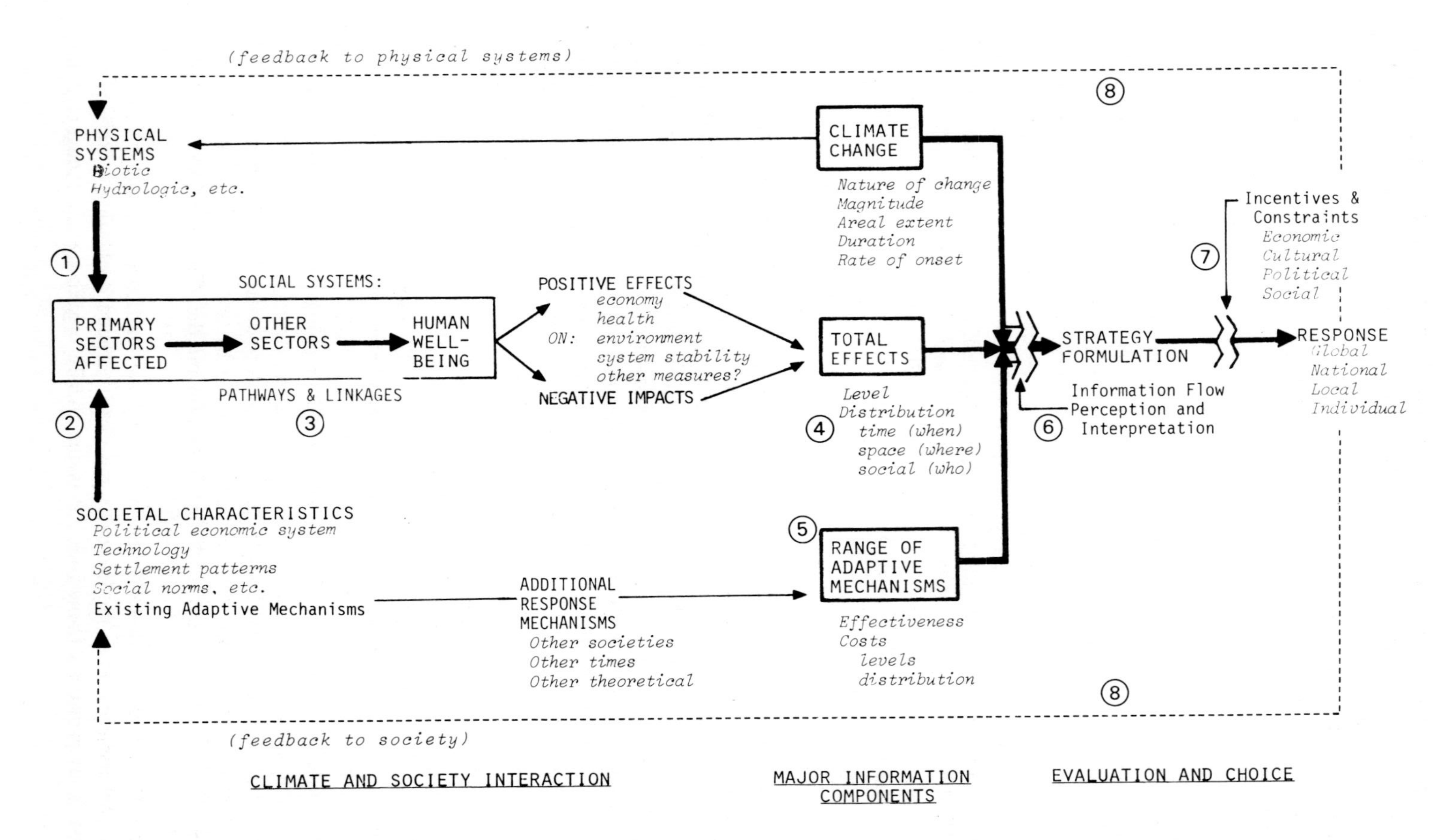

Fig. 1. A framework for identification of research on the societal effects of CO_2-induced climate change, with an emphasis on decision information and response.

CO_2 research needs. In this respect, the framework emphasizes strongly the role of information, decision-making, and response. This is purposeful. A major assumption is that the research results are intended for key decision-makers; therefore, it is critical not only to understand the processes of society and climate interaction, but also to know about the role of information and the process by which it leads to effective response.

0.2. The Nature of Climate Change and Social Adaptation

Implicit to this framework are also assumptions about the nature of climate variation and change and about the ways in which societies adapt. As Hare (1979; personal communication) points out, climate is an intellectual construct composed of an ensemble of central tendencies, variabilities about these tendencies, and rare extreme events, all derived from a span of time long enough to give the averages, spectra, and extremes some statistical meaning. Although there is the possibility of a sudden "flip" to a new climatic ensemble from CO_2 buildup, in all likelihood the shifts will be slow and continuous. The best understanding of the atmospheric effects of CO_2 buildup is in terms of their influence on central tendencies, as in gradually increasing average global temperatures. However, climate variabilities and frequencies of extreme events may also be affected, though we are currently not sure how; there is not always a clearcut relationship between trends in central tendencies and variance, as recently demonstrated by van Loon and Williams (1978). Nevertheless, it is sobering to note, as did Bryson and Padoch (1980), that " . . . an often overlooked fact about climatic data is that small differences in the mean temperature may mean significant differences in the frequency of occurrence of extreme values" (p. 589); this is particularly true for certain regions of the world, which might be termed climatically marginal environments. Finally, actual regional climatic changes could prove to be quite disparate, and may actually differ in sign from one region to the next. In short, the global effects of a CO_2 buildup will be anything but uniform (Kellogg and Schware, 1981).

The critical issue for us is the process by which society adapts to climatic change. One possibility is that society gradually adapts to a slow cumulative change in central tendencies. If that is simply the case, society can be optimistic since human adaptability appears rapid in comparison to the slow rates of climatic change due to CO_2 (Margolis, 1979). However, it is also possible that society does not respond to slow changes in averages but rather reacts to discrete events which represent departures from average conditions; and that the accumulation of a series of step-like social changes in the short term leads to adaptation to climate change in the long term. If this is the case, the *rate of societal adaptation* would then change with the *change in frequency* of such initiating climate events.

For example, to a farmer the knowledge that average temperature may increase by 2°C over the next 50 years means little. Farmers react to good and poor crop years. When droughts or freezes are too extreme and/or frequent in relation to farm operations, changes will be made in land use, technology, or even location. Similarly, alterations in water supply systems tend to be made on the heels of precipitation failures severe enough to jar communities into action; perceived changes in the frequency of such events may even prompt system redesign.

The notion that the changing frequencies of disruptive (or beneficial) climatic occurrences may be the key factor in societal adaptation to climate change finds support in several recent case studies. For instance, Whyte (1981), in a study of climate influences on agriculture in early Scotland, argues that " . . . climatic deterioration in marginal areas is more likely to have manifested itself to farmers and their landlords in terms of increasing frequencies of short-term difficulties rather than in declining average productivity over extended periods" (p. 25). Similarly, Parry (1978) concludes that the increased incidence of harvest failure in upland fields of Little Ice Age Scotland was of greater significance in changing land use and settlement patterns than the gradual changes in mean yields. Moreover, instances exist in which societies may have failed to perceive even rapid climate changes, thus making them extremely vulnerable to disruptive climatic occurrences; this may have been the case in the disappearance of the Norse Greenlanders (McGovern, 1981).

If, indeed, peoples and societies do adapt to climate change in a discrete, step-like manner, then the occurrences which stimulate societal adaptation are ones for which society has plenty of experience – droughts, frosts, cold spells, etc. Included are opportunities as well as adversities, like increases in growing seasons with favorable weather. So, although we may not know exactly how the frequency distributions of socially relevant climatic phenomena will change with CO_2 enrichment of the atmosphere, at least we can learn about the adaptation process by examining societal response within our existing socio-technical systems under present climatic conditions.

Regardless of whether society adapts slowly to changes in central tendencies or in a step-like fashion to changes in variability – or some combination of both – the research community has its challenge. In the absence of better theory, we assume that both modes of adaptation may be at work. In fact, the most pressing research need overall is the deveopment of theory regarding societal adaptation to climate change. We remain hopeful that many of the research opportunities described herein will lead to improved theoretical understanding.

0.3. Key Questions for Research

Our framework suggests eight key questions which pertain directly to the issue of CO_2-induced climate change.

(1) What are the relationships between *climate variations and sectors of society*?

(2) What are the characteristics of society that determine differential *vulnerability* to climatic change?

(3) What are the *pathways* and *linkages* within social systems through which the effects of climatic variations are transmitted?

(4) What is the *level and distribution of total effects* arising from climatic change?

(5) What is the range of *mechanisms* by which societies adjust to climatic variations? And what are their costs and effects?

(6) How is information (scientific or otherwise) on climate change and its effects *perceived, interpreted, valued, and channeled* into strategy formulation?

(7) What are the *conditions of choice* which guide societal response to climatic change?

(8) What are the *dynamic feedback effects* to nature, society, and subsequent response?

The circled numbers in Figure 1 indicate how each of the eight key questions relates to the framework. In the body of the text we will ask what is known about each question, what are the important gaps in knowledge, and what research should be conducted to fill those gaps. Our thinking partly reflects a geographic perspective; at this point, the reader who is unfamiliar with the discipline of geography may wish to divert his or her attention to the appendix before reading further.

1. What Are the Relationships Between Climate Variations and Sectors of Society?

Although climate change or fluctuation may have direct implications for human well-being through physiological effects, the more important adverse effects stem from disruption of the socio-economic sectors upon which society relies for normal functioning. Aside from the possibility of a dramatic sea level rise and inundation of coastal settlements (Schneider and Chen, 1980), it is probably safe to assume that the major impacts (either positive or negative) from CO_2-induced climate change will depend largely on the ways in which sectors such as agriculture, energy, transportation, water supply, or fisheries are affected.

1.1. The Climate-Sector Connection

Much of the work carried out in climate-sector research is concerned with establishing *climate transfer functions* (for example, to translate winter weather conditions into heating fuel demand for energy, precipitation and temperature into crop yield for agriculture, or precipitation into streamflow for water supply). Although establishing the direct functional *relationship* is central to such efforts, another important element for planning and design is related to *timing* – that is, to frequency, probability, persistence, and/or duration. The two go hand-in-hand. For instance, while the direct effects of growing season precipitation and temperature on wheat yields are of prime interest, the frequency with which particular climatic conditions lead to agricultural impacts can be equally important. Climate transfer functions provide basic building blocks upon which more complex models of climate-sectoral systems can be constructed.

For purposes of identifying those sectors for which further research would prove most useful, it is instructive to separate sectors roughly into three groups: socio-technical, physico-ecologic, and intermediate sectors. *Socio-technical* sectors are those which are imbedded primarily in human systems and are largely human-created. Examples include energy, transportation and the construction industry. If one adheres to the arguments presented in Bennett and Chorley (1978), socio-technical systems can be characterized as flexible, adaptable, and resilient in the face of perturbation. With respect to time-scales comparable to CO_2-induced climate change, rates of adjustment within these systems appear rapid. For example, the socio-technical changes necessary to meet soaring electricity demands in the United States, or to meet new standards of air pollution in the United Kingdom (Burton *et al.*, 1974), transpired relatively quickly over a few decades. Rough estimates of the time required for *major* structural alterations in technological systems vary from 20 to 50 years – about the time required for climate changes from CO_2. Incremental adjustments (similar to what we are seeing with respect to world oil supplies) can take place concurrently with those major alterations, especially if currently available climate information is used in planning supply and demand (Quirk, 1980).

The conclusion is that with respect to a slow-onset CO_2 problem, climate-related research on the socio-technical sectors is less urgent than research on other sectors.[1] Two additional reasons support this argument. First, we already possess a fairly good understanding of the relationship of weather and climate to these sectors. For example, the National Oceanic and Atmospheric Administration (NOAA) has been developing climate transfer functions to relate seasonal weather forecasts to heating fuel demands in the United States (Mitchell *et al.*, 1973; NOAA, 1974), while similar work has been conducted in the United Kingdom on heating fuel and electricity (Craddock, 1965; Barnett, 1972). Furthermore, knowledge about short-term weather effects on the outdoor construction industry is fairly extensive (see major works by Russo *et al.*, 1965, and Environmental Science Services Administration, 1966; also Roberts, 1960, or Maunder *et al.*, 1971). The same holds true for the transportation sector (see McQuigg, 1975, for an overview of economic impact of weather and climate). Second, the understanding of these specific relationships is only useful as it applies to the particular socio-technical context within which it is developed, and these contexts evolve quickly over time. In hindsight, would it have proven worthwhile to have initiated major research 40 years ago on the consequences of a long-term climate change on steam locomotive transportation?

On the other hand, we face a rather different situation with respect to *physico-ecologic* sectors. These sectors are imbedded firmly in the natural systems, and include fisheries and forestry. Inherently, these sectors may not contain the flexibility and resiliency characteristic of social systems. In contrast to social systems, climatic perturbations encounter a restricted number of routes through natural ecosystems. Instead of leading to new equilibria, these perturbations may cause qualitative change, as in the extinction of a population (Holling, 1973). Many such systems may be fragile and sensitive to local environmental alteration brought about by climate change. The case of "El Niño" is illustrative: The sudden depletion of anchovies induced by the recurrence of the stronger (and longer) than normal outbreak of warm surface waters and a temporary failure of cold upwelling (perhaps related to regional wind patterns) brought the Peruvian anchovy industry to a standstill. Similarly, the California coast sardine industry totally collapsed with a combination of environmental fluctuation and a human perturbation, overfishing (Radovich, 1981).

While fishery resources may indeed be sensitive to climatic changes, we have limited knowledge upon which to construct climate transfer functions (cf., Cushing, 1979; Murray *et al.*, 1980). We perhaps possess better understanding of the related ecological, economic, and policy aspects (Rettig, 1978; Caviedes, 1975). The same holds true of forestry, in which little work has been done on climate/yield relationships.[2] In short, with the prospect of shifts in climate from increased CO_2 on the far horizon, we have limited knowledge of the major consequences to key physico-ecologic sectors.

Perhaps the sectors of greatest social concern are those *intermediate*, "hybrid" sectors, like agriculture and water supply (plus certain energy systems, such as hydro-electric power). It is probably within these sectors that most of the research on climate-sector relationships has been conducted and where further work is still most needed. In the agricultural sector, for example, a great deal of effort has been expended in developing crop-yield models (e.g., see reviews by McQuigg, 1975; Baier, 1977; Biswas, 1980b; or U.S. Dept. of Transportation, 1975). Empirically derived statistical models represent one approach. These models are based largely on historical records of precipitation, temperature, and crop yields; the multiple regression models of U.S. grain yields by Thompson (1962, 1969a, b) are among the most widely recognized.

One major problem encountered in these models is that of separating out the influence of technological trends in order to ascertain the sensitivity of crop yields to weather, or vice versa. McQuigg *et al.* (1973) and Haigh (1977) built upon the work of Thompson and others in attempting to separate technological and climatic influences on U.S. crop yields. Their main conclusion is that yields are still subject to climate variability as in the past, despite technology. Others (Newman, 1978; U.S. Department of Agriculture, 1974), using similar data, reach opposite conclusions. Together these studies reflect the uncertainty and controversy surrounding the issue (Warrick, 1980).

Experimental research, in which attempts are made to control systematically for selected weather and technology variables, is another approach to the same problem (McQuigg, 1975). Physiological studies (e.g., Gupta, 1975) and simulation models address the problem by modeling the processes by which environmental stresses actually affect plant growth (Baier, 1977); thus, they deal directly with the "black box" problems inherent to regression modeling (Katz, 1977).

While climate-yield research has been extensive in the United States, Canada (Baier and Williams, 1974) and other industrialized countries, many other places and crops have received little attention. Preliminary crop-yield models have been developed for parts of India, Korea, Bangladesh, Burma, Sri Lanka, Thailand, and Japan (cf. Takahashi and Yoshino, 1978), and similar efforts are underway for Latin America (Food and Climate Forum, 1981, p. 68). But on the whole, from a global perspective, the knowledge about climate-yield relationships remains meager. For example, the most ambitious global yield prediction scheme – NASA's Large Area Crop Inventory Experiment (LACIE) project, which is based on Thompson's early crop-yield regression models (Environmental Data Information Service [EDIS], 1979) – only managed to make reasonable crop estimates for three countries: Canada, the United States, and the Soviet Union. Efforts to assess the likely effects of a CO_2-induced climate change on global food systems are constrained by the lack of knowledge of climate-yield relationships.[3]

In the water supply sector, climate transfer functions describe climate-hydrologic relationships (say, between weather variables and streamflow) needed for water resource planning and design. Although models abound, there is some question of their credibility. For instance, one study tested ten models in up to six watersheds and found the forecasted streamflows were often greatly in error (World Meteorological Organization [WMO], 1974, as reported in National Academy of Science [NAS], 1977, p. 28). As far as information on "timing" (e.g., probability of high or low streamflows), the same caution should be advanced. Typically, values of central tendency or variability are determined from climatic or hydrologic data derived from a historical period of record (often a short one). They serve as the basis for evaluating water project design performance, on the assumption that the future will be a replication of the past (Schaake and Kaczmarek, 1979). If the period of record is too short or unrepresentative of long-term conditions, the results may lead to difficulties in water resource management. This is exemplified in the familiar case of the Colorado River, in which hydrological studies of streamflow were based on a rather short span of wet years, which led to overestimation of water availability and overextension of demands in later years (Dracup, 1977).

In short, there is often inadequate information about agricultural and water supply sectors, with respect both to how they are affected by climatic variables and to the likely distributions of disruptive events in time. For the CO_2 problem, there are several implications. First, careful assessment of the effects of climate change on key sectors is obstructed by insufficient development of (or uncertainties about) climate transfer functions themselves. Second, since prevailing evaluation and planning methods assume a constant trend based on historic data and a future which will be like the past, existing water supply designs or food production policies may become outmoded in the event of climate change due to CO_2. Four opportunities for research can help rectify this situation.

1.2. Research Opportunities

In order to anticipate the possible effects of a CO_2-induced climate change, the following research activities are suggested: a systematic review of basic research on climate-resource linkages; further research on crop-yield modeling; sector sensitivity analyses; and sector resiliency analyses.

Particularly for sectors like fisheries and forestry, existing studies of climate-resource linkages have not been compiled in a way that allows the development of climate transfer functions for making meaningful estimations of the sectoral impacts of CO_2-induced climate changes. A number of studies address the physical mechanisms linking climate, upwelling, biological productivity, and commerical fisheries for particular coastal regions and fish populations (e.g., the Peruvian anchoveta industry and the occurrence of "El Niño"; the climatic mechanisms involved in the upwelling region and its fishery off the Oregon coast; the relationships between air and sea temperatures, extent of ice cover, and the migration routes of Icelandic herring. See, for example, Thompson, 1977, 1981, or J. Johnson, 1976, for reviews). But most often these linkages are forged by those whose interests lie firmly in marine biology or in oceanography, not in atmospheric sciences or in the impacts of climate change. As a result, the research has not been organized in a manner conductive to tracing, from the top down, the likely shifts in fishery resources or utilization which might accompany climatic changes resulting from CO_2 buildup (e.g., would global warming result in a permanent shift in the average position of subtropical high-pressure systems and cause a spatial displacement of coastal upwelling fisheries?).

Therefore, *systematic reviews are needed of basic research on climate-fishery relationships*. Such reviews should be conducted *from the perspective of climate and climate change* in order to integrate the wide-ranging array of studies which address the linkages between climate variation and living marine resources. This kind of activity would help to assemble and organize existing knowledge and to identify potentially interesting and important areas for further research.

For forestry resources, Reifsnyder (1976), in a National Academy of Sciences review, emphasized the need for expanded research to develop greater understanding of climate-forestry linkages.

Second, refinement of existing transfer functions appears warranted in water supply and agriculture. In particular, research on *crop-yield modelling* should be expanded, not only to sharpen existing models but to develop climate transfer functions for other areas of the globe. But rather than models based solely on empirical-statistical analyses (i.e., multiple-regression models), we need analytic models that simulate the actual processes linking crop production to climate. Eventually, empirical-statistical models must be supplemented by simulation models of the determinant linkages, particularly if CO_2-induced climate changes transcend the bounds of recent experience. The prospect that CO_2-induced climate changes will involve notable alterations (whether desirable or undesirable) in global patterns of agricultural production is very likely. The importance of building an understanding or climate-yield relationships in an international context becomes increasingly urgent in light of the trend toward global interconnectiveness in food production and consumption. Currently, work in this field is being carried forth by NOAA (EDIS, 1979) and others, and is deserving of solid support.

Third, there is need for exploratory research to assess the *sensitivity of various sectors to climate change*. Research could proceed in several ways. One approach is to re-evaluate the design adequacy of water or energy systems, replacing assumptions of constant trends with assumptions of CO_2-induced trend changes. For example, how might an altered hydrologic regime that results in lower average streamflows in the Colorado River Basin affect the supplies and demands for water and energy in the semi-arid western United States where serious problems already loom on the horizon? Existing studies (by the Corps of Engineers, Bureau of Reclamation, and others) that have evaluated the design adequacy of the Colorado water management system under projections of future agricultural, industrial, and municipal demands could be re-worked in order to ascertain if (or to what degree) shortages might arise with differing assumptions about streamflow distributions in a CO_2-enriched future. Another approach would be to run existing models of various sectors under particular sets of climate conditions. For example, in a recent study the National Defense University (NDU, 1980) estimated the impacts on world grain production from hypothetical changes in climate, and then traced the societal consequences through the use of a complex global agricultural trade model. For both approaches, the objective would be to identify the tolerances of existing sectors to changes in physical parameters. It is quite possible that under certain circumstances, some sectors may prove to be rather insensitive to climate changes of the order of magnitude likely under CO_2 global warming (a conclusion reached by the NDU study). Obviously, this sort of research is tied to, and would profit from, refinement of the climate transfer functions noted above.

A fourth opportunity for research is to examine the *resiliency of sectors to potentially disruptive perturbations*. (The term "resiliency", as used here, is the ability of systems to absorb perturbation through multiple pathways, whereby new equilibria may be achieved but basic structural relationships remain unaltered [Holling, 1973]). We know relatively little about the degree of resiliency of many societal sectors or the mechanisms by which it is attained. However, we do know that some sectors of society are surprisingly resilient under conditions of stress which exceed planned design. For example, Russell, Arey, and Kates (1970) suggest that many New England water supply systems can easily accommodate 10–20% shortage through emergency conservation mechanisms. It has been shown that some western localities in the United States (e.g., Marin County, California) were able to cope with up to 40% reductions in water supply during the mid-seventies drought (Jackson, 1979). Case studies of sectoral resiliencies under perturbations such as droughts, floods, cold spells, or heat waves would greatly increase our understanding of their capacity to handle a CO_2-induced change. Such case studies should be selected globally from a range of socio-economic systems and levels of development.

2. What Are the Characteristics of Society That Determine Vulnerability to Climatic Variation?

Why is it that some nations, regions, or communities exhibit a remarkable degree of resistance to climatic perturbations, while others are subject to chronic disruption? A large number of historical case studies have touched upon this question (for example, see the recent climate and history volumes edited by Rotberg and Rabb, 1980; Wigley *et al.*, 1981; Smith and Parry, 1981.) Hardly a discipline is unrepresented. Historians, geographers, social psychologists, climatologists, anthropologists, sociologists, political scientists, ecologists, and so on, have drawn upon the diversity of their experience to develop an understanding of the bases of societal vulnerability. An underlying premise of these studies is that in order to fashion effective strategies for dealing with the societal impacts of climate variation in the future, we must understand the reasons for vulnerability.

However, wide-ranging excursions into the past to develop lessons for the future have not led to common theoretical understanding. Instead, the field has been characterized by applications of several different theoretical perspectives to the problem, with the result that a number of explanations of vulnerability – some conflicting, some overlapping – have emerged. We label these the perceptual, environmental/ecological, economic, and social structural. Because of the critical implications these explanations hold for strategy choice, we will briefly discuss them below.

2.1. Explanations of Vulnerability

From a *perceptual* perspective, the explanation for vulnerability is conceived in terms of the ways in which individuals perceive their environment and how they make decisions. The major underlying assumption is that societal patterns of vulnerability represent the aggregate of individual decisions and, therefore, that in order to understand vulnerability we must learn about the process of individual choice.

One major thrust of geographic research on natural hazards and resource management has followed this line of reasoning (Sewell and Burton, 1971; White, 1973). Much of this research was aimed at constructing and testing descriptive models of choice from empirical case studies; Simon's (1956) notion of "bounded rationality" served as one theoretical basis for perception models of resource decision-making (White, 1961). For example, early research on flood harzards sought to explore the perceptions held by flood plain residents of the nature and likelihood of flood events and of the range of possible adjustments to them (Burton *et al.*, 1968; Kates, 1962). A number of case studies of other natural hazards – e.g., droughts, earthquakes, hurricanes – ensued. Perhaps one of the best known is Saarinen's (1966) study of the perceptions of recurring droughts by Great Plains agriculturalists. In 1973, White summarized the research of that early work, indicating that such variables as managerial role, experience, personality, and decision situation appear to be significant factors in determining the ways in which choices about hazards or resources are made. Later geographic studies expanded into cross-cultural comparisons of hazard perception and, despite some methodological difficulties, provided useful insights into hazard vulnerability in an international context (Saarinen, 1974; White, 1974).

Similar approaches can be found scattered throughout the literature of other disciplines. A few deal with long-term climate change. For example, an interesting study of archaeological evidence from Greenland led McGovern (1981) to conclude that poor short-term decisions by elite elders on the eve of the Little Ice Age resulted in the disappearance of the Norse colonies. The common theme of these studies is that societal vulnerability is a function of human perception. What other explanations for vulnerability could be offered?

Environmental ecological explanations for societal vulnerability borrow theoretical frameworks of the biological sciences and apply them to human societies, as in 'cultural ecology' (e.g., see Grossman, 1977, for a comparison of approaches in anthropology and geography). The emphasis is on the interrelatedness of components within ecosystems of which man is a part. The major assumption is that the concepts which describe the processes of natural ecosystems are applicable to human systems: Vulnerability is a function of the degree to which humans violate ecological principles which maintain stability within the system. Such ecological reasoning was applied widely in explaining resource degradation during the environmental movement of the 1960s and early 1970s. The ecological lessons were recited in a multitude of social contexts: Human population growth, if unchecked, eventually exceeds carrying capacity, in turn degrading man's resource base; ecosystems simplified by human intervention lack diversity and are susceptible to disruption; etc. (e.g., see Ehrlich and Ehrlich, 1970).

The ecological perspective has been applied to a number of case studies which relate societal vulnerability to environmental variation. For example, Deevey *et al.* (1979) have attributed the decline of the Mayans to accumulated vulnerability resulting from a long-term trend in soil fertility depletion. Similarly, collapse of Mesopotamian culture has been portrayed as a consequence of salinization and siltation in irrigated agricultural systems along the Tigris-Euphrates river valley (Jacobsen, 1958). In today's agricultural systems, many researchers have warned about the dangers of eliminating ecological diversity through monoculture and pesticides (e.g., Harris, 1969; Ehrenfeld, 1972; Manners, 1974) or through narrowing of the genetic crop base (National Academy of Sciences, 1972). The consequences of violating such ecological principles have been reported in a number of case study accounts; a well-known study is the Food and Agriculture Organization (1974) report of the pest population explosion following the use of pesticides on cotton in Peru, and the return of stable populations upon the cessation of its use. In an extreme case, Norwine (1978) goes so far as to claim that "our species is more vulnerable to slight climate variations than ever before in the history of mankind." He cites the following reasons for this state of affairs: dependence upon a small number of hybrids, costly technological innovations, and large total population.

The ecological explanation for vulnerability eventually blends into a technological explanation. Specific technologies, or technological systems, are seen to affect vulnerability, usually through ecological disruption or violation of ecological principles.

Technological homogeneity, specialization, and centralization replace diversity, multiplicity, and redundancy – elements of ecosystem stability. Some studies are specifically concerned with technology assessment, as for example, Biswas (1979), who assesses the potential for Green Revolution technology to increase societal vulnerability to climatic variation. There are numerous studies of this nature. The point is that environment and technology are often put forth as the determinants of societal vulnerability to climatic fluctuations.

Probably the most theoretically well-developed paradigm within the social sciences is economics. *Economic explanations* for climatic impacts rest on a bed of neo-classical and welfare economic approaches to resources and to environmental quality. In a resource context, climate could be viewed as another factor of production. One could argue theoretically (as per Barnett and Morse, 1963) that long-term CO_2-induced climate change is analogous to a resource "scarcity" and presents no problem that could not be remedied in the market place: as a resource (e.g., precipitation and hence irrigation water) becomes scarce, prices rise, consumption drops, substitutes are found, and new technologies developed.

In the environmental quality context, anthropogenic climate problems are viewed as "market failures". CO_2 is seen as an "externality" problem – an effect (detrimental or beneficial) of production which is not borne by the producers nor reflected in market price – deriving from the exploitation of a "common property resource", the atmosphere. Because the atmosphere is not privately owned and no one can be excluded from its use, its service as a waste receptacle for CO_2 or other by-products conflicts with its service as an environmental quality control mechanism (Kneese *et al.*, 1970). Preferred strategies to correct the problems, then, often involve "internalizing" the externalities – pollution charges, subsidies, taxation, etc. We are slowing moving toward an understanding of the atmosphere as a resource to be managed by economic tools (Ausubel, 1980).

For both resources and environment, the economic approach is to allocate goods and services to their most efficient use, as valued by individuals within society. Thus, there is an underlying belief in optimality, equilibrium, and order in society. Theoretically, economists would argue that there is an optimal mix of climate change and production pursuits: that is, where the marginal costs of correcting CO_2 emissions equal the marginal benefits of damages prevented. In short, vulnerability is a form of cost deriving from market failure and is theoretically correctable to an optimal level which maximizes social welfare.

Finally, the *social-structural* explanation turns to the organization of society for understanding the bases of vulnerability. For example, a summary of one collaborative set of global case studies of natural hazards (which, incidentally, also contains strong perceptual and ecological components) discusses differential vulnerabilities of traditional, developed, and developing societies (Burton *et al.*, 1978). The authors observed that folk or pre-industrial societies are characterized by high absorptive capacity and resiliency, based on a large number of adaptive adjustments to environmental perturbations. On the other hand, the industrialized societies possess a large array of technological mechanisms to manage the environment (e.g., dams, irrigation schemes, warning systems) and loss sharing mechanisms (e.g., relief programs). It is argued that the most vulnerable societies are those in which social structures are undergoing transition from traditional to industrialized. It is the transitional, developing society which has dismantled traditional mechanisms for coping with perturbations like climatic fluctuations, but has not yet incorporated the technological prowess of the developed world.

In contrast, another set of case studies views vulnerability more as a problem of underdevelopment: the most vulnerable countries are often the poorest, least developed. Explanation lies in the process of imposing one political economic system upon another, particularly the forcing of capitalist market systems (which emphasize profit maximization) upon traditional systems (which tend to emphasize risk minimization) (see O'Keefe and Wisner, 1975). One result of this process is to reduce drastically the range of adjustive mechanisms. This effect was described by Wisner (1978), for example, who claims that in Eastern Kenya the imposition of modern agricultural systems on traditional systems leaves only migration to urban areas for wage labor as a strategy against drought. K. Johnson (1976) argues convincingly in a case study of the Otomi of Mexico that ethnoscientific methods are forgotten in the process of adopting new technologies, which leads to "de-skilling" of the populace and greater vulnerability to drought. Various studies purport to show that the same process promotes a greater reliance upon marginal lands for subsistence while better lands are utilized for cash crops, thus intensifying land deterioration and the ability to withstand climatic fluctuations – a "marginalization" process (Waddell, 1975; Regan, 1980; Hewitt, 1980; Spitz, 1980).

In contrast to the developing world, Worster (1979) provides a socio-structural explanation for vulnerability in a highly developed nation in his examination of the 1930s Dust Bowl in the Great Plains of the United States. He argues that the reason for one of the nation's greatest environmental crises was not the environment itself nor even the agriculturalists' misperception of it. Rather the explanation lay within the cultural-economic system with its emphasis on economic efficiency. In the 1930s, agriculturalists responded exactly as the system demanded. Problems of drought vulnerability were, and are, chronic, and reflect the structure of our socio-economic system. Thus, the future may see a repeat of the tragedy.

2.2. *The Explanations in Perspective*

To a large extent the initial theoretical perspective – whether perceptual, environmental/ecological, economic, socio-structural, or some variant thereof – that the researcher brings to bear on the problem predetermines conclusions as to the reasons for societal vulnerability to climate variation. This is entirely understandable (if not somewhat counter to the beliefs we often express about the objectivity of science). Furthermore, a corollary to this observation raises some very important issues: namely, that the particular explanation one holds for societal vulnerability largely predetermines the *strategies* that are offered for coping with problems like climatic fluctuation.

For example, if one believes that drought hazard in the U.S. Great Plains is largely a problem of misperception of environment and poor decisions by agriculturalists, then the appropriate strategy is to provide better information – as through county extention agents – in order to elicit more appropriate behavior (a "cognitive fix" in the words of Heberlein, 1973). On the other hand, if one holds an environmental/ecological view (either explicitly or implicitly) that drought is a problem of moisture stress, one is more prone to invest in cloud seeding, drought resistant crops, or irrigation schemes (a "technological fix"). The economic explanation might lead to price supports or land set-aside programs. Finally, if Great Plains droughts are seen as symptomatic of broader inadequacies in the socio-economic structure of societal systems, then perhaps strategies such as legislating land-use regulation are put forth (a "structural" fix).

The point to be emphasized here is the critical importance of gaining a sound theoretical understanding of the reasons for societal vulnerability. For it is this understanding which strongly influences the pattern of strategies perceived as efficacious. The thread of theoretical development vis-à-vis climate-society interaction initiated at the SCOPE Workshop on the Climate/Society Interface (SCOPE, 1978), and carried on by Kates (1980a) and Timmerman (1981), must be pursued further. If society is to respond effectively to CO_2-induced climate change in the future, knowledge of the processes which promote vulnerability is vital.

The above overview of the major themes in explanations for societal vulnerability to environmental variation is intended to convey the general thrusts of research and the levels of understanding developing therefrom. We reach four conclusions. First, there is *no common conceptual framework* for the study of human effects of climate change, but rather a number of loosely connected (sometimes overlapping, often conflicting) perspectives brought to bear on the questions of societal vulnerability. Certainly, all of the perspectives have some merit; the problem is to combine them sensibly. Second, as a consequence, the *methodological approaches vary greatly* from one study to the next, often within one perspective. This presents serious difficulties in comparability of results and in theory development. Third, *researchers often demonstrate limited awareness of other case study work* which bears directly on their own work. Fourth, the above problems are exacerbated by the *lack of common research questions*. If case studies such as the ones described above are to build upon one another to contribute to theoretical understanding, then at a minimum there must be a sense of a shared research focus. Otherwise, case studies tend to remain idiosyncratic and unreplicable. The research community can help in four ways.

2.3. Research Opportunities

First, there is immediate need for a *critical review of conceptual and methodological approaches to climate impact assessment*. This is an urgent task, since we appear to be on the brink of a whole new round of case studies, stimulated in part by the formulation of the climate impact phase of the World Climate Programme. The Scientific Committee on Problems of the Environment (SCOPE) for one, has recently sponsored a research project of this nature. The critical review promises to generate new ideas about concepts and methods, and to enlarge and strengthen the international research community interested in such work (Kates, 1980a, and personal communication). This sort of research deserves solid support and encouragement as a high-priority topic.

The SCOPE project also takes an initial step toward meeting a second need, that is, a *networking of research and researchers* concerned with the societal effects of climate change. As used here, the term "networking" connotes a blend of coordination and feedback among researchers, and might encompass the following kinds of activities:

(1) The establishment of a coordinating body, or clearinghouse, whose function it is to keep researchers informed of other research endeavors and research findings (as, for example, through a climate and society newsletter).

(2) A series of workshops in which researchers with common research interests could exchange ideas, share findings, and sharpen methods. Perhaps the greatest benefit could be the formulation of, and agreement upon, a set of priority research questions which could then be pursued systematically by those involved.

We are presently hampered in drawing sound conclusions about the societal effects of climate change because of the disparate nature of climate-society research, especially case study work. A networking of researchers and research activity would enhance greatly the usefulness of past and prospective work in this field, at relatively low cost.

Third, the research community could profit considerably from the assemblage of a systematic *inventory of potential case studies*. More often than not, the selection of particular case studies has been rather haphazard, frequently depending upon researchers' peripheral interests, expertise, or ongoing research. A deliberate selection of case studies, made possible by a thorough inventory of possibilities, could go far in fashioning a set of studies which are comparable and replicable.

There are at least two ways to go about constructing a useful inventory. One is to select cases on the basis of objective identification of periods of climate change in a region, in terms of both long-term climate change or short-term fluctuations.[4] These periods could then be differentiated according to effect: those with apparent negative societal impact and those with little, or positive, effects. Much of the research to date dwells only on the adversities of climate change; yet, examining both positive and negative effects is necessary in order to gain full understanding of societal vulnerability to environmental change.

A second way of constructing an inventory of potential case studies involves the identification of historical situations *analogous* to changes in climate. One such situation is the case in which people, for one reason or another, misperceive the climate. For example, the agricultural settlers who moved into the semi-arid Mallee area of South Australia around 1900 faced both climatic adversities and opportunities for which they were largely unaware and unprepared; over time, experience with their new environment led to the development of adaptive capabilities necessary to fashion a livelihood (Heathcote, 1974). For all intents and purposes, the problems and opportunities encountered often parallel those presented by actual climate change. History is full of many such examples.

Moreover, the number of possible analogues increases substantially if the CO_2-induced climate change is considered synonymous to many problems in resource management. In this light, shifts in resource supply or demand may be equivalent to alterations in climate. For instance, unexpected increases in demands for hydroelectric power over time could be considered analogous to climate-related changes in streamflow. An upsurge in crop yields resulting from technological change (such as occurred in the United States since the 1940s) and its side effects (national overproduction, adequate global food supplies, depressed grain prices, land retirement, etc.) may be analogous to a regional improvement in climate for crop production. In many ways tightening of oil imports is analogous to exceptionally cold winters, and so on. Such problems in resource supply and demand can shed light on the problems of climate change.

The last research opportunity builds directly on the previous two: we need *actual case study investigations of the effects of climatic change (or analogous events) on society*. But unlike our present stock of case studies, it is critical that similar research questions are asked of each (realizing, of course, that new questions will emerge in the process), that the methods employed allow for comparability of results, and that the case studies themselves be parsimoniously and systematically selected. One useful way of selecting case studies is suggested by Table I, below:

TABLE I: Schema for selection of case studies of climatic changes or fluctuation of similar magnitude

	Historical period	
Cultural context	T_1	T_2
Case study area # 1	Major societal effects	Minor societal effects
Case study area # 2	Minor societal effects	Major societal effects

Following this schema, one is led to ask two questions: (1) Over time, why do we find differences in climate impacts within a society, given similar climatic changes? If, say, negative impacts increased, can we identify the processes of change within a society which explain the shifts in vulnerability? (2) At any given time, why do we observe differences in societal effects of climate variation from one cultural context to another? Can we identify the particular characteristics of societies which explain the differential vulnerabilities?

For carefully selected study areas, a combination of temporal, cross-sectional, spatial, and cross-cultural approaches – cast within common conceptual and methodological frameworks – promises to lead to more generalizable results than heretofore available. One would hope that the case study investigations could build upon one another in such a way as to contribute to the development of sound theories of climate and society interaction.

3. What Are the *Pathways and Linkages* Within Social Systems Through Which the Effects of Climatic Variations Are Transmitted?

The human consequences of climatic variation depend largely on the ways in which direct sectoral effects filter through the socio-economic-political fabric of society. As described under questions # 1 and # 2 above, societal sectors like energy, transportation, fisheries, forestry, agriculture, and water supply are subject to disruption, as determined jointly by the characteristics of society and the nature of climatic change. However, reduced or increased crop yields, altered water supply, or modified hydroelectric output are not the ultimate concerns; rather, it is the degree to which people are actually affected, as expressed in changes in income, human health, community stability, settlement patterns, or the like. It is the pathways taken by climatic variations through space and societal organizations which largely determine the patterns of human consequences.

For example, Figure 2 was constructed for the purpose of providing a framework for tracing the impacts of drought occurrence in the U.S. Great Plains (Warrick and Bowden, 1981). The diagram depicts a variety of hypothetical pathways that drought impacts could take, spanning spatial scales (from local to global) and systems affected (from agricultural to social). From the lower left to the upper right quadrants in Figure 2, the drift of impact is from the more physical, direct, and immediate to the more social, indirect, and remote. The initial perturbation originates as meteorological (or hydrological) drought, that is, a decline in available moisture. This phenomenon becomes agricultural drought when yields fall below some perceived threshold, and it translates into drought impact if and when the stress is detected in the economic and social sectors. The degree to which the initial climatic event is transformed into stress is influenced by several factors, including individual and social perceptions of drought, market prices, government policy, and farm stability.

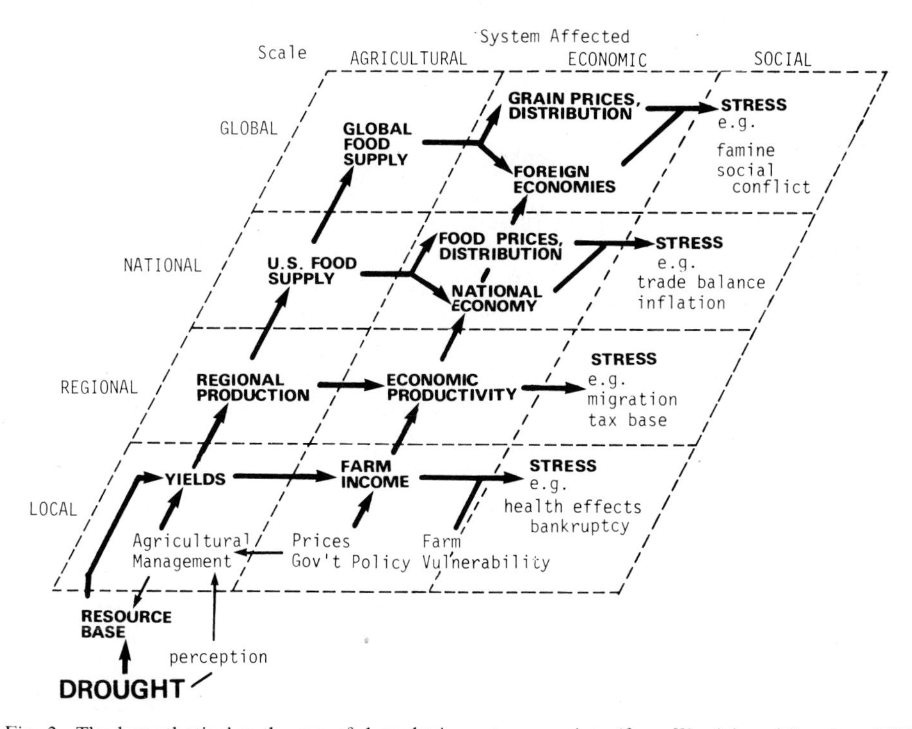

Fig. 2. The hypothetical pathways of drought impacts on society (from Warrick and Bowden, 1981).

Let us make two additional comments about the drought example. First, while Figure 2 displays a hypothetical set of pathways and linkages, the actual routes of drought impact can vary over time. For instance, in the Great Plains during the 1930s, drought impacts took an essentially horizontal pathway to local stress (Warrick *et al.*, 1975). However, there is concern today that future droughts may follow a more vertical pathway. This pathway leads from yield declines to lowered national and global food reserves, with resultant increases in food prices and severe health effects among the poorest nations. Whether such a pathway would actually evolve following the onset of a rare, major drought in the Great Plains (or elsewhere) is a matter of debate.

The second point is that as pathways of drought shift, so does the range of strategies for drought management. Theoretically, opportunities for preventing drought impacts exist at any box or arrow in Figure 2. For instance, if future drought assumed a global pathway, then strategies for preventing stress from Great Plains droughts could include maintenance of grain reserves (Schneider and Mesirow, 1976) or assistance to developing countries for development of food self-sufficiency. However, the usual response is to pursue conventional drought strategies (such as irrigation expansion or conservation practices) which may be more appropriate to drought impacts of the past than to those of the present or future.

The above example serves to illustrate the following point: in order to anticipate the full range of consequences from, and possible responses to, climatic variation, it is vital to understand the pathways and linkages by which such occurrences are transmitted through society. Furthermore, given the dynamic nature of these pathways, one needs to obtain some understanding of how they are shifting over time. This applies to the spectrum of societal sectors which may be affected by CO_2-induced climate changes.

3.1. Models for Tracing the Effects

There exists a broad range of atmospheric, hydrologic, ecological, agricultural, urban, regional, and even global models for examining environmental problems (Holcomb Research Institute, 1976). Many of these have potential for tracing climatic variations through socio-economic systems. For example, one well-developed approach, designed to calculate the ripple effects of changes in regional or national economies, is economic input-output analysis. Input-output models are, essentially, models of pathways and linkages within the economic sector, as between extractive industry, agriculture, manufacturing, etc. Generally, they are used to estimate the indirect economic benefits of increased production or services within a geographic area. In a few instances, however, input-output models have been employed to estimate the pattern of indirect losses from economically disruptive events. For instance, Cochrane *et al.* (1974) used input-output analysis to estimate the potential economic losses from a hypothetical recurrence of the 1906 earthquake in San Francisco. Conceivably, similar analyses could be conducted of the regional indirect economic benefits and losses from CO_2-induced climate change.

Although input-output models are probably among the most widely used models available, they comprise only a limited set of relationships – all within the economic sector. Components such as public policy or climate differences are exogenous. As such, they can only treat selected portions of the pathways and linkages potentially implicated in the flow of effects from climatic change. This is one fundmental drawback of the entire arsenal of models currently at our disposal: they tend to be directed toward particular disciplinary interests, the modelers themselves having difficulty crossing interdisciplinary boundaries (Holcomb Research Institute, 1976).

Thus, referring back to the example of drought pathways laid out in Figure 2, there exists a plethora of models to fill virtually every individual "box" in the diagram: national economic models, regional input-output models, crop-yield models, local agricultural operations models, etc. A few purport to model entire "rows" of boxes (as, for example, the global agricultural trade models). But by and large, the linkages – or "arrows" – in Figure 2 remain unforged. Clearly, these need specification if the full range of consequences of climate change are to be explored. Several important lines of research are suggested.

3.2. Research Opportunities

A very promising research endeavor involves the *linking of major models*. As discussed above, we have models of climate and environmental systems, of agriculture and other resource systems, and of socio-economic systems, at a variety of scales. But we do not have them interconnected in a manner which allows us to estimate easily the full effects of climate fluctuations or to explore the results of alternative policy options.

To illustrate, for the agriculture and food sector, a number of global models have been developed during the last decade which simulate grain production, trade flows, and consumption at the world scale: for example, the Model of International Relation in Agriculture (MOIRA; Linnemann *et al.*, 1979) or the interactive global model of population, food, and policy constructed at the University of Southern California (Enzer *et al.*, 1978). However, neither MOIRA nor the USC model accommodates a direct link to climate and its effects on yields. One possibility is to construct a model that does incorporate a climate-yield component; this approach is currently being pursued at the University of California at Los Angeles in which analytic yield models are being developed for this very purpose (Terjung, 1981). Another approach is simply to use existing crop-yield models and global food models in combination, whereby the results of one serve as input to the other. The latter approach has been used, for instance, in determining the international consequences of a global climate change (National Defense University, 1980) and in a preliminary study aimed at tracing the global impacts of a recurrence of a severe drought in the Great Plains of the United States (Warrick and Kates, 1981). The linking of climate and environmental models to several key sectors (like agriculture or energy) at several scales of analysis (regional, national, or global) promises a large payoff in exploring the societal effects of climate change.

A second opportunity for research involves *tracing the actual pathways of societal effects of climate variation through empirical case studies*. In part, this work is needed in order to build and refine the models described above and to "keep them honest". One important aspect of such research would be identifying how linkages and pathways have shifted over time. The objective of such research would be to gain understanding of the trends in societal sensitivity to climate variation. One might ask, for example, how the economic and social effects of droughts on grain production in the Soviet Union have changed over time. As argued above, this information is pivotal in attempting to estimate how the full range of societal impacts might unfold given the occurrence of major climate change in the future.

4. What Is the *Level and Distribution of Effects* from Climatic Change?

The effects, both positive and negative, of climatic variation can be viewed as one major informational component to ameliorative strategy formulation, as we depicted in Figure 1. As society faces policy decisions about CO_2, at least three major questions arise: What is the total *level* of benefits or losses? How are the effects *distributed* in time, space, and society? What are the *trends*?

4.1. Level of Effects

There are two very good reasons for compiling information on the level of effects. First, it is an essential ingredient for *evaluation* of alternative strategies. Estimates of aggregate positive and negative impacts provide baseline information by which the benefits of strategies or policies can be assessed. For example, one measure of the agricultural benefits of weather modification for hail suppression is the dollar value of damages prevented. Conventional evaluative methods like benefit-cost analysis or cost-effectiveness analysis demand such data (in commensurable dollar values) for purposes of comparing the costs of strategy implementation to the expected benefits accruing to society.

Second, information on the level of total effects is important in placing societal problems into perspective: How does CO_2 fit among the list of *priorities* of problems to be addressed? As Kenneth Boulding (1980) suggests, society in general, and science in particular, often expends a good deal of effort on seemingly less pressing problems, while the more critical problems (like preventing nuclear war) are neglected. How the CO_2 problem matches up to other societal problems is a fundamental issue at both national and global scales.

With respect to negative impacts arising from short-term climatic episodes, such as droughts or other extreme events, the literature on natural hazards contains fairly comprehensive assessments. Nationally and globally, society does a fair job of estimating losses from natural hazards (White and Haas, 1975; White, 1974), at least in terms of direct dollar losses and lives lost (Table II). We do a worse job of calculating indirect

TABLE II: Selected Estimates of Natural Hazard Losses (from Kates 1980b)

					Losses and costs per capita at risk			
Harzard	Country	Total pop.	Pop. of risk	Annual death rate/ million at risk	Damages losses	Costs of loss reduction	Total costs	Total costs at % of GNP
Drought	Tanzania	13	12	40	$ 0.70	$ 0.80	$ 1.50	1.84
	Australia	13	1	0	24.00	19.00	43.00	0.10
Floods	Sri Lanka	13	3	5	13.40	1.60	15.00	2.13
	United States	207	25	2	40.00	8.00	48.00	0.11
Tropical cyclone	Bangladesh	72	10	3000	3.00	0.40	3.40	0.73
	United States	207	30	2	13.30	1.20	14.50	0.40

economic losses, environmental effects, and social disruption. Some of these other impacts may be quite significant, but often we remain unsure of their correct measures or indicators. Displacement of families, abandonment of livelihood, community disruption, and malnutrition are difficult effects to place on a common yardstick. Yet, these are the appropriate descriptions of societal impact in most parts of the world.

For this reason, we often face a rather difficult task in tabulating or monitoring the impacts of climatic fluctuations, particularly in areas of the world in which slow-onset, climate-related processes exact an insidious, debilitating toll on human populations. This is the case with the problem of desertification in the Sahel, for example. In this region, the ability to recognize clearly the "early-warning" signals through monitoring of human well-being and livelihood systems would help in combatting the effects of desertification. Development of sound indicators of human impact would be a major contribution toward this goal.

4.2. Distribution of Effects

Who carries the burden of loss from climatic fluctuations? And who benefits? The question can be asked about groups within nations or about nations throughout the global system. If we had better understanding of the social, temporal, and spatial distributions of effects, society would be in a better position to target programs and policies to the most vulnerable parties and to address explicitly the inequities promoted by our social systems. In addition, a solid base of data on social distribution of effects would likely shed light on the reasons for vulnerability. What do we already know?

For a few empirical case studies we have some well-documented research of the *social* distribution of effects, especially with regard to climate episodes like drought or other extreme events. For example, it has been argued convincingly that the major share of impacts of the 1930s Dust Bowl in the U.S. Great Plains was suffered by the poorest – the tenant farmers, sharecroppers, and migrant laborers (Great Plains Committee, 1937). Some of the major government programs designed to relieve drought stress actually exacerbated the impacts on the hardest hit (Stein, 1973). In contrast, many of the well-to-do farmers benefited from the drought in the long run by being able to clear their debts for as little as ten cents on the dollar with hard-pressed banks and by acquiring bankrupt farms for little or nothing (Worster, 1979; Fossey, 1977). Of course, the fact that someone actually benefited is not surprising; the bulk of climate-society interaction benefits mankind. The question of how the benefits and costs are distributed is the issue.

Similar studies at, say, a national level are few. Dacy and Kunreuther (1969) extend several case studies into an assessment of federal relief policies for disasters. One study by Cochrane (1975) of the distributive efforts of natural hazards in the United States concluded that " . . . lower income groups consistently bear a disproportionate share of the losses . . . " incurred from natural hazards. These groups also receive the least disaster relief, have less insurance, and are more prone to damages. The recent, extended heat wave over the south-central United States exhibited clear-cut differences in impacts vis-à-vis social class. Health effects were severest among the poor and elderly. Monetary losses (i.e., added cooling costs) were borne by middle- and upper-level income groups (Center for Environmental Assessment Services, 1980).

These studies suggest that there may be fundamental social difference in benefits and loss-sharing within societies. Such studies, however, encounter tough sledding with respect to data availability. Presently, in order to associate effects with income groups in the United States, for instance, one is forced to delve into the files of the Small Business Administration, Red Cross, insurance organizations, etc. As a matter of course, data on impacts from hazards in the United States are not collected systematically with respect to income class (or in any way, for that matter). We expect the situation to be no better – probably worse – in the international context. This lack of data on social distribution of effects could be a major constraint to effective strategy formulation with respect to CO_2-induced climate change.

The *temporal* distribution of effects is another subject about which we have relatively little understanding. In the United States, it is often assumed that lingering secondary impacts of natural disasters are severe enough to warrant post-disaster programs of recovery. However, two separate studies of long-term disaster impacts in the United States, one by Friesema *et al.* (1979) and the other by Wright *et al.* (1979), show that such impacts are minimal. The latter study, at census tract level, explored both economic (e.g., retail sales) and social (e.g., population change, income level) indicators and found no evidence of persistent post-disaster disruption. On the other hand, in other social settings, one finds examples of enduring effects long after the initial disaster impact. For instance, the social and cultural impacts of the eruption of Paricutin Volcano in Mexico persisted for decades (Rees, 1970). Why the difference? In which circumstances and socio-cultural settings should we anticipate chronic effects of disruptive climatic (or other environmental) perturbations? And how do we make intertemporal comparisons of impacts (at what social discount rate) for purposes of evaluating the benefits and costs of climate changes in the future (d'Arge, 1979)? These questions relate directly to CO_2 strategy formulation but remain largely unanswered.

A final issue concerns *spatial* distribution of impacts. The most pressing question lies at a global scale: Will CO_2-induced climate change exacerbate the differences between rich nations and poor nations, between the high latitudes and equatorial regions? Unfortunately, this issue is often obscured by focusing solely on the notions of average regional warming, drying, etc. To repeat a point made earlier, it is probably not only the slow average changes in central tendencies that matter to society, but also the changes in frequency of discrete atmospheric events – events which are occurring today. It is relevant to the CO_2 issue to ask: Do major recurring events like droughts bring consistently greater impacts to struggling nations than to developed nations, and further divide the world into "haves" and "have-nots"? If, as Garcia and Escudero (1980) argue, the explanation for these impacts lies in the political and economic relationships of nations instead of in the physical systems, the room for remedying the inequities of spatial distribution of impacts is large indeed. So while it is true that CO_2-induced climate change may slowly alter the regional frequencies of such events in the future, the basic processes by which the benefits and costs of climate fluctuations are distributed *are observable today*. We do not have to wait for better information from atmospheric scientists on the exact nature of regional climate shifts. The global laboratory is open for research now. The opportunity to understand the links between climatic events and spatial inequities should be grasped enthusiastically – CO_2-induced climate change or not – if we wish to avoid the pessimistic scenario painted by Heilbroner

(1974) of a self-destructing, conflict-ridden world divided and aligned over resources and wealth.

4.3. Research Opportunities

Research can contribute in four ways: First, there is a need to develop a comprehensive set of *indicators of societal impact* from climate-related events. Conventional measures of dollar damages, while useful, are inappropriate for many purposes and in many socio-cultural contexts. Sensitive indicators which detect alterations in economic, cultural, and human well-being would allow careful monitoring of societies in order to identify signs of societal stress from climate fluctuations – an early-warning capability.

Second, a need exists for the development of *methodological frameworks for systematic collection of data on the social distribution of climate effects* (as, for example, by income class). Once applied, knowledge about who shoulders the burdens and who reaps the benefits would provide a solid data base from which to address the issue of inequities – of the processes by which they are maintained and of the strategies or policies which might alleviate them. Consistent, comparable information is not currently available. The need exists at both national and international levels.

Third, a very worthwhile research effort is the *investigation of the global distribution of effects arising from climatic fluctuations*. Do climate-related occurrences such as droughts or "El Niños" act to increase further the discrepancies between rich and poor nations, given the present global economic system? CO_2-induced climate change may change the frequencies of such events in the future, but many of the events themselves – the nature of their effects and their implications for global equity – are available for study now.

Despite all research efforts expended on the CO_2 problem, a large amount of uncertainty will remain. There is a good chance that atmospheric scientists may never be able to predict precisely the average climate changes or the changed frequencies of climatic events in key regions of the world. Moreover, it is almost certain that social scientists will fail to predict accurately the nature of social change fifty years hence. Perhaps the alternative to prediction is to *assess the broad trends in climate-society interaction* in order to ascertain the likely direction of effects. For example, two case studies of drought impacts at a time scale of a century, one in the Great Plains and one in the Sahel-Sudanic zone of West Africa, conclude that the local impacts of recurring droughts of similar magnitudes have lessened over time (Bowden *et al.*, 1981). Historical case studies of the trends, using time-series or cross-sectional data on socio-economic effects, could provide valuable clues as to the societal situations in which future climatic alterations may be handled well and those in which the potential for catastrophe may be building.

5. What is the Range of *Mechanisms* by Which Societies Adjust to Climatic Fluctuations? And What Are Their Costs and Effects?

As society faces the prospect of a possible climate change from CO_2 or other causes, the issue of what mechanisms or strategies could be employed to cope with the effects has more than just academic interest. Decision-makers will require information on the full range of possible mechanisms and their effects. In part, researchers can help provide this knowledge by drawing upon investigations already conducted on human response to natural hazards and to long-term climate change. Opportunities for new research endeavors also exist, particularly in developing an understanding of the effectiveness of adjustive mechanisms and of the social costs associated with them, as described below.

5.1. The Range of Adjustments

The resiliency and adaptability of a social system to environmental fluctuation or change is linked inextricably to its repertoire of adaptive mechanisms, or adjustments, which can be called upon to buffer society against adverse impacts (or to take advantage of new opportunities). To date, a substantial amount of interdisciplinary research has been devoted to identifying and understanding human adjustments to extreme events in nature. Much less attention has been directed toward learning about adaptation to slow environmental change.

One central goal of research on extreme events in nature is to assess how "man adjusts to risk and uncertainty in natural systems" (White, 1973). Empirical studies have focused largely on identifying the range of adjustments theoretically possible, delineating the particular adjustments actually adopted by individuals and groups, and understanding the processes of adjustment choice and adoption. For example, one collaborative research project sought explicitly to identify adjustments to natural hazards in an international context (White, 1974). Twenty-four hazards were investigated, twenty-one of which were atmospheric in origin. Collective and individual adjustments adopted at local, national, and global levels were described. Similarly, a study of 15 natural hazards in the United States provides an exhaustive list of adjustments (White and Haas, 1975). These studies, along with anthropological (see Torry, 1979, for review) and sociological (Dynes, 1970) studies, provide a wealth of information on discrete mechanisms for adjusting to environmental fluctuations.

A body of literature on human impacts of, and response to, short-term climate fluctuations is accumulating rapidly, as evidenced by a recent extensive bibliography compiled by Rabb (1983) for a DOE-AAAS panel on CO_2 effects. The studies of harsh climate conditions of 1816–19 in Europe (Post, 1977) or the western U.S. drought in the 1970s (Jackson, 1979) are illustrative. These studies deal with environmental changes which have direct applicability to the CO_2 issue. The literature concerning human adaptation to slow environmental change is scanty in comparison. One recent example is a study by McGovern (1981), who examined the relationship between a cooling of Greenland's climate over several centuries and adjustment decisions by Norse Greenlanders. Another example is the work of Baerreis and Bryson (1968) who studied the response of Mill Greek Indians to a steady desiccation of the mid-west prairies of North America in the twelfth century A.D. (also, see Bryson, 1975). But, on the whole, such studies are few in number.

It is unclear at this time whether enough studies of hazards or climate change have been conducted to provide an exhaustive inventory of potential adjustments. But it becomes increasingly clear that we have enough of a sample to offer classifications of adjustments which could serve as common frameworks for systematically assessing the range of possible adjustments to CO_2-induced climate change. For example, Table III draws upon the classificatory scheme offered by Burton *et al.* (1978) and compares human adjustments to a spectrum of climate variations, culled from four studies.

TABLE III: Adjustments to climate variation in four case studies

Type of adjustment	Climate variability		Climate change	
	Drought (U.S.: Warrick *et al.*, 1975)	Frost (U.S.: Huszar, 1975)	Regional Cooling (Norse Greenland, c. 1300: McGovern 1980)	Regional Drying (Central North America, c. 1100: Baerreis and Bryson, 1968)
Modify the physical event	Cloud seeding Irrigation Water storage Evapotranspiration suppressants	Heaters Wind machines Irrigation Artificial clouds		
Modify the loss potential	Resistant crop varieties Diversification	Change planting and harvesting times	Improve sealing techniques Storage of fodder and food Decrease 'cash hunt' for trade Reduce reliance on domesticates	Change diet (hunt)
Change location	Land use regulation Abandon sensitive areas	Change topographic setting	Alter grazing pressure Shift location to the coast	Abandon maize cropping Abandon sensitive areas
Accept or share losses	Accept loss Crop insurance Relief and rehabilitation	Accept loss Crop insurance	Reduce expenditures on ceremonial structures	

Finally, while we have an abundance of studies which identify such adjustments, we know relatively little about the hidden social costs associated with them. Although our society performs an adequate job of tabulating the direct, economic costs of adjustments, there is a category of insidious social costs which remain largely unspecified and unmeasured. Yet these may be of critical importance. For example, the apparent lessening of drought impact on the U.S. Great Plains has occurred with the advent of high technology, larger farms, and fewer farmers – along with disintegration of Plains communities, dislocated families, and altered rural life-style (Warrick, 1980). As CO_2 strategy decisions are faced in the future, the issue of whether similar sorts of social costs could accompany particular adjustive mechanisms may arise.

Two opportunities for research are suggested by the above discussion.

5.2. Research Opportunities

First, existing research on natural hazards and societal effects of climate change should be canvassed extensively to construct a comprehensive *roster of adjustive mechanisms*. This compilation of mechanisms would benefit from the application of a common classificatory framework appropriate to anticipated CO_2-induced climate impact problems, as noted above. Although some additional new case study work could be carried out, it is felt that such a roster of mechanisms could be mined from the backlog of completed case studies.

Second, *studies of the long-term social costs of adjustment strategies* may allow society to anticipate the possible undesirable side effects associated with strategies contemplated for CO_2-induced climate change. Such studies could focus on climate fluctuations which are potentially related to the CO_2 issue, like droughts, floods, frosts, cooling, or warming. The value of this information would lie in avoiding situations in which unanticipated insidious costs eventually outweigh the social benefits of adjustment strategies.

6. How Is *Information* (Scientific or Otherwise) on Climate Change *Perceived, Interpreted, Valued, and Channeled* Into Strategy Evaluation?

Information on climate change, societal effects, and adjustive mechanisms gets filtered in a variety of ways into strategy and policy formulations. Scientific information on CO_2 and its effects will ultimately leave the hands of the scientific community and will be digested by the media, decision-makers, and the public at large. Unfortunately, we have little understanding of how that information will be interpreted and of the resulting social and economic consequences of its dissemination. The ways in which that information is incorporated into the decision process has enormous implications for eventual strategy or policy choice regarding CO_2-induced climate change.

Within this context, two specific problems arise. First, there is the danger that the information on CO_2 (which, from the societal viewpoint, may be seen as a prediction or forecast) might set in motion a set of behaviors that will lead to unexpected, adverse impacts on society. Clearly, this problem is not specific to CO_2. Within the field of resource management in general, the complications arising from expert information, lay interpretation, and public response are well documented (Sewell and Burton, 1971; White, 1966). In many instances, the development of scientific information progresses at a rate which exceeds society's ability to make best use of it. For example, while enormous efforts have been directed at developing a predictive capability for major U.S. droughts, we have scant knowledge of how the forecast could, or would, be ultilized by agriculture and industry (Warrick *et al.*, 1975; Yevjevich *et al.*, 1978). Similarly, if we had prior understanding of the likely perceptions of key decision-makers to state-of-the-art CO_2 information, the scientific community and the political system would be in a much better position to fashion the context, mode of presentation, and timing of such information in order to elicit a more desirable response.

A second problem is that scientific research on climate change, societal effects, and mechanisms for effective response proceeds with little feedback from user groups. It is possible that as atmospheric scientists produce information on midlatitude average temperature changes, for example, the users – policy decision-makers, farmers, water supply managers, etc. – may desire information on other parameters, such as rate of onset, likelihood of catastrophic events, frequency and duration of certain events, and so on. The discrepancy between what the scientific community produces and what the users need could be large indeed.

6.1. Research Opportunities

Two research opportunities flow from the above discussion. The first is an *investigation of the socio-economic consequences of dissemination of scientific information on CO_2*. This assumes that increasing knowledge of the climate effects of CO_2 enrichment will be available at some future date, and that the information will be regarded publicly as a forecast or prediction – whether intended so or not. What will be the likely responses of key decision-makers to the information? Within varying social and economic contexts, what consequences (both beneficial and detrimental) might arise from those response patterns? The most appropriate scale at which to address these questions is international.

A second research opportunity is to *assess user needs with respect to information concerning CO_2*. Without getting critical feedback from those who will eventually formulate and adopt strategies and policies, the research community runs the danger of expending tremendous effort on producing inappropriate information. This research endeavor, along with the first, could go far in assuring that the best use is made of the scientific community's role in societal adjustment to a possible CO_2-induced climate change.

7. What Are the *Conditions of Choice* Which Guide Societal Response to Climate Change?

Despite input from scientific investigators and careful scrutiny of all alternative strategies by decision-makers, the eventual responses to CO_2-climate change will be guided at all levels by a complex array of economic, political, cultural, and legal factors. Together they constitute what we call here the "conditions of choice" and comprise the context of decision-making which encourages or inhibits particular modes of responses to environmental or resource problems. Such constraints on decisions occur at all levels of social organization.

For example, for decades American communities facing choices about response to flood hazard were enticed into choosing from a narrow range of engineering works (dams, levees, channelization) in lieu of alternative strategies like flood plain regulation or warning systems (White, 1969). In part, this was due to bureaucratic bias and to federal cost-sharing policies. At the national level, efforts to institute effective national land use regulations in geologic hazard areas are severely constrained by cultural values about the rights of private land ownership and by legal entanglements (Baker and McPhee, 1975). Internationally, agreements on strategies for preventing depletion of important ocean fisheries are hampered by the common-property nature of the resource and by the lack of international institutional arrangements (Christy and Scott, 1965; McKernan, 1972). The 1972 Stockholm Conference on the Environment discovered that agreement on cooperative strategies for handling global environmental problems is stifled by differences in priorities of national aims, as between economic development and environmental protection.

At all levels, the conditions of choice often whittle away at a theoretically large number of alternative strategies, leaving pitifully few perceived practicable choices. It would be overly sanguine to expect that research which simply delimits the range of alternatives and their consequences will automatically lead to optimal choice. Experience dictates otherwise. Understanding of the ways in which decisions are guided by the conditions of choice is necessary in order to anticipate how societies are apt to select among possible strategies for handling the CO_2-induced climate change.

One source of major influences on choices is found in national policy – in broad national response modes and the priorities of national aims. For instance, global comparison of natural hazards management shows that national *response modes* fall largely into four categories: provision of relief, control of the physical event, comprehensive damage reduction, and multi-hazard management (Burton *et al.*, 1978). In the past, the United States relied heavily on the first two, then switched to the third, and now is taking tentative steps toward the fourth. Other nations display different patterns. Comparative cross-cultural analyses can discover the relative advantages and disadvantages of particular hazard response modes within differing societal contexts. This kind of knowledge can provide clues as to the patterns of response most applicable to CO_2-induced climate change.

Moreover, adjustment decisions are further guided by the relative emphases placed upon particular *national aims*. In the United States, several national aims enter into decisions regarding natural resources in general and water resource management in particular: economic efficiency, human health, environmental protection, equity in income distribution, and regional development (Water Resources Council, 1976; White, 1969). Choice of adjustment strategies varies as the relative weights attached to particular aims shift over time. For example, the massive river basin development schemes involving large dams and other engineering structures (prevalent in the 1950s and justified on the grounds of economic efficiency and regional development) began to fall into disfavor in the 1960s as greater consideration was given to the national aim of environmental protection.

The point to be emphasized here is that a given set of national aims can be considered the common denominator which underlies all resource-related issues and decisions –CO_2-induced climate change included. This is an important, fundamental connection between responses to CO_2 and responses to other societal problems like air pollution, drought, or energy. In essence, the desirability of any adjustment strategy – to climate change or otherwise – can be judged according to the degree to which it contributes to national income, leads to equitable distributions of wealth, meets goals of environmental protection, and so forth. In this way, particular strategies for CO_2 differ only in specific purpose from other resource strategies, but ultimately they contribute to the same set of national aims.

This suggests that we need to step back and view societal response to CO_2 in its larger resource context. This means ceasing to contemplate response to CO_2-induced climate change as a separate problem (or opportunity). We must begin to establish the associations with other societal issues. There are two practical reasons for doing so. First, to the degree that CO_2 strategies can be tactically piggybacked onto other problems or projects which are perceived to be of a more urgent nature (like energy or global food supplies), the greater the likelihood of successful adoption. CO_2 by itself is unlikely to rise to the top of national priority lists. Jointly-designed or dual-purpose strategies promise greater ease of implementation.

Second, CO_2 response strategies might achieve greater benefits at less cost if their adoption is hooked directly into the attainment of specific national aims. For instance, in developing countries CO_2 response could be tied directly to the goal of economic development. Similarly, response to CO_2 in the developed world might be linked intentionally with the broader aim of protecting our environment, as, for example, in reducing both local air pollution and CO_2 emissions by switching away from coal burning power plants. Such linkages could help to sharpen the issues, to guide strategy choice, and to clarify the streams of benefits to be derived nationally from purposeful societal response to CO_2.

In sum, the message of this section is as follows: knowledge of the ways in which particular conditions of choice (i.e., prevailing response modes or national aims) guide, or could guide, societal response to CO_2 is necessary in order to allow careful fashioning of strategies which promise the greatest net benefits for society and which have a high likelihood of implementation. Several research opportunities are suggested.

7.1. Research Opportunities

First, research leading to fuller understanding of the likely *constraints and incentives to CO_2 strategy adoption* is basic. Better descriptive models of choice in resource management would shed light on the CO_2 issue. This includes all levels, from local to global. For some resources and regions, we have already a fairly good understanding of the processes involved and a wealth of existing studies. This is the case with water resource management in the western United States, for example. But for other areas knowledge is scarce. Perhaps the greatest research payoff would come from studies within the context of the global political economy: how does the capitalist market economy constrain or otherwise guide resource decisions in developing countries at individual, local, or national levels? The global study of the 1972 world food situation by Garcia and Escudero (1980), for example, is enlightening in this regard. A major assumption, of course, is that the same incentives and constraints which are at work to guide societal strategies to droughts, energy shortages, food problems, water management, or the like, are directly applicable to the CO_2 issue. We believe they are.

A second opportunity builds on the first: based on understanding the conditions of choice, research directed towards identifying viable *connections between CO_2 strategies and other existing resource problems* could facilitate adoption of strategies and more explicitly meet the priorities of national aims. How might strategies for adjusting to CO_2 climate change, reducing agricultural drought hazard, and achieving economic development aims be integrated and implemented in the tier of sub-Saharan developing countries? Inasmuch as CO_2 will be perceived as a remote problem in many parts of the world, the chances of formulating patterns of response to benefit the entire global community would be enhanced considerably by addressing such questions.

8. What Are the *Dynamic Feedback Effects* to Nature, Society, and Subsequent Response?

Once adopted and implemented, strategies for adjusting to a situation of climatic change feed back to alter the physical systems or the social systems (refer to Figure 1). The process of response is continuous, changing as social systems themselves change, as thresholds of perception shift, as newly conceived (or perceived) strategies arise, and as conditions of choice evolve. Thus, we are portraying a dynamic system of climate-society interaction, one which must be studied ultimately in a dynamic – not static – fashion.

Because of the dynamic complexities, every study which attempts to capture the whole set of relationships in one setting over time appears unique and highly descriptive. As Kates (1980a) notes, this is partly why many holistic studies of climate impact seem so idiosyncratic and "fuzzy" to non-social scientists. On the other hand, the seemingly "rigorous" studies often focus on isolated aspects of the problem. Thus, we face a dilemma of sorts. Ultimately, we would like to understand, theoretically, the entire pattern and process of human response to climatic variation, but our research procedures usually guide us to specific pieces of the problem (and we are not completely certain that the whole is equal to the sum of its parts). Limited attempts have been made to overcome the dilemma through the use of systems modelling, but the potential for wider applicability of the method to climate impact assessment is unclear at this time. There is a need to investigate further systems modelling and related tools for this purpose.

In terms of actual experience, how effective have various strategies been in helping society adjust to climatic variations? Surprisingly little research has been conducted on this question. In the United States, most research efforts provide prescriptive answers to normative question (should cloud seeding be widely adopted or water conservation techniques encouraged?). Very few post-audit analyses of past adjustment strategies have been conducted to ascertain actual performance (White and Haas, 1975). For example, for nearly 40 years the federal government has encouraged certain drought adjustment strategies for the U.S. Great Plains, which include a mix of river basin-wide and groundwater irrigation, soil and water conservation techniques, and so on. Yet, we have little idea if, or to what degree, these practices have achieved their intended aims. One finds many studies which assess such adjustments on an experimental basis (e.g. agricultural experiment station research) but very few which attempt hindsight-evaluation in the real-world socio-economic, political, and environmental setting (as did Riebsame, 1981) This represents a major knowledge gap, since many of these same strategies are potentially applicable to problems engendered by CO_2 -induced climate change.

8.1. Research Opportunities

Several research areas emerge from the above discussion. First, there needs to be an assessment of the *potential of simulation modelling* as a method for analyzing the effects of and responses to climatic change on society. Simulation models might prove to be a useful means of integrating a number of climate-society relationships, of exploring their interaction over time, and of assessing the possible effects of strategy or policy changes. This is clearly an interdisciplinary endeavor, and incidentally, one which might serve as a useful vehicle for bringing together those with differing viewpoints.

A second research opportunity lies in conducting empirical *post-audit analyses of past adjustment strategies*. The purpose would be to gain understanding of their effectiveness in actual use beyond theoretical or normative evaluations. The focus should be on studies of both long-term climate change and shorter-term fluctuations which relate directly to the CO_2 issue. The experience in natural resource management or natural hazard management should prove helpful. Knowledge of the efficacy of alternative adjustive mechanisms is a critical informational component to development of strategies for coping with climate change, and should thus receive strong research support.

9. Summary and Conclusions

Within the time scale of human existence, climate changes are not unusual. The Neolithic age was well underway when the Altithermal period of warmer average temperatures prevailed over the globe. Ancient civilizations witnessed the decline of the Altithermal with the change to a cooler global climate about 4500 years ago. The Little Ice Age occurred but a minute ago in the lifetime of mankind. Superimposed upon longer-term climate changes were numerous medium- and short-term fluctuations. Some of these apparently coincided with growing, healthy societies, while others found societies troubled and declining. Climate changes presumably can bring boom to some and bust to others, although in hindsight the cause-and-effect nature of the relationship is extremely difficult to establish.

The possibility of a climate change within the next 50 years, then, is not so unusual, given the long-term perspective. What *is* unique about the prospect of a CO_2-induced climate change, however, is that: (1) it would be generated largely by human activity through massive combustion of fossil fuels, and (2) it is the first time in human history that we have *prior* scientific information that a climate change is likely. Heretofore, climate changes were phenomena recognized after the fact. The implications are that not only does society (in the global sense of the word) have the option to implement preventative measures if it so chooses, but it also has the opportunity to anticipate the possible outcomes (both benefits and costs) of climate changes and to fashion strategies accordingly. For an environmental change of such a large magnitude and scale, this is unprecedented in the history of human affairs.

The challenges to both the physical and social sciences are enormous. The particular climate impact assessments that are required are extremely demanding because of the far-reaching consequences of a CO_2-climate change. Moreover, society is now requesting answers to direct, applied questions which the scientific community is unprepared (and often reluctant) to address and which ultimately require interdisciplinary cooperation – a formidable hurdle given the tradition and strength of disciplinary barriers. As a result, the scientific community often finds itself lacking the theoretical and conceptual frameworks and the methodologies it needs to answer the socially and economically relevant questions about the shifts in climate. This is especially true for the social sciences. Existing methods have to be re-examined, and new methods and approaches developed and tested. Above all, there is an urgent need to ask what we already *do* know about important aspects of the CO_2 issue and in what broad directions research should now be directed.

As we reflect on the set of research questions and opportunities outlined above (as summarized in Table IV), we reach three broad conclusions. First, opportunities for studying the societal effects of a potential climate change lie only secondarily in specifying the likely direct impacts (as in, say, estimating wheat yield declines given a reduction in average precipitation). Although such estimates are useful and practical, the greater opportunities lie in gaining a fuller understanding of the nature and process of societal vulnerability, adaptability, sector sensitivities and resiliencies, and decision-making and response to environmental variation. It is this fundamental knowledge which, in the long run, will prove more valuable than mechanistic delineations of direct sectoral impacts.

Second, because of the complexity and holistic character of the CO_2 issue, disciplinary

TABLE IV: Opportunities for research

1. What are the relationships between *climate variations and sectors* of society?
   (a) Review of basic research on climate-resource sector linkages.
   (b) Crop-yield modelling.
   (c) Sector sensitivity analyses.
   (d) Sector resiliency analyses.

2. What are the characteristics of society that determine *vulnerability* to climatic variation?
   (a) Critical review of conceptual and methodological approaches to climate impact assessment.
   (b) Networking of research and researchers concerned with societal effects of climate change.
   (c) Systematic inventory of potential case studies.
   (d) Case study investigation of effects of climate change (or analogous events) on society.

3. What are the *pathways and linkages* within social systems through which the effects of climate change are transmitted?
   (a) Linking of major climate-related models with sectoral models.
   (b) Tracing actual pathways of societal effects of climatic variations through case study research.

4. What is the *level and distribution of effects* from climatic change?
   (a) Develop indicators of societal impacts.
   (b) Construct methodologies for systematic collection of data on social distribution of climate effects.
   (c) Investigations of global distributions of effects arising from climatic fluctuations.
   (d) Assess trends in climate-society interactions.

5. What is the range of *mechanisms* by which societies adjust to climatic fluctuations? And what are their costs and effects?
   (a) Construct roster of adjustive mechanisms.
   (b) Studies of long-term social costs of adjustment strategies to short-term climate fluctuations.

6. How is *information* (scientific or otherwise) on climate change *perceived, interpreted, valued, and channeled* into strategy evaluation?
   (a) Investigation of the socio-economic consequences of the dissemination of scientific information on CO_2.
   (b) Assess user needs with respect to CO_2 information.

7. What are the *conditions of choice* which guide societal response to Climate Change?
   (a) Determine potential constraints and incentives to CO_2 strategy adoption.
   (b) Determine connections between CO_2 strategies and other existing resource problems.

8. What are the *dynamic feedback effects* to nature, society, and subsequent response?
   (a) Investigate potentials of simulation modelling for climate impact assessment.
   (b) Post-audit analyses of past adjustment strategies.

research foci have only limited application. Interdisciplinary communication and research are central to progress in understanding climate-society interaction in general and CO_2-induced climate change in particular. The research opportunities suggested herein reflect this view.

Finally, from a societal standpoint, CO_2-induced climate change is not an isolated problem. Many of the basic processes – of vulnerability, of effects and their distribution, of strategy formulation, etc. – are common to a broad spectrum of resource, environmental, and natural hazards contexts which exist today. The implications are twofold. In order to conduct meaningful research, we do *not* have to wait until reasonably precise forecasts of regional climate changes are available; research related to the societal effects of, and response to, CO_2-induced climate change can proceed now. Also, the eventual strategies which society adopts to deal with climate change will be, and should be, intertwined with a whole set of related environmental and social problems. Many of the issues are similar and amenable to common societal responses. What we learn about a range of social, climatic, and environmental problems will apply directly to an understanding of the societal effects of CO_2-induced climate change. And vice versa.

Acknowledgements

Early drafts were reviewed by F. Kenneth Hare, Ian Burton, and the members of the AAAS/DOE Panel.

Appendix: The Role of Geography

Probably more than any other discipline, geography is extremely wide-ranging in subject matter and approach. Geographers have spent considerable time arguing among themselves as to what constitutes the distinctive claims of geography within academia. On the basis of content, geographers have claimed nothing less than the entire surface of the earth, physical and human systems included. With respect to scale, geographic studies have ranged from investigations of spatial behavior in hospital rooms to global resource patterns. At one time or another, geographers have offered their preoccupation with space as their calling card (but what discipline is without a spatial context in its focus of study?). In terms of content, scale, or dimension, geography always eludes clear definitions.

Perhaps more illuminating is to ask what geographers do. One view (Pattison, 1964; Taafe, 1974) identifies four traditions in geography: Physical, regional, spatial, and man-environment. Within these traditions, geographers consistently have dipped heavily into related disciplines to acquire theory and methodology, and in the process have spread themselves thinly over broad intellectual ground. One result is that geographers have assumed hybrid identities: cultural geographers, social geographers, behavioral geographers, regional geographers, physical geographers, historical geographers, Marxist geographers, economic geographers, geomorphologists, climatologists, human ecologists, and on and on – and all within the same bailiwick.

Perhaps it is this characteristic of crossing disciplinary boundaries which gives geography its uniqueness, in two ways. First, operationally, geography seems to come closer to an interdisciplinary field than others. Geographers often provide much-needed synthesis when integration is most needed. (Not that geography has special claim to that role; often, it just turns out that way.) Second, because geographers historically have always kept one foot in the physical sciences (climatology has been a traditional focus in geography), they are noted for their attempts to bridge the gap between the environmental and human realms, between physical science and social science. Our assessment here reflects both an interdisciplinary synthesis and a man-environment perspective on the issue of CO_2-induced climate change.

National Center for Atmospheric Research, Boulder, CO 80307, U.S.A.

Department of Geography,
University of Wyoming,
Laramie, WY 82070 U.S.A.

Notes

[1] Which is not to say that research on alternative energy systems or energy conservation as substitutes for fossil fuel burning should not be pursued vigorously.
[2] However, Baumgartner (1979) tends to downplay the potential impacts of climate shifts on forest ecosystems, despite the fact that forest shifts are traditional indicators of past climate fluctuations.
[3] This was cited as a major reason why better results were not forthcoming from LACIE (EDIS, 1979).
[4] A difficulty which might be encountered here is in being "objective". Often climate "changes" are inferred from apparent social or ecological impacts, a procedure which can lead to tautological reasoning if caution is not exercised.

References

Ausubel, J.: 1980, 'Economics in the Air', in A. K. Biswas (ed.), *Climatic Constraints and Human Activities*, 13–59. Oxford, Pergamon Press.

Baerreis, D. A. and Bryson, R. A.: 1968, 'Climatic Change and the Mill Creek Culture of Iowa', *J. Iowa Archaeological Society* **15**, 1.

Baker, E. J., and McPhee, J. G.: 1975, *Land Use Management and Regulation in Hazardous Areas: A Research Assessment*. Institute of Behavioral Science, University of Colorado, Boulder, Colorado.

Baier, W.: 1977, 'Crop-Weather Models and Their Use in Yield Assessments', World Metorological Organization Technical Note No. 151. Geneva, WMO.

Baier, W. and Williams, G. D. V.: 1974, 'Regional Wheat Yield Predictions From Weather Data in Canada', *Proceedings of the WMO Symposium Agrometeorology of the Wheat Crop*, Braunschwg, Federal Republic of Germany, 22 October, 1973. WMO No. 396, 265–283. Geneva, WMO.

Barnett, C. V.: 1972, 'Weather and the Short-Term Forecasting of Electricity Demand', in J. A. Taylor (ed.), *Weather Forecasting for Agriculture and Industry*, Rutherford, Farleigh, Dickinson University Press.

Barnett, H. J. and Morse, C. M.: 1963, *Scarcity and Growth: The Economics of Natural Resource Availability*, Baltimore, Johns Hopkins Press.

Baumgartner, A.: 1979, 'Climate Variability and Forestry', in World Meteorological Organization (ed.), *World Climate Conferences: A Conference of Experts on Climate and Mankind*, Geneva, WMO.

Bennett, R. J. and Chorley, R. J.: 1978, *Environmental Systems*. Princeton, New Jersey, Princeton University Press.

Biswas, A. K.: 1980a, 'Climate and Economic Development', *The Ecologist* **9**, 188.

Biswas, A. K.: 1980b, 'Crop-Climate Models: A Review of the State of the Art', in J. Ausubel and A. K. Biswas (eds.), *Climatic Constraints and Human Activities*, 75–92. Oxford, Pergamon Press.

Biswas, M. R.: 1979, 'Energy and Food Production', in A. K. Biswas and M. R. Biswas (eds.), *Food, Climate and Man*, J. Wiley and Sons.

Biswas, M. R., and Biswas, A. K., eds,' 1979, *Food, Climate and Man*, New York, J. Wiley and Sons.

Boulding, K.: 1980, 'Science: Our Common Heritage', *Science* **207**, 831.

Bowden, M. J., Kates, R., Kay, P., Riebsame, W., Gould, H., Johnson, D., Warrick, R., and Weiner,

D.: 1981, 'The Effect of Climate Fluctuations on Human Populations: Two Hypotheses', in M. Ingram, G. Farmer, and T. M. L. Wigley (eds.), *Climate and History*, Cambridge University Press.

Bryson, R. A.: 1975, 'Some Cultural and Economic Consequences of Climatic Change', Report No. 60. Institute of Environmental Studies, University of Wisconsin, Madison.

Bryson, R. A., Lamb, H. H., and Donley, D. L.: 1974, 'Drought and the Decline of Mycenae', *Antiquity* **48**, 46.

Bryson, R. A. and Padoch, C.: 1980, 'On the Climates of History', *J. of Interdisciplinary History* **10**, 583.

Burton, I., Kates, R. W., and White, G. F.: 1968, 'The Human Ecology of Extreme Geophysical Events', Natural Hazards Research Working Paper No. 1, Department of Geography, University of Toronto.

Burton, I., Kates, R. W., and White, G. F.: 1978, *The Environment as Hazard*, New York, Oxford University Press.

Burton, I., Billingsley, D., Blacksell, M., Chapman, V., Kirkby, A., Foster, L., and Wall, G.: 1974, 'Public Response to a Successful Air Pollution Control Program', in J. A. Taylor (ed.), *Climatic Resources*, London, David and Charles.

Carpenter, R.: 1968, *Discontinuity in Greek Civilization*, New York, W. W. Norton.

Caviedes, C. N.: 1975, 'El Niño 1972: Its Climatic, Ecological, Human, and Economic Implications', *Geogr. Rev.* **65**, 493.

Center for Environmental Assessment Services: 1980, 'Climate Impact Assessment: Annual Summary'. Washington, D.C., U.S. Department of Commerce.

Christy, F. T. and Scott, A.: 1965, *The Common Wealth in Ocean Fisheries*. Johns Hopkins, Baltimore.

Cochrane, H. C.: 1975, 'Natural Hazards: Their Distributive Impacts', Boulder, Colorado, Institute of Behavioral Science, University of Colorado.

Cochrane, H. C., Haas, J. E., and Kates, R. W.: 1974, 'Social Science Perspectives on the Coming San Francisco Earthquake: Economic, Prediction, and Reconstruction', Natural Hazards Research Working Paper No. 25. Institute of Behavioral Science, University of Colorado, Boulder, Colorado.

Craddock, J. M.: 1965, 'Domestic Fuel Consumption and Winter Temperatures in London', *Weather* **20**, 257.

Cushing, D. H.: 1979, 'Climatic Variation and Marine Fisheries', in World Meteorological Organization (ed.), *World Climate Conference: A Conference of Experts on Climate and Mankind*, Geneva, WMO.

d'Arge, R.: 1979, 'Climate and Economic Activity', in World Meteorological Organization (ed.), *World Climate Conference: A Conference of Experts on Climate and Mankind*, Geneva.

Dacy, D. C. and Kunreuther, H.: 1969, *The Economics of Natural Disasters: Implications for Federal Policy*. New York, Free Press.

Deevey, E. S., Rice, D., Rice, P., Vaughan, H., Brenner, M., and Flannery, M.: 1979, 'Mayan Urbanism: Impact on a Tropical Karst Environment', *Science* **206**, 298.

Dracup, J. A.: 1977, 'Impact on the Colorado River Basin And Southwest Water Supply', in National Academy of Sciences (ed.), *Climate, Climatic Change and Water Supply*, Washington, D.C., National Academy of Sciences.

Dynes, R. R.: 1970, *Organized Behavior in Disaster*. Lexington, Massachusetts, D.C. Heath.

Ehrenfeld, D.: 1972, *Conserving Life on Earth*. New York, Oxford.

Ehrlich, P. R. and Ehrlich, A. H.: 1970, *Population, Resources, Environment: Issues in Human Ecology*, San Francisco, Freeman.

Environmental Data Information Service (EDIS): 1979, 'The Large-Area Crop Inventory Experiment', Environmental Data Service, National Oceanic and Atmospheric Administration, Washington, D.C.

Environmental Science Services Administration (ESSA): 1966, 'Weather and the Construction Industry', ESSA/PI660013. Washington, D.C., GPO.

Enzer, S., Drobnick, R., and Alter, L.: 1978, *Neither Feast nor Famine*, Lexington, Massachusetts, Lexington Books.

Food and Climate Forum: 1981, *Food and Climate Review 1980–81*, Aspen Institute for Humanistic Studies. Boulder, Colorado.

Food and Agriculture Organization (FAO): 1974, *Environment and Development*, Thirteenth FAO Regional Conference for Latin America, Panama City, Panama. Rome FAO.

Fossey, W. R.: 1977, 'Talkin' Dust Dowl Blues: A Study of Oklahoma's Cultural Identity During the Great Depression', *Chronicles of Oklahoma* **55**, 12.

Friesema, H. P., Caporaso, J., Goldstein, G., Lineberry, R., and McCleary, R.: 1979, *Aftermath; Communities After National Disasters*, Beverly Hills, Sage.

Garcia, R. and Escudero, J.: 1980 (draft), 'The Constant Catastrophe', Vol. 2; An IFIAS Report on 'Drought and Man: The 1972 Case History'.

Great Plains Committee: 1937, *The Future of the Great Plains*, 75th Congress, First Session, Document No. 144, U.S. House of Representatives. Washington, D.C., GPO.

Grossman, L.: 1977, 'Man-Environment Relationships in Anthropology and Geography'. *Annals. Assoc. Amer. Geogr.* **67**, 126.

Grove, A. T.: 1973, 'Desertification in the African Environment', in D. Dalby and R. H. Church (eds.), *Drought in Africa*, London, University of London School of Oriental and African Studies.

Gupta, U. S. (ed.): 1975, *Physiological Aspects of Dryland Farming*, New Delhi, Oxford and IBH Publishers.

Haigh, P.: 1977, 'Separating the Effects of Weather and Management on Crop Production', Contract Report, C. F. Kettering Foundation, Grant # ST–77–4. Columbia, Missouri.

Hare, F. K.: 1979, 'Climatic Variation and Variability: Empirical Evidence From Meteorological and Other Sources', in World Meteorological Organization (ed.), *World Climate Conference: A Conference of Experts on Climate and Mankind*, Geneva, WMO.

Harris, D. R.: 1969, 'The Ecology of Agricultural Systems', in R. U. Cooke and J. H. Johnson (ed.), *Trends in Geography*, Oxford, Pergamon Press.

Harrison, P.: 1979, 'The Curse of the Tropics', *New Scientist* **22**, 602.

Heathcote, R. L.: 1974, 'Drought in South Australia', in G. F. White (ed.), *Natural Hazard: Local, National, Global*, New York, Oxford University Press.

Heberlein, T. A.: 1973, 'The Three Fixes: Technological, Cognitive and Structural', paper presented at the Water Community Conference. Seattle, Washington.

Heilbroner, R. L.: 1974, *An Inquiry into the Human Prospect*, New York, Norton and Co.

Hewitt, K.: 1980, 'Perspectives on Development and Small Settlements in the Third World With Special Reference to Pakistan', in P. O'Keefe and K. Johnson (eds.), *Environment and Development: Community Perspectives*, Program on International Development, Clark University, Worcester, Massachusetts.

Holcomb Research Institute: 1976, *Environmental Modeling and Decision-Making: The United States Experience*, New York, Praeger Publishers.

Holling, C. S.: 1973, 'Resilience and Stablility in Ecological Systems', *Ann. Rev. Eco. and Systematics* **4**, 1.

Huntington, E.: 1915, *Civilization and Climate*, Yale University Press, New Haven, Connecticut.

Huszar, P.: 1975, *Frost Hazard in the United States: A Research Assessment*, Boulder, Colorado, Institute of Behavioral Science, University of Colorado.

Jacobsen, T.: 1958, 'Salt and Silt in Ancient Mesopotamian Agriculture', *Science* **128**, 1251.

Jackson, R. H.: 1979, 'Drought in the Western United States: Perception and Adjustment 1976–1977', Paper presented at the Annual Meeting, Association of American Geographers. Philadelphia, PA, April 22–25.

Johnson, J. H.: 1976, 'Effects of Climate Changes on Marine Food Production', Appendix A in National Academy of Sciences, *Climate and Food: Climatic Fluctuation and U.S. Agricultural Production*, Washington, D.C., NAS.

Johnson, K.: 1976, 'Do As the Land Bids: A Study of Otomi Resource-Use on the Eve of Irrigation', Ph.D. dissertation, Graduate School of Geography, Clark University, Worcester, MA.

Kates, R. W.: 1962, *Hazard and Choice Perception in Flood Plain Management*, Research Paper No. 78. Chicago, University of Chicago, Department of Geography.

Kates, R. W.: 1980a, 'The Methodology of Climate Impact Assessment: A Note', unpublished MS, Clark University, Worcester, MA.

Kates, R. W.: 1980b, 'Climate and Society: Lessons From Recent Events', *Weather* **35**, 17.

Katz, R.: 1977, 'Assessing the Impact of Climatic Change on Food Production', *Climatic Change* **1**, 85.

Kellogg, W. W. and Schware, R.: 1981, *Climate Change and Society: The Consequences of Increasing Atmospheric Carbon Dioxide*, Boulder, Colorado, Westview Press.

Kneese, A. V., Ayres, R. U., and d'Arge, R. C.: 1970, *Economics and the Environment: A Materials Balance Approach*, Washington, D.C., Resources for the Future.
Lambert, L. D.: 1975, 'The Role of Climate in the Economic Development of Nations', *Land Economics* **47**, 339.
Linnemann, H., DeHoogh, J., Keyzer, M. A., and van Heemst, H. D. J.: 1979, *MOIRA: Model of International Relations in Agriculture*. Amsterdam, North-Holland Publishing Company.
Manners, I. R.: 1974, 'The Environmental Impact of Modern Agricultural Technologies', in I. R. Manners and M. W. Mikesell (eds.), *Perspectives on Environment*, Washington, D.C., Assoc. Amer. Geographers.
Margolis, H.: 1979, 'Estimating Social Impacts of Climate Change: What Might Be Done vs. What Is Likely to Be Done', Paper present to AAAS/DOE Workshop on Environmental and Societal Consequences of a Possible CO_2-Induced Climate Change, Annapolis, Maryland.
Maunder, W. J., Johnson, S. R., and McQuigg, J. D.: 1971, 'The Effect of Weather on Road Construction: Application of a Simulation Model'. *Monthly Weather Rev.* **99**, 946.
McGovern, T. H.: 1981, 'The Economics of Extinction in Norse Greenland', in M. Farmer, M. Ingram, and T. M. Wigley (eds.), *Climate and History*, Cambridge, University of Cambridge Press.
McKernan, D.: 1972, 'World Fisheries World Concern', in B. Rothschild (ed.), *World Fisheries Policy: Multidisciplinary Views*, University of Washington Press, Seattle.
McQuigg, J. D.: 1975, *Economic Impacts of Weather Variability*, Columbia, Missouri, Atmospheric Science Department, University of Missouri-Columbia.
McQuigg, J. D., Thompson, L., LeDuc, S., Lockard, M., McKay, G.: 1973, *The Influence of Weather and Climate on United States Grain Yields: Bumper Crops or Droughts*, Washington, D.C., U.S. Department of Commerce.
Mitchell, J. M. Jr., Felch, R. E., Gilman, D. L., Quinlan, F. T., and Rotty, R. M.: 1973, 'Variability of Seasonal Total Heating Fuel Demand in the United States', National Oceanic and Atmospheric Administration. Washington, D.C.
Murray, T., Ingham, M., LeDuc, S. K., and Steyaert, L. T.: 1980, 'Impact of Climatic Factors on Early Life Stages of Atlantic Mackerel', paper presented at the Conference on Climatic Impacts and Societal Response of the American Meteorological Society, August. Milwaukee, WI.
National Academy of Sciences (NAS): 1972, *Genetic Vulnerability of Major Crops*, Washington, D.C., NAS.
National Academy of Sciences (NAS): 1977, *Climate, Climatic Change and Water Supply*, Washington, D.C,., NAS.
National Defense University: 1978, *Climate Change to the Year 2000: A Survey of Expert Opinion*, Washington, D.C., GPO.
National Defense University: 1980, *Crop Yields and Climate Change to the Year 2000*, Washington, D.C., GPO.
National Oceanic and Atmospheric Administration (NOAA): 1974, 'Variability of Seasonal Total Fuel Demand in the United States', *Environmental Data Service*, January, 5–9.
Newman, J. E.: 1978, 'Drought Impacts on American Agricultural Productivity', in N. E. Rosenberg (ed.), *North American Droughts*, Boulder, Colorado, Westview Press.
Norwine, J.: 1978, *Climate and Human Ecology*, Houston, D. Armstrong.
O'Keefe, P., and Wisner, B.: 1975, 'African Drought: The State of the Game', in P. Richards (ed.), *African Environment*, London, International African Institute.
Parry, M. L.: 1978, *Climatic Change, Agriculture and Settlement*, Kent, UK, Dawson.
Pattison, W. D., 1964: 'The Four Traditions of Geography', *Journal of Geography* **63**, 211.
Post, J. D.: 1977, *The Last Great Subsistence Crisis in the Western World*, Baltimore, Johns Hopkins University Press.
Quirk, W. J.: 1980, 'Using Improved Climate Information to Better Manage Energy Systems', Paper presented at the Conference on Climatic Impacts and Societal Response of the American Meteorological Society, August.
Rabb, T.: 1983, 'Climate and Society in History: A Research Agenda', in Chen, R. S., Boulding, E., and Schneider, S. H. (eds.), *Social Science Research and Climate Change: An Interdisciplinary Appraisal*, D. Reidel Publ. Co., Dordrecht, Holland, pp. 62–76 (this volume).

Radovich, J.: 1981, 'The Collapse of the California Sardine Industry: What Have We Learned?', in M. Glantz and J. D. Thompson (eds.), *Resource Management and Environmental Uncertainty:*
Rees, J. D.; 1970, 'Paricutin Revisited: A Review of Man's Attempts to Adapt to Ecological Changes Resulting From Volcanic Catastrophe', *Geoforum*, April, 7–26.
Reifsnyder, W. E.: 1976, 'Forest Lands', Appendix B in National Academy of Sciences, *Climate and Food: Climatic Fluctuation and U.S. Agricultural Production*, Washington, D.C., National Academy of Sciences.
Regan, C.: 1980: 'Underdevelopment, Vulnerability and Starvation: Ireland, 1845–48', in P. O'Keefe and K. Johnson (eds.), *Environment and Development: Community Perspectives*, Worcester, Massachusetts: Program on International Development, Clark University.
Rettig, R. B.: 1978, 'Managing Fisheries for Optimum Yield: The Need for Integrated Research in Physical, Life and Social Sciences', in M. Glantz, H. van Loon, and E. Armstrong (eds.), *Multidisciplinary Research Related to the Atmospheric Sciences*, Boulder, Colorado, National Center for Atmospheric Research.
Riebsame, W.: 1981, 'Adjustment to Drought in the Spring Wheat Area of North Dakota: A Case Study of Climate Impacts on Agriculture', Ph.D. Dissertation, Clark University, Worcester, MA.
Roberts, P. R.: 1960, 'Adverse Weather – Its Effects on Engineering Design and Construction', *Civil Engineering* **30**, 35.
Rotberg, R. I. and Rabb, T. K. (eds.): 1980, *Journal of Interdisciplinary History* **10**, Cambridge, Massachusetts, MIT Press.
Russell, C. S., Arey, D. G., and Kates, R. W.: 1970, *Drought and Water Supply*, Baltimore, Johns Hopkins University Press.
Russo, J. A., *et al.*: 1965, 'The Operational and Economic Impact of Weather on the Construction Industry', The Travelers Research Center, Inc., Hartford, CT.
Saarinen, T.: 1966, *Perception of the Drought Hazard on the Great Plains*, Research Paper No. 106. Chicago, Department of Geography, University of Chicago.
Saarinen, T. F.: 1974, 'Problems in the Use of a Standardized Questionnaire for Cross-Cultural Research on Perception of Natural Hazards', in G. F. White (ed.), *Natural Hazards: Local, National, Global*, New York, Oxford University Press.
Schaake, J. D., Jr. and Kaczmarek, Z.: 1979, 'Climate Variability and the Design and Operation of Water Resource Systems', in World Meteorological Organization (ed.), *World Climate Conference: A Conference of Experts on Climate and Mankind*, Geneva, WMO.
Schneider, S. H. and Chen, R. S.: 1980, 'Carbon Dioxide Warming and Coastline Flooding: Physical Factors and Climatic Impact', *Annual Review of Energy* **5**, 107.
Schneider, S. H. and Mesirow, L. E.: 1976, *The Genesis Strategy*, New York, Plenum.
Scientific Committee on Problems of the Environment (SCOPE): 1978, *Workshop on Climate/Society Interface*, Paris, SCOPE Secretariat.
Sewell, W. R. D. and Burton, I. (eds): 1971, *Perceptions and Attitudes in Research Management*. Ottawa, Information Canada.
Simon, H. A.: 1956, 'Rational Choice and the Structure of the Environment', *Psych. Rev.* **63**, 129.
Smith, C. D. and Parry, M.: 1981, *Consequences of Climatic Change*, Department of Geography, University of Nottingham, UK.
Spitz, P.: 1980, 'Drought and Self-Provisioning', in J. Ausubel and A. K. Biswas (eds.), *Climatic Constraints and Human Activities*, 125–147. Oxford, Pergamon Press.
Stein, W. J.: 1973, *California and the Dust Bowl Migration*, Westport, Connecticut: Greenwood Press.
Taaffe, E. J.: 1974, 'The Spatial View in Context', *Annals Assoc. Amer. Geogr.* **64**, 1.
Takahashi, K. and Yoshino, M. M.: 1978, *Climatic Change and Food Production*, University of Tokyo Press.
Terjung, W. E.: 1981, 'Toward Level-5 Geographical Modeling: A Working Structure', Paper presented to 77th AAG Meeting. Los Angeles.
Thompson, J. D.: 1977, 'Ocean Deserts and Ocean Oases', Chapter 6 in M. Glantz (ed.), *Desertification: Environmental Degradation in and around Arid Lands*, Boulder, Westview Press.
Thompson, J. D.: 1981, 'Climate, Upwelling, and Biological Productivity: Some Primary Relationships',

Chapter 2 in M. Glantz and J. D. Thompson (eds.), *Resource Management and Environmental Uncertainty: Lessons from Coastal Upwelling Fisheries*, New York, John Wiley and Sons.

Thompson, L. M.: 1962, 'Evaluation of Weather Factors in the Production of Wheat in the United States', *Journal of Soil and Water Conservation* **17**, 149.

Thompson, L. M.: 1969a, 'Weather and Technology in the Production of Wheat in the United States', *Journal of Soil and Water Conservation* **23**, 219.

Thompson, L. M.: 1969b, 'Weather and Technology in the Production of Corn in the U.S. Corn Belt', *Agronomy Journal* **61**, 453.

Timmerman, P.: 1981, 'Vulnerability, Resilience and the Collapse of Society', Environmental Monograph No. 1. Institute for Environmental Studies, University of Toronto.

Torry, W. I.: 1979, 'Anthropological Studies in Hazardous Environments: Past Trends and New Horizons', *Current Anthropology*, Vol. 20, No. 3.

United States Department of Agriculture (USDA): 1974, *The World Food Situation and Prospects to 1980*, Foreign Agriculture Report No. 98. Washington, D.C., GPO.

U.S. Department of Transportation: 1975, 'Impacts of Climatic Change on the Biosphere', CIAP Monograph 5, Part 2, *Climatic Effects*. Washington, D.C., GPO.

van Loon, H. and Williams, J.: 1978, 'The Association Between Mean Temperature and Interannual Variability', *Monthly Weather Rev.* **106**, 1012.

Waddell, E.: 1975, 'How the Enga Cope With Frosts: Responses to Climatic Perturbations in the Central Highlands of New Guinea', *Human Ecology* **3**, 249.

Warrick, R. A.: 1980, 'Drought in the Great Plains: A Case Study of Research on Climate and Society in the U.S.A.', in J. Ausubel and A. K. Biswas (eds.), *Climatic Constraints and Human Activities*, Oxford: Pergamon Press.

Warrick, R. A. (with Trainer, P. B., Baker, E. J., and Brinkman, W.): 1975, *Drought Hazard in the United States: A Research Assessment*. Boulder, Colorado, Institute of Behavioral Science, University of Colorado.

Warrick, R. A. and Bowden, M. J.: 1981, 'Changing Impacts of Drought in the Great Plains', in M. Lawson and M. Baker (eds.), *The Great Plains, Perspectives and Prospects*, Lincoln, Nebraska, Center for Great Plains Studies.

Warrick, R. A. and Kates, R.: 1981, 'Testing Hypotheses About the Effects of Climate Fluctuations on Human Populations', Paper presented at the Annual Meeting of the American Association for the Advancement of Science, Toronto.

Water Resources Council (WRC): 1976, *Unified National Program for Flood Plain Management*. Washington, D.C., GPO.

White, G. F.: 1961, 'The Choice of Use in Resource Management', *Natural Resources Journal* **1**, 23.

White, G. F.: 1966, 'Formation and Role of Public Attitudes', in H. Jarrett (ed.), *Environmental Quality in a Growing Economy*, Baltimore, Johns Hopkins Press.

White, G. F.: 1969, *Strategies of American Water Management*, Ann Arbor, University of Michigan Press.

White, G. F.: 1973, 'Natural Hazards Research', in R. J. Chorley (ed.), *Directions in Geography*, London, Methuen.

White, G. F. (ed.): 1974, *Natural Hazards: Local, National, Global*, New York, Oxford University Press.

White, G. F. and Haas, J. E.: 1975, *Assessment of Research on Natural Hazards*. Cambridge, MIT Press.

Whyte, I.: 1981, 'Human Response to Short- and Long-Term Climatic Fluctuations: The Example of Early Scotland', in C. D. Smith and M. Parry (eds.), *Consequences of Climatic Change*, 17–29. Department of Geography, University of Nottingham.

Wigley, T. M. L., Ingram, M. J., and Farmer, G. (eds.): 1981, *Climate and History*, Cambridge University Press.

Wisner, B.: 1978, 'The Human Ecology of Drought in Eastern Kenya', Ph.D. Dissertation, Graduate School of Geography, Clark University.

World Meteorological Organization: 1974, 'Intercomparison of Conceptual Models Used in Operational Hydrological Forecasting', Operational Hydrology Report No. 7. Geneva: WMO.

Worster, D : 1979, *Dust Bowl: The Southern Great Plains in the 1930s*, New York, Oxford University Press.
Wright, J. D., Rossi, P. H., Wright, S. R., and Weber-Burdin, E.: 1979, *After the Clean-up: Long-Range Effects of Natural Disasters*, Beverly Hills, California, Sage.
Yevjevich, V., Hall, W. A., and Salas, J. D.: 1978, *Drought Research Needs*, Fort Collins, Water Resources Publications.

INTRODUCTION TO: CLIMATE AND SOCIETY IN HISTORY: A RESEARCH AGENDA

In the spring of 1979 Princeton University historian Theodore Rabb and his Massachusetts Institute of Technology colleague Robert Rotberg co-convened a workshop of some two dozen climatologists and historians on the subject of climate and history. Rabb appealed to the group to take the next interdisciplinary step, albeit cautiously, by beginning to collaborate on projects which could "enrich both climatology and history" (Rabb, 1980). At first this requires that "all parties are clear about (1) the context, (2) the method, and (3) the final aim". While this modest statement suggests no bold or daring initiative, it revealed that Rabb recognized that the process of interdisciplinary cooperation between historian and climatologist could best begin with small, solid steps. He knew that grand conclusions or general theories on the impacts of climate on history would have to be built on a solid base of data. The recognition that there are interdisciplinary prerequisites to fruitful climate and history research led directly to Professor Rabb's participation in Panel IV of the AAAS/DOE climate project. He proposes here to encourage "a major cooperative and interdisciplinary effort, linking historians with scientists from various disciplines so as to produce a uniform, standard data base" of both climatic and historical events, concentrating at first where the data are richest: from the 13th century to the present and in Europe. (He also offers a "data base" of his own: a very extensive bibliography of climate and history works through mid-1982.)

After the data have been assembled, we can begin to addess the more basic intellectual issue of CO_2 /climate impact assessment: how has climate change modified human societies and what societal conditions or responses provided greater or lesser adaptive ability to deal with the climatic stresses? Although one could argue about the organizational details of the very specific research project charted by Rabb on his Figure 1, I suspect that any successful future climate and history research project would likely have the same basic elements that he has outlined in his paper in this volume. In the meanwhile, the climate and history bibliography that Rabb assembled with the help of research assistant Lewis Steinberg will undoubtedly prove to be a major resource for the growing field of climate and history.

S. H. SCHNEIDER

Reference

Rabb, Theodore K.: 'The Historian and the Climatologist'. *Journal of Interdisciplinary History* **10**, 4, Spring 1980, 831–38; reprinted in *Climate and History*, R. I. Rotberg and T. K. Rabb, (eds.), (Princeton Univ. Press, Princeton NJ, 1981), pp. 251–7.

R. S. Chen, E. Boulding, and S. H. Schneider (eds.), Social Science Research and Climate Change: An Interdisciplinary Appraisal, 61.

THEODORE K. RABB

CLIMATE AND SOCIETY IN HISTORY: A RESEARCH AGENDA

1. Introduction: Problems, Needs, and Possibilities

A historical component will be essential to any research plan which seeks to evaluate the likely impact of CO_2-induced climatic changes on human society. Information from the past provides us with the only possible laboratory for generating and testing hypotheses about the effects of, and responses to, climatic or other large-scale physical shifts. The problems we face over the next fifty years will be unique primarily because they will be appearing at this particular juncture, the late twentieth and early twenty-first centuries. The attitudes, governmental structures, and economic and technical means that will be drawn on to confront the problems will be products of that period, and therefore are not available for study. But analogies, proxy data which can reveal how the processes of change and adaptation function, are abundant if we turn to the past. Here we can uncover the continuities, the general (perhaps even theoretical) relationships between variations in the environment, variations in behavior, and social vulnerability, that can help guide policymakers in the future.

Historians have probably expended less effort than other social scientists on these issues. Accordingly, their research is at a more tentative stage, full of enormous gaps in knowledge. They have given considerable attention to major social upheavals, but less to physical changes *per se*. A great deal of basic groundwork will therefore be required, and it might be useful to think in terms of a need for "basic" as well as "applied" research. Much fundamental information still has to be recovered, and this is an urgent "basic" task, because it will serve as a resource for all investigators, not just historians; but at the same time a number of specific historical projects directly related to the CO_2 problem – in other words, "applied" research – are already possible. A combination of both types of work should be a high priority in studies of the effects of climate over the next few years.

The most important contributions historians can make to public policy dealing with the consequences of CO_2-induced climatic change are, in essence, threefold.

First, they can determine how the climate has in fact changed in the past, with special reference to situations that may parallel the conditions that are now predicted.

Second, they can assess the impact of those changes on human societies – on agriculture, on settlement, on demographic behavior, and on attitudes.

Third, they can evaluate the different ways in which societies have responded or adapted. The last question is the most difficult, and is predicted on a reasonably confident set of answers to the first two. For it is this problem that raises the normative issues on which actual policy recommendations must be based – what has been successful, what has failed (and by what criteria), and how does climate differ in its effects from other physical and non-physical phenomena (such as warfare).

R. S. Chen, E. Boulding, and S. H. Schneider (eds.), Social Science Research and Climate Change: An Interdisciplinary Appriasal, 62–76.

This paper will outline the work that has been done on these three questions, specify the most important gaps in our knowledge, and make recommendations for further research. The major aim will be to suggest what kinds of inquiry will enable us to define:

(a) the ways that climatic changes have affected societies,

(b) the types of adaptation that have protected or strengthened those societies, and

(c) the lessons that can be drawn by policy makers from the experiences of recent centuries.

To the extent possible, the paper will indicate the specific projects that could be undertaken, the available resources, and the scholars who have a particular interest in these areas.

In an ideal world, all of history would be our province. We would study every instance of climatic change for which evidence survives, regardless of country or period. From such comprehensive data universal conclusions could be drawn. To be more practical, one must assume that a sampling of a few significant cases, taken primarily from the most thoroughly studied periods of western history, will produce enough varieties of material to support the understanding we seek. Accordingly, we will concentrate on the so-called "Little Ice Age" in Europe, the history of North America, and a few comparative examples of major events in other regions, with particular attention to Japan (which is well documented). The Little Ice Age is particularly relevant to the CO_2 problem, because it also included slow, long-term shifts, often masked by short-term trends. North America is appropriate because of the audience for this project; moreover, like the Little Ice Age, it enjoys the advantage of considerable previous research, which allows suggestions for the future to be more specific. Some well-studied instances of climatic change in other places or times will broaden the range of reference, but will not be the subject of specific recommendations for further investigation.

A few comments on what follows may be in order. Although climatologists have paid considerable attention to the general trends of the past thousand years and earlier, not many historians have been interested in centuries-long developments. Rather, their investigations of physical conditions have tended to foucs on relatively short-term and immediate influences -- demographic shifts, warfare, famines, disease, and weather. The links between such events and climate have not been too thoroughly explored. Therefore, although these studies do not necessarily provide comprehensive information about the ways that human societies respond to alterations in their physical environment, they can to some extent serve as proxies for research on the historical results of climatic change. Adaptations vary in kind enormously, and such problems as soil exhaustion or salination, irrigation, and deforestation may offer close parallels to climate, but the central need is to understand better the process of adaptation itself. Our chief concern, therefore, must be to see how and when and in what different ways it operates, and topics like demography, war, famine, plague, and weather, which are often more amply documented than climate, will have to form part of the research proposals.

Also by way of introduction, it should be noted that both of the key terms, "climate" and "adaptation", retain some ambiguity. Except in a few dramatic instances, such as floods or droughts, climate itself always has a mediated effect on society. Its immediate manifestation, for example, may be in the onset and duration of the growing season, which *in turn* has consequences for crops and agriculture, and only *then* makes itself felt in prices or mortality rates. So many other variables may affect prices and mortality that it is often difficult to isolate the specific influence of the climate. For the very process of mediation is shaped by social, economic, political, and cultural circumstances. Behavior depends on perception; the latter is therefore often the key to the question "has something changed?" This makes it all the more important, in the historical context, to accept proximate events, such as a rise in deaths, as evidence, and indeed to use their variation as a means of tracing how vulnerability to climate can change. One of the outcomes of this research effort may indeed be to establish the limits of interpretation: we may be able to specify much more clearly what can and cannot be defined as the effects of climate.

Adaptation is also a nebulous concept. It can take place at many different levels – the individual, the group, the locality, the region, the government, the continent; it can be deliberate, a planned response to a perceived need, or accidental, a spontaneous decision not specifically aimed at climatic conditions; or it can occur without any real awareness of the conditions that require it. And the forms it takes change not only with the problem – potato blight vs. longer winters, for example – but also by time and place – for instance, Ireland vs. the Netherlands in the 1840's. Again, perception is often the crucial element. Where this paper is concerned, the level, the self-consciousness, and the form of adaptation will be regarded as the best means of understanding comparatively the way it functions and the effectiveness of its outcome.

It is important, finally, to remember that precision in historical research is impossible. At best we can hope to examine the range within which climate affects societies (if at all), the ways societies respond, and the degree to which the process is influenced by time and place.

2. The Questions Historians Can Answer

The following are the main categories into which historians' research can be divided. They are presented in three groups, each relating to the major subjects listed in the first paragraph of this paper.

(1) What are the climatic changes that have taken place in the past?
 (a) What have the long-term changes been?
 (b) What have the short-term changes been?
 (c) How have the short-term related to, or masked, the long-term?
 (d) What have the different types of change been?
  (i) Warmer.
  (ii) Colder.
  (iii) Wetter.
  (iv) Dryer.
  (v) More windy or stormy.
  (vi) Less windy or stormy.
  (vii) Increased variation.
  (viii) Less variation.
  (ix) Change in some seasons but not in others.
(2) What have been the effects of climatic change on human societies?
 (a) How have they affected agriculture (including fishing and forestry)?
 (b) How have they affected patterns of settlement?
 (c) How have they affected demographic trends (including health)?

(d) How have these effects altered, in turn, political institutions, economic development, and social relationships?
(e) How have they affected mentalities?

(3) How have societies responded?
(a) How has awareness of change developed, and how has it led to action?
(b) How have *ad hoc* adaptations operated?
(c) Have deliberate, planned responses differed sigmificantly in their results from more spontaneous responses?
(d) How have localities responded -- villages or cities?
(e) How have regional or national authorities responded?
(f) What are the ingredients of successful response?

The remainder of the paper will address each of questions in order, indicating the principal areas of research by reference to the appended bibliography of approximately one thousand items, which is intended as an introduction to the field. Only a few of the most important findings will be described, however; the main emphasis will be on ways to answer more firmly the questions that have just been outlined. In that endeavor the Bibliography will be a substantial resource, for it offers both some answers and a guide to the materials that will be needed for further investigations. Yet the primary concern will be to suggest more specifically what ought to be done by historians over the next few years.

3. The Areas of Needed Research

3.1. What Are the Climatic Changes That Have Taken Place in the Past?

Before any estimate of an answer can be made, we must have the basic data from which to work. A considerable amount of research has been done in the area, though little of it by historians. Documentary evidence and observation are important sources, but thus far they have been used only slightly to confirm the results derived from glaciers, tree rings, lake sediments, mineral isotopes, phenological data, and other physical analysis. Moreover, the literature is at present uneven and inconsistent, though ample, because little effort has been made to integrate all possible sources into a single comprehensive survey of climatic change. [Bibliography Sections A-G.]

What is needed is a major cooperative and interdisciplinary effort, linking historians with scientists from various disciplines so as to produce a uniform, standard data base. The main literature that would provide guidelines for such an enterprise are the works in the Bibliography, Sections B through F, by Alexandre, Bergthorssen, Bryson, Catchpole, Craddock, Fritts, Ingram, Ladurie, Lamb, Landsberg, Manley, Moodie, Pfister, and Schweingruber. The sources should be world-wide, and as eclectic as possible. But the nature of the information to be collected should be limited, to keep it within feasible bounds. The form and range of the data bank should be as follows:

(i) *Chronology*: From the 13th century to the present, with additional data for the previous four hundred years, if possible, so as to provide a context for the change around 1200 proposed by Bryson. But the main reason for beginning in the thirteenth century is itself to provide a context for the Little Ice Age, which has been posited as beginning anywhere from the fourteenth to the seventeenth centuries. It is to encompass the

Little Ice Age that the data should go so far back, in order to determine which periods witnessed slow, and which rapid climatic shifts, and how the two are to be distinguished. It is especially important to define the slow warming that apparently ended the Little Ice Age because that period may be the best analogue for the predicted pattern of CO_2-induced change.

(ii) *Time Scale*: The unit of analysis may have to change as we come forward in time and the information becomes richer. A final decision would have to be made by those undertaking the research. For the present, it seems plausible that through the fifteenth century sufficient resolution could be achieved to permit seasonal or perhaps monthly figures; from the sixteenth until the eighteenth centuries, ten-day or weekly figures should be recoverable; and by the year 1800 (and in some cases earlier) daily measures should be sought. For these purposes techniques like low-resolution pollen analysis and dendrochronology may be less useful, but they will be important as checks on the more detailed figures derived from documentary materials. The latter will include such information as the freezing and thawing of rivers and canals and the dates of harvests as well as the direct testimony of letters, diaries, and government records.

(iii) *Indices*: Although not all data may be recoverable for all periods, divided into equally precise time-scales, the following should be measured:

(1) Temperature
(2) Precipitation
(3) Length of snow cover
(4) Windiness
(5) Incidence of extreme conditions, such as frost, hail, floods, and drought.

(iv) *Geography*: Since the main focus of the historical project will be on developments in Europe and North America, the data collection should emphasize these two continents. Depending on the sources available, it will probably be advisable to separate the results according to regions, possible three or four on each continent, with comparative data as a control from Asia and other continents.

(v) *Presentation*: The raw data should of course be assembled in time series, but the research must provide information that is organized and synthesized to aid analysis as well as description. It is important, first, that the data bank include calculations of the proportional changes in the indices from time period to time period, and provide geographic comparisons. It will also be essential to specify the rate and range of the changes, because there is reason to believe that the speed and inconsistency of change may have more impact on human societies than the change itself.

This research should be organized by an international, interdisciplinary team. Two related efforts that are now under way, at the Climatic Research Unit at the University of East Anglia in England, and at the CLIMHIST Project at the University of Bern in Switzerland, should be linked with a central American effort, perhaps under the aegis of the National Center for Atmospheric Research in Boulder, Colorado. With a director, a twenty-person interdisciplinary board of scholarly advisors and contributors at related centers like the University of Maine, and a research team of two full-time assistants, the basic data bank could be assembled in a period of five to seven years. This team will also assist the research projects described further below.

The remainder of this section will outline the substantive questions the compilers of the climatic data bank ought to address, beyond the basic collection, synthesis, and organization of information. Providing answers to the four questions listed in Section 2, above, under 1. (a)–(c) and (d) (i)–(ix), should be their prime objectives. Behind those questions, however, lie difficult issues of conceptualization.

First, how is one to define long-term and short-term? This is a matter to which we will return when discussing perceptions of change. But it is significant even at this first stage of laying out the data because the present project has arisen out of the very expectation that CO_2-induced change will be gradual and long-term. One of the problems the compilers of the data bank will have to address is how to distinguish cycles of different lengths, because their impacts may differ. They will also have to try to weigh the significance of brief counter-trends. Three clear-cut examples can be taken from the Little Ice Age, and they should be studied in depth: the periods 1500–1550, 1650–1688, and the 1730's, which were warm interludes within a general cooling. From American data, the exceptional warm spell of the 1820's should also help reveal how the short-term can mask the long-term, and what we need to know in order to understand such "deceptions" better. (See Smith *et al.* in Section L of the Bibliography.)

Second, how can conflicting interpretations or results be resolved? Two major conflicts, in particular, would repay the close attention of the compilers of the data bank. One concerns the bewildering variety of opinions about the Little Ice Age: how is it to be defined, when did it begin and end, and is this formulation a useful heuristic device? As the literature in Sections A and B of the Bibliography reveal, there are almost as many chronologies and definitions as there are scholars who have written about the problem. A merging of different data, and a mediation between different interpretations, will be necessary as the project accumulates enough data to reach relatively firm conclusions. Another conflict, which highlights the difficulty of reconciling contradictory evidence, has to do with the climate of the Great Plains after 1850. On this subject one set of evidence, represented by Lawson, indicates that the Plains became drier; other evidence, drawn from free-rings by Fritts, shows that they became wetter. (See Section F of the Bibliography.) Results from different sources and methods are unlikely ever to coincide perfectly, but this is an example of a problem that will recur whenever research findings prove to be diametrically opposed to one another. Without the kind of overarching effort that the compilers of the data bank will undertake, it will be impossible to create a standard, accepted chronology that will supersede such uncertainties.

This, then, should be the first task of the next few years. Beyond the purposes just surveyed, it will have three other notable advantages. First, it will be a research resource for all investigators, scientists as well as social scientists. Second, it can be pursued simultaneously with, and indeed can lend valuable support to, the research (discussed below) which will examine the effects of climate on society. Third, it will provide a much more solid basis for extrapolations into the future, for it will enable policy planners to follow trends over a longer period than is now possible and thus to make more confident predictions.

3.2. What Have Been the Effects of Climatic Changes on Human Societies?

In Section 2, above, it was suggested that this large and difficult problem should be divided into five separate questions, dealing in turn with agriculture; settlement; demography;

changes in institutions, economies, and social relations; and mentalities (or attitudes). Those five areas derive from various fields within historical research, but they are of course closely interrelated. Methods and conclusions from one spill over into others, and thus, although the questions can be kept distinct, the research projects will overlap.

Within each area, moreover, there are different mechanisms at work. To be specific, four sub-categories of the larger questions must always be kept in mind:

(i) the effects of sudden disasters,
(ii) the effects of short-term but sharp climatic shifts,
(iii) the effects of slow and gradual changes in climatic regimes, and
(iv) the ways in which perceptions and social, political, and economic conditions mediate the effects of both short-term and long-term changes.

Which is the most important consideration may vary from area to area, and from time to time, but a prime purpose of the research must be to define how these influences operated and to judge their relative significance as transformers of human societies.

3.2.1. *How Has Climate Affected Agriculture*? As in all areas of research dealing with the consequences of climatic phenomena, the bulk of work on agricultural questions has focussed on short-term and usually catastrophic effects. Moreover, the emphasis has been on the physical rather than the social environment. For example, there is a much larger literature on drought, notably the American Dust Bowls of the 1930's and the Sahel Drought of the 1960's, than there is on the consequences of, say, general rainfall or temperature patterns for agricultural regimes and the geographic distribution of specific farming techniques (Bibliography, Sections H, J, K, and O in comparison with Sections L, S, and T).

Nevertheless, a considerable amount of work has been done on harvest yields and crop failures, though not necessarily with agreement about the relationship with climate. (Compare Pfister, de Vries, Smith *et al.*, and McCloskey in the Bibliography, Sections L and S.)

What is need now is a large-scale systematic and comparative study of the research that has been completed. From the articles and books written by Appleby, Ladurie and Goy, Pfister, Titow, van Bath, and the others listed in the Bibliography, Sections L and S (which is itself only a sampling of the main landmarks in the field), a statistical series parallel to the one on climatic indices should be prepared. This would not require as much research in original sources as the climatic data would, because a lot of the material for the series is available in print. There are problems with the most abundant source, tithe records, while other sources, such as data on output and yields, are often fragmentary, but it should be possible to construct a basic series.

The agricultural series should also cover the period from the year 1200 to the present, using the growing season as the unit of analysis wherever possible, and specifying the impact of technological advances on the figures. The data should be divided by crops – at least grains and vines should be included, though it would also be useful to differentiate such food for the poor as oats, barley, and potatoes – and by major geographic regions in Europe and North America. Among the information to be included should be:

(i) yields,
(ii) length of growing season,
(iii) date of harvest,

(iv) failures,
(v) changes in crops, and
(vi) changes in farming techniques.

The specific contents of the data, however, will depend on what the compilers feel can be recovered with some accuracy. Their ultimate purpose will be to provide a series that can be compared directly with the climatic series, so that correlations can be made between short-term and long-term changes in the two indices. Special attention should be given to the effects on marginal lands, which are sensitive indicators, and to the consequences of irrigation, which is a proxy for rainfall. Attention should also be given to fishing and forestry.

This would be a smaller-scale enterprise than the creation of the climatic data bank. It would require a five-year effort by a director, a research assistant, and a board of five scholarly advisors, linked with the research team working on climate, though it should have a separate director.

3.2.2. *How Has the Climate Affected Patterns of Settlement*? This issue is closely related both to agriculture and to demography, because much of the scholarly literature has concentrated on the utilization of marginal lands and on migratory movements as functions of agricultural and demographic conditions. The research can therefore be undertaken by the group dealing with demography, but the question should be kept distinct, even though this is a subject on which the evidence is often ambiguous, and arguments have to be made by inference. For it is of special interest to the CO_2 project, since the need to relocate populations is a possible outcome of future climatic change. Current findings should therefore be examined to some extent in connection with agriculture, but especially in conjunction with the demographic study described in the next section. The aim should be to determine if broad generalizations are possible in the following areas:

(i) how do slow changes (in temperature and precipitation) affect settlement,
(ii) what are the consequences of disasters (single and multiple),
(iii) does the population of towns (over 3000 inhabitants) or cities (over 10 000 inhabitants) react differently from rural populations,
(iv) have the reactions changed in the past 1000 years, and if so, to what extent because of technological advances, and
(v) in all these areas are migrations and patterns of settlement influenced by different forces?

In this study, as in the agricultural study, some attention has to be paid to major non-climatic influences on migration and settlement patterns, especially demographic, economic, attitudinal, and political change. But the chief purpose should be to assemble sufficient examples, from enough different cultures and periods, to specify the range within which certain consequences, such as the inability to maintain marginal lands, follow. The principal literature is listed in Sections L and M of the Bibliography, notably the works by Barreis and Bryson, Bell, Carpenter, McGhee, Parry, and Murphy, which, together with Bryson and Murphy's book (Section A) also offer guides to other research that has been done.

As has been indicated, although settlement is a distinct subject, it should be linked in a research plan with demography, which follows.

3.2.3. *How Has the Climate Affected Demography*? Demography has been one of the liveliest fields in historical research for the past twenty years. Where climate is concerned, it is also one of the most unsettled. Few scholars would suggest that climate affects demographic behavior directly, but a number have sought to find links via disease and nutrition. At this stage, however, one has to say that research is still inconclusive. The strongest case has been made in England, where mortality and fertility rates have been connected with temperature levels by Lee (Bibliography, Section L), who has also investigated the change in the relationship over time. Others, too, notably Appleby, Menken *et al.*, and Post (Bibliography, Section Q) have made important contributions to an understanding of the interaction between climate, and thus nutrition and disease, on the one hand, and fertility and mortality on the other hand.

What is needed here is a major effort, taking off from the enormous body of demographic research that has been done (including the compilation of major data banks), to bring current findings together, link them with the information gathered in the climatic data bank, and produce rigorous statistical analyses, similar to those compiled by Lee for England, which can establish the relationship between climatic and demographic variations. The following are among the questions that should be considered:

(i) How do temperature and precipitation relate to mortality, fertility, and the marriage rate: Is there a time lag, a seasonal relationship, a geographic pattern?

(ii) To what extent is the connection mediated by disease and the availability of food?

(iii) Can one trace a direct effect on health at all?

The research team dealing with both settlement and demography ought to be about the same size as the one working on agriculture, and also linked closely to the climatic history data bank. Again a director who is experienced in this field should work with a small advisory group and a research assistant.

3.2.4. *How Have the Agricultural and Demographic Effects of Climate Influenced Political Institutions, Economic Development, and Social Relationships (in Other Words, the Main Contours of History)?* For this question a separate director, with two research assistants, will be needed, to work closely with both the argricultural and the demographic research teams. Seeking answers will be a task which, guided by a board of scholarly advisors, should take about five to seven years. As a more general "historical" undertaking, it should be led by a broad-gauged historian, experienced in a wide variety of fields, who will have to undertake a number of case studies to define the impact of climate, disease, harvests, migrations, and demography (as outlined by the other research teams) on the standard events of political, economic, and social history.

Among the events that might be studied are:

(i) The so-called "general crisis" of the seventeenth century, an era of political upheavals and revolts.

(ii) The expansion of Europe, especially the major overseas migrations.

(iii) The development of new governmental powers to meet the new situations caused by physical changes (e.g., welfare systems).

(iv) Differential economic growth, especially in the rate of industrialization, due to constraints on the allocation of resources and capital/labor relations.

(v) The comparative growth of cities (across time and space) and intranational migrations.
(vi) The changing levels, incidence, and seasonality of poor relief.

In each case the attempt will be made to trace the links, if any, between physical change, attitudes, and the historical events. The purpose will be to define the way the process can work, the range within which it operates, and how it has changed over time. There has been some scholarship on these issues (see Bibliography, Sections M, N, O, R, and S), but largely in terms of disaster and decline. The effort now must be to see if there is a more regular, less dramatic connection as well; to understand why and when certain climatic and social changes cannot be linked with one another; and to try to devise and test broad hypotheses about the relationship.

3.2.5. *How Has Climate Affected Mentalities (or Attitudes)*? This is an almost entirely uncharted area, and will require a considerable research effort, because attitudes are the vital intermediary between physical change and a change in behavior. Method is not a problem. What is required is the classic historian's endeavor of studying documents, books, paintings, and other such expressions of opinion, in search of comments about weather, climate, and its consequences. There have been a few efforts in this direction, notably the works in Section N of the Bibliography. But a systematic confrontation of the main issues involved in this question has not been attempted.

A historian of mentalities, with experience in studying attitudes, would be an excellent director. He would need two research assistants, and could expect to take at least five years to arrive at his results. Again, though, he should be closely linked to the compilers of the main climatic data bank, and to the agricutlural, demographic, and historical research teams. The particular questions that will have to be answered will include:

(i) What kinds, degrees, and directions of change make people aware of alteration in the physical environment?
(ii) When they refer to weather or climate, do they think of it as affecting their lives, and if so, in what ways?
(iii) Can climate induce optimism about people's control over their own lives, or only pessimism? Why do responses differ, and how have they changed over time?
(iv) Is there a difference in the perception of different kinds of climatic effect – e.g., the recurrent drought cycle, warming or cooling trends, and increasing or lessening variability in weather conditions?
(v) Has the impact of long-term change, in contrast with short-term change, altered by time and place?

The sources are almost endless – letters, diaries, government documents, books, and paintings. The research should concentrate on well-defined periods and territories which experienced climatic change and for which considerable published materials are available. It may well be that this investigation will simply form a part of, or will be subsumed by, the other inquiries that historians will undertake. For example, governmental responses (to be discussed below) may prove to be our best "proxy" evidence for an awareness of the effects of climate. Moreover, in contrast to the other investigations, this one is unlikely to yield quantitative results. Nevertheless, the subject of mentalities deserves a distinct place in the research effort, because our findings will carry useful implications for future action only if they take into account people's consciousness of change as well as the change itself.

3.3. How Have Societies Responded?

The set of questions that bears on this issue is the crux of the historical project. Thus far we have envisioned a central research effort, producing a data bank on the climate of the past, that is linked with four teams specializing in the effects of climate on agriculture, demography (including settlement patterns), historical events, and mentalities. Their function will be to assemble the evidence that can document how climate has changed and how it has affected human societies. This might be called the "basic" research. We now move to the reciprocal question, at the heart of the CO_2 project: how have human societies reacted to these influences, and in particular, how have they adapted?

There are a number of major categories under which answers should be sought. This should, however, be the responsibility of a single research group, cutting across and interacting with the other five, but always emphasizing the *response* to physical change rather than its causes. The literature on this issue is scattered throughout the Bibliography, though it is particularly the concern of the works in Sections U and W. See, too, the studies by de Vries, Patch, Pfister, Smith *et al*, and Swan in Section L, and to a lesser extent by Brooks, Glantz and Katz, Heathcote, and Saarinen in Section H. Most of the research on disasters (Sections H-K, O, and Q) also considers the nature of a society's response, as does the parallel literature on the effects of war (Section V). This is a subject which must include disruptions that are analogues to climatic change, such as plague and warfare, because the parallels and contrasts in the mechanism of response will allow the investigators to determine the ingredients of success more precisely. It may be that adaptations to environmental change (such as the studies in soil exhaustion, deforestation, resource scarcity, and pollution mentioned in Section W of the Bibliography) provide the closest parallel, but the diversity of the cases obviously should be as wide as possible. One needs to remember, too, that some human activities, such as irrigation, deforestation, and industralization, themselves affect the climate and require further responses.

Given the range of the more specific questions which follow, one has to recommend a research team as large as the one which will be compiling the climate data bank: a director, two research assistants, and an advisory board of up to twnety scholars from all fields. This, too, should be thought of as at least a five-year effort.

The remainder of this section will discuss in somewhat more detail the six major categories into which this inquiry ought to be divided. None should be regarded as entirely distinct, because all will relate to one another. But a separation of the main topics seems advisable.

3.3.1. *How Has Awareness of Change Developed, and How Has It Led to Action*? Although it was the last topic in the previous section, where it followed the more tangible developments that affect attitudes, in this section it must come first. Response depends on awareness, and therefore one must begin with perception, using as much as possible the findings of the mentalities research team. Among the distinctions that will have to be made are the following:

(i) Is the determination to take action nurtured by the severity, by the duration, or by the direction of physical change (i.e., is a cooler trend more galvanizing than a warmer, a drier than a wetter)?
(ii) Can one differentiate among the effects of a sudden, short-term dislocation; a medium-term development like an unusual regime that lasts a number of years (such as the warm and stable climate between 1715 and 1740); and a long-term shift more than half a century long? Is one more important than the others?
(iii) Do these influences vary by country, region, or period?
(iv) How much credence is given to public statements and/or warnings?
(v) How has the understanding of climate, and the possibility of controlling its effects, changed?
(vi) What attitudes lead to action, and how are they generated?

A few specific case studies should be chosen which can illuminate these issues. Periods when action was taken, or policy was formulated, such as the famines in France in the 1690's, the high mortality of the 1740's, or the disasters of the 1810's or 1930's, should offer the best evidence, and permit comparisons across time, place, and conditions. An example of change in the absence of disaster, such as the development of new agricultural techniques by the first American colonists, would provide a useful contrasting case study. It would also be interesting to see how the mix of crops in an area changes (for instance, a growing reliance on potatoes) and what effects this has.

3.3.2. *How Have Ad Hoc Adaptations Operated*? The most general form of adaptation is the rule-of-thumb, almost unconscious change in behavior as circumstances change. The abandoment of marginal lands as population declines, the improvisation of building techniques when people move or materials become unavailable, the revision of fishing patterns as the fish change traditional routes, and the development of new avenues of communication or new sources of heat, are all examples of the slow, almost imperceptible transformations that have taken place constantly throughout history. They are usually thought of as unpremeditated and accidental, like the discovery of gunpowder. What needs to be learned is exactly how spontaneous they were, how they were transmitted from group to group, and the sorts of pressures – social, environmental, or cultural – that helped them spread.

To this end, it would again be useful to concentrate on a few case studies. Because of the interest in the distinction between long-term and short-term change, the case studies should include both gradual and rapid adaptations, and indeed an attempt should be made to distinguish the two types. Among other possibilities,

(i) the search for additional sources of heat during the Little Ice Age,
(ii) changes in work habits among the first settlers of North America,
(iii) the adoption of new building styles and new forms of family life by the pioneers on the American frontier, and
(iv) a comparative analysis of the different responses to a few selected natural disasters,

should offer enough subject matter to resolve some of the major issues. Throughout the research, the aim should be to find out, not only how people adapted and whether they felt they were responding to climatic conditions, but also what made them adapt as they did, which were the most successful adaptations, and why. A central concern should be to determine whether societies have become more insulated from climatic effects, and in what ways, so that the present-day context can be better understood.

3.3.3. *Have Deliberate, Planned Responses Differed Significantly in Their Results from More Spontaneous Responses*? On this subject the case studies will have to range across the entire spectrum of deliberate policy making in the face of material change. They will have to examine legislation dealing with agriculture, settlement, and population, as well as regulations designed to alleviate the consequences of alterations in the physical environment, such as the organization of charities to handle vagrancy or food disribution to handle famine. Both national governments and regional authorities should be investigated, and it would probably be best to concentrate on four or five countries and periods, perhaps linked to the times and places being studied for *ad hoc* adaptation. Thus comparative research on

(i) England, France, and the Netherlands in the period 1550–1750,
(ii) colonial and early national North America,
(iii) England in the 1810's and 1820's, and
(iv) the United States in the 1930's,

would run parallel to the work being done on *ad hoc* responses to change.

The goal should be to show how policies (and the influences shaping them) have changed over time and place, how they have differed by culture and according to the problem at hand, how they have related to *ad hoc* adaptations, what has constituted success, and how it has been achieved. In all these questions, the way problems are perceived, and the way people have been (or have not been) persuaded to accept official solutions – for example, the Poor Law in England, grain distribution in France, population resettlement and changes in crops in Prussia, disaster relief in the United States – must be part of the inquiry. The difference between long-term and short-term change, and between local and national or regional response, is also essential to the research.

3.3.4. *How Have Localities Responded – Villages or Cities*? This is part of the inquiry just outlined, but a distinct set of answers should be provided about the geographic unit of response as well as the nature of the response. From the case studies listed above, it should be possible to tell how different kinds of localities have adapted – by location, size, and period. There will doubtless be a significant variation between rural and urban areas, and it will be important to indicate where adaptation is easier, and for what reasons. Does a village change more easily, for example, because it is compact and united, or a city, because it is ruled by a more sophisticated government apparatus? In order to address these issues, the case studies will have to include detailed local analyses – for which some literature already exists (Bibliography, Sections P, U, and W).

3.3.5. *How Have Regional or National Authorities Responded*? Here again the inquiry is not basically different from the one outlined above, under question (3.3.3). But once more it is the unit making the decision rather than the type of decision that ought to receive distinct attention here. The same case studies that explore the nature of deliberate, planned responses should be used to understand the ways in which adaptations devised by national or regional authorities differ from those developed by localities. Is one more effective than the other? Which works more efficiently for a situation involving long-term change, and which for short-term change? How are the two linked? By what process are decisions implemented? How have the answers to these questions changed over time and by place? The definition of "success" and "effectiveness" will be an essential part of this inquiry.

3.3.6. *What Are the Ingredients of Successful Responses*? Taking off from the issue posed at the end of the last section, a general concern with the meaning of "success" and "effectiveness" will have to pervade the work of this research team. How are these terms to be defined – by the numbers of people affected, by the speed of the response, by the opinions of those involved, by the balance between benefits (agricultural, economic, demographic, political, and/or cultural) and losses? The criteria will have to be specified, as well as the means of judging them. For unless the project can reach conclusions about the relative success of various adaptations to physical change, it cannot make substantive policy recommendations.

4. Summary: An Agenda for Research

Given the questions that should be asked, and the research enterprises that will be necessary to answer them, the historical project ought to be organized in accordance with the chart, below.

ORGANIZATION OF THE RESEARCH PROJECT

PHASE 1:
Data Bank on Climatic History
Director and Two Assistants
Board of Advisors

PHASE 2:
Agriculture Research Project
Director
One Assistant
Board of Advisors

Demography Research Project
Director
One Assistant
Board of Advisors

Central Coordinating Committee
The Six Directors
Representatives from the Boards of Advisors

History Research Project
Director
Two Assistants
Board of Advisors

Mentalities Research Project
Director
Two Assistants
Board of Advisors

PHASE 3:
Societal Response Research Project
Director and Two Assistants
Board of Advisors

Interval between phases: one to two years. Once started, each activity continues until the end of the project. Expected total duration: at least five to seven years. *Note*: This is a *minimum* estimate. The phases may in fact require three or more years each.

The structure outlined (above) summarizes the recommendations made in the course of this paper. It also suggests a chronological as well as an organizational framework. The Data Bank on Climatic History should begin its work first, in Phase 1, and remain in existence, as a central resource, throughout the undertaking. The four Research Projects dealing with the effects of climate should start operations a few years afterwards, in Phase 2, as soon as some of the preliminary work on the data bank is complete, so that its findings can be used as a guide in their own research. A Central Coordinating Committee, linking the different groups on matters of common concern, should be formed at the same time. Finally, the Societal Response Research Project should get under way only in Phase 3, a few years later, when the four other Research Projects have produced sufficient results to create a basis for its work. The entire effort should take something on the order of seven or more years. It would require part-time commitments from the six Directors, and fulltime employment of the ten Research Assistants.

The outcome of all this research will enable policy makers to understand for the first time how, historically, societies have been affected by climatic and physical changes, and how they have been induced to take appropriate and effective measures of responses.

Princeton University

Research Assistant:
LEWIS STEINBERG

THEODORE K. RABB

BIBLIOGRAPHY

to accompany

Climate and Society in History: A Research Agenda

Note

This bibliography is intended, not as a definitive survey, but as a starting point for the research into the historical effects of climate that is described in the accompanying paper. It is by no means exhaustive, but the approximately one thousand items do represent the major landmarks in the fields, as well as the most important recent research. A few annotations have been added, where appropriate, and a specific mention has been made of the excellent bibliographies which appear in many of the works. The bulk of the list was assembled in the spring of 1980, with the assistance of Lewis Steinberg. Since that time, additional entries have been made, bringing the list up to date as of the end of 1981. At the end of the bibliography are appended the names of the sixty journals which were consulted – and from some of which the 1980/81 articles were taken; this may be useful as a guide to the main patrons of publications in this field. A final point: although some attempt has been made to follow the standard style of a bibliography, there has not been time to create the perfect consistency that a professional bibliographer would expect. There has been an effort to make sure that a publication can be found, and its date given; but there has not been the exhaustive checking of spelling, wording, and style of reference to ensure absolute consistency. Thus in some cases only the year of a journal is given, sometimes the volume, sometimes the issue, and sometimes the page numbers of an article, but not always. Foreign accents have been omitted. The purpose has been to create a useful tool, rather than a polished work.

Table of Contents

R. S. Chen, E. Boulding, and S. H. Schneider (eds.), Social Science Research and Climate Change: An Interdisciplinary Appraisal, 77–114.

Abbreviations

Annales: *Annales Economies Societes Civilisations.*
C&H I: *Climate and History: Studies in Interdisciplinary History*. (See Rabb, p. 80.)
C&H II: *Climate and History: Studies in Past Climates and Their Impact on Man.*
EHR: *Economic History Review*. (See Wigley, p. 80.)
ICCH: International Conference on Climate and History, 1979.
J. Ec. Hist.: *Journal of Economic History.*
J. Eur. Ec. Hist.: *Journal of European Economic History.*
JIH: *Journal of Interdisciplinary History.*
JM: *Journal of Meteorology.*
MM: *Meteorological Magazine.*
PPP: *Palaeogeography, Palaeoclimatology, Palaeoecology.*
P&P: *Past and Present.*
QJRMS: *Quarterly Journal of the Royal Meteorological Society.*
QR: *Quaternary Research.*

A. Climatic Change and Its Social Impacts: General Views

Anderson, J. L: 1981, 'Climatic Change in European Economic History', unpub. ms., presented at the ICCH, University of East Anglia, July 1979. An important critique of the existing literature on the influence of climate and climatic change on human affairs. To be published in Paul Uselding, (ed.), *Research in Economic History*, Vol. 6.

Barry, Roger: 1977, 'Short-Term Climatic Fluctuations', *Progress in Physical Geography* **1**, 114–25.

Beckinsale, R. 1965, 'Climatic Change, a Critique of Modern Theories', in Whittow, J. and Wood, P. (eds.), *Essays in Geography for Austin Miller*, Reading, 1–38.

Bowden, M. J., Kates, R. W., Kay, P. A., Riebsame, W. E., Warrick, R. A., Johnson, D. L., Gould, H. A., and Wiener, D.: 'The Effect of Climate Fluctuations on Human Populations: Two Hypotheses', *C&H:II*.

Bryson, R. A.: 1968, 'A Reconciliation of Several Theories of Climatic Change', *Weatherwise*.

Bryson, R. A.: 1974, 'A Perspective on Climatic Change', *Science* **184**, 753–60.

Bryson, R. A.: 1975, *Some Cultural and Economic Consequences of Climatic Change*, Madison, Wisconsin.

Bryson, R. A.: 1975, 'The Lessons of Climatic History', *Environmental Conservation*, **2**, **3**, 1963–79.

Bryson, R. A.: 1976, 'The Lessons of Climatic History', *Ecologist* **6**, 205–211.

Bryson, R. A. and Murphy, Thomas: 1977, *Climates of Hunger: Mankind and the World's Changing Weather*, Madison, Wisconsin.

Bryson, R. A. and Padoch, Christine: 1980, 'On The Climates of History', *JIH* **10**, 583ff.

Chappel, John: 1970, 'Climatic Change Reconsidered: Another Look at "The Pulse of Asia",' *Geographical Review* **60**, 347–73.

Chaunu, P.: 1967, 'Le Climat et l'Histoire, a Propos d'un Livre Recent', *Revue Historique*.

Clark University Climate and Society Research Group: 1979, *The Effect of Climate Fluctuations on Human Populations*, Worcester, Mass. Excellent bibliography.

Claxton, Robert M. and Hecht, Alan D.: 1978, 'Climatic and Human History in Europe and Latin America: An Opportunity for Comparative Study', *Climatic Change* **1**, 195–203.

Dunbar, M.: 1976, 'Climatic Change and Northern Development', *Arctic*. **29**, 183–93.

Dury, George H.: 1967, 'Climatic Change as a Geographical Backdrop', *Australian Geography* **10**, 231–42.

Fischer, David H.: 1980, 'Climate and History: Priorities for Research', *JIH* **10**, 821–30. Reprinted in *C&H* I.

Gunnarsson, Gisli: 1979, 'Some Interdisciplinary Problems of Climate and History: A Study in Causal Relations and Other Questions of Methodology', unpub. paper, ICCH, University of East Anglia.

Huntington, E.: 1907, *The Pulse of Asia*, Boston. Huntington's work has been continuously criticized for its simple-minded climatic determinism and its emphasis upon speculation (rather than empirical data-gathering) in reaching its conclusions.

Huntington, E.: 1924, *Civilization and Climate*. New Haven, 3rd ed.

Ingram, M. J. and Lamb, H. H.: 1980, 'Climate and History', Conference Report *P&P* **88**, 136ff.

Johnson, C. G. and Smith. L. P. (eds.): 1965, *Biological Significance of Climatic Changes in Britain*, New York.

Journal of Interdisciplinary History: 1980, Vol. **X**, No. 4, *History and Climate: Interdisciplinary Exporations*. Reprinted in *C&H* I.

Ladurie, E. LeRoy: 1959, 'Histoire et Climat', *Annales* pp. 3–24. Translated in Peter Burke (ed.), *Economy and Society in Early Modern Europe: Essays from Annales*. New York, 1972, pp. 134–169.

Ladurie, E. LeRoy: 1961, 'Aspects Historiques de la Nouvelle Climatologie', *Revue Historique*.

Lamb, H. H.: 1969, 'Climatic Fluctuations', in H. Flohn (ed.), *World Climate of Climatology* 173–249.

Lamb, H. H.: 1970, 'Volcanic Dust in the Atmosphere; with a Chronology and Assessment of its Meteorological Significance', *Philosophical Transactions of the Royal Society of London: Mathematical and Physical Sciences* **266**, 1178, 425–533.

Lamb, H. H.: 1972, *Climate: Past, Present and Future*, I, London, and II, London, 1977.

Lamb, H. H.: 1973, 'Climatic Change and Foresight in Agriculture: The Possibilities of Long-Term Weather Advice', *Outlook in Agriculture* 7, 203–10.

Lamb, H. H.: 1978, 'The Variability of Climate: Observation and Understanding', *Proceedings of the Nordic Symposium on Climatic Changes and Related Problems*, Copenhagen, pp. 144–55.

Lamb, H. H.: 1979, 'An Approach to the Study of the Development of Climate and Its Impact in Human Affairs'. ICCH, University of East Anglia, Published in *C&H* II.

Lawrence, E. N.: 1969, 'Effects of Urbanization on Long-Term Changes of Weather Temperature in the London Region', *MM*. **98**.

Lockwood, J.: 1977, 'Long-Term Climatic Changes', *Progress in Physical Geography* **1**, 104–113.

Mason, B. J.: 1976, 'Towards the Understanding and Prediction of Climatic Variations', *QJRMS* **102**, 476–98.

Mitchell, J. M.: 1968, 'Causes of Climatic Change', *Meteorological Monthly* **8**, 155–9.

Mitchell, Murray: 1976, 'An Overview of Climatic Variability and Its Causal Mechanisms', *QR* **6**, 481–94.

Morth, Umbert Hans: 1978, 'Climatological Research: An Interdisciplinary Study', *New Scientist* 691–3.

Pittock, A. B.: 1972, 'How Important Are Climatic Changes?', *Weather* **27**, 262–71.

Pittock, A. B., Frakes, L. A., Jenssen, D., Peterson, J. A., and Zillman, J. (eds.): 1978, *Climatic Change and Variability – A Southern Perspective*, Cambridge. See especially pp. 294–338: 'The Effect of Climatic Change and Variability on Mankind'.

Post, John D.: 1973, 'Meteorological Historiography', *JIH* **3**, 721–32.

Post, John D.: 1979, 'Climatic Change and Historical Explanation', *JIH* **10**, 2, 291–301.

Rabb, Theodore K. and Rotberg, Robert I.: 1981, *Climate and History: Studies in Interdisciplinary History*, Princeton.

Rabb, Theodore K.: 1980, 'The Historian and the Climatologist', *JIH* **10**, 4, 831–38. Reprinted in *C&H* II.

Schneider, S. H. and Temkin, R. L.: 1978, 'Climatic Changes and Human Affairs', in J. Gribbin (ed.), *Climatic Change*, Cambridge, Massachusetts, pp. 228–46.

Smith, L. P.: 1963, 'The Significance of Climate Variation in Britain', in *Changes of Climate, Proceedings of the UNESCO/WMO 1961 Rome Symposium*. Paris, pp. 455–63.

Smith. L. P.: 1965, 'Possible Changes in Seasonal Weather', in Johnson, C. G. and Smith, L. P. (eds.), *The Biological Significance of Climatic Changes in Britain*, London, pp. 187–91.

Smith, L. P.: 1970, 'The Changing Climate', *Agricultural Meteorology* 7, 361–2.

Warrick, R. and Riebsame, William: 1981, 'Societal Response to CO_2-Induced Climate Change: Opportunities for Research'. Paper prepared for the AAAS-DOE Panel on Social and Institutional Responses to CO_2-Induced Climate Change. *Climatic Change* **3**, 387. A major statement on the interaction between climate and society.

Wigley, T. M. L., Ingram, M. J., and Farmer, G.: 1981, *Climate and History: Studies in Past Climates and Their Impact on Man*. Cambridge.

Winstanley, D.: 1973, 'Rainfall Patterns and General Atmospheric Circulation', *Nature* **245**, 90–4.

B. Reconstructions of Past Climate: General and Methodological Works

Adem, Julian: 1981, 'Numerical Expenments on Ice Age Climates', *Climatic Change* **3**, 156ff.

Baker, Donald G.: 1980, "Botanical and Chemical Evidence of Climatic Change: A Comment," *JIH* **10**, 4, 813–820. Reprinted in *C&H* I.

Barron, E. J., Sloan II, J. L., and Harrison, C. G. A.: 1980, 'Potential Significance of Land-Sea Distribution and Surface Albedo Variations as a Climatic Factor; 180 m.y. to the Present', *PPP* **30**, 17ff.

Bell, Barbara: 1980, "Analysis of Viticultural Data by Cumulative Deviations', *JIH.* **10**, 4, 851–8. Reprinted in *C&H* I.

Birks, H. J. B.: 'The Use of Pollen Analysis in the Reconstruction of Past Climate: A Review'. Paper delivered at the ICCH. As indicated, a review essay with an extensive bibliography. Published in *C&H* II.

Brett, D. W.: 1978, 'Elm Tree Rings as a Rainfall Record', *Weather* **33**, 87–94.

Brinkman, A. R.: 1976, 'Surface Temperature Trend for the Northern Hemisphere – Updated', *QR* **6**, 355–8.

Britton, C. E.: 1937, 'A Meteorological Chronology to A. D. 1450', *Geographical Monthly* **8**.

Bryson, R. A.: 1966, 'Airmasses, Streamlines, and the Boreal Forest', *Geographical Bulletin* **VIII**, 228–69.

Bryson, R. A.: 1978, 'Cultural, Economic and Climatic Records', in A. B. Pittock, *et al.* (eds.), *Climatic Change and Variability – A Southern Perspective*. Cambridge, Massachusetts.

Bryson, R. A. and Padoch, Christine: 1980, 'On the Climates of History', *JIH* **10**, 4, 583–98. Reprinted in *C&H* I.

Catchpole, A. J. W. and Moodie, D. W.: 1976, 'Valid Climatological Data from Historical Sources by Content Analysis', *Science* **193**, 51–3.

Catchpole, A. J. W. and Moodie, D. W.: 1978, 'Archives and the Environmental Scientist', *Archivaria* **6**, 113–36.

Claxton, Robert: 1977–1980, *A Bibliography of Recent Works Regarding Climatic Variations and Its Effects in Historic Times*, 4 vols. Published by Claxton at West Georgia College, Carollton, Geogia, 30118, USA. An indispensable reference tool.

Cornish, P.: 1977, 'Changes in Seasonal and Annual Rainfall in New South Wales', *Search* **8**, 38–40.

Craddock, J. M.: 1974, 'Phenological Indicators and Climates', *Weather* **29**, 332–43.

Dansgaard, W.: 1981, 'Ice Core Studies: Dating the Past to Find the Future', *Nature* **90**, 360ff.

Denton, G. H. and Karlen, W.: 1973, 'Holocene Climatic Variations – Their Pattern and Possible Cause', *QR* **3**, 2, 155–205.

Dyer, Thomas G. J.: 1975, 'Secular Variation in South African Rainfall', University of Witwaterstrand; unpublished dissertation.

Dyer, Thomas G. J. and Curtis, B.: 1978, 'Analysis of Tree Rings and Climatic Changes', *South African Journal of Science* **74**, 176–8.

Eddy, J. A.: 1980, 'Climate and the Role of the Sun', *JIH* **10**, Reprinted in *C&H* I.

Flohn, H.: 1979, 'On Time Scales and Causes of Abrupt Paleoclimatic Events', *Quaternary Research* **12**, 135ff.

Fritts, H. C.: 1972, 'Tree Rings and Climate', *Scientific American*, **226**, 16, 91–100.

Fritts, H. C.: 1976, *Tree Rings and Climate.*

Fritts, H. C.: 1978, 'Tree Rings, A Record of Seasonal Variations in Past Climate', *Naturwissenschaften* **65**, 48–56.

Fritts, H. C., Lofgren, G. R., and Gordon, G. A.: 'Reconstructing Seasons to Century Time Scale Variations in Climate from Tree-Ring Evidence', *C&H* II.

Fritts, H. C., Lofgren, G. R., and Gordon, Geoffrey: 1979, 'Variations in Climate Since 1602 as Reconstructed from Tree Rings', *QR* **12**, 18ff.

Gray, J.: 'The Use of Stable-Isotope Data in Climate Reconstruction'. Unpublished paper presented at the ICCH, University of East Anglia. Includes lengthy bibliography. Published in *C&H* II.

Grove, J. M.: 1979, 'The Glacial History of the Holocene', *Progress of Physical Geography* **3**, 1–54. A major review essay: see for additional bibliography.

Groverman, Brian and Landsberg, H. E.: 1979, *Reconstruction of Northern Hemisphere Tempeature: 1579–1880.* College Park, Maryland.

Hage, Keith: 1977, 'Local History as a Source of Climatic Information', in J. Powell, (ed.), *Applications of Climatology* 11–28.

Harvey, L. D.: 1980, 'Solar Variability as a Contributing factor to Holocene Climatic Change', *Progress in Physical Geography* **4**, 487ff.

Hecht, Alan *et al.*: 1979, 'Paleoclimatic Research: Status and Opportunities', *QR* **12**, 6–17.

Herlihy, David: 1980, 'Climate and Documentary Sources: A Comment,' *JIH* **10**, 4, 713–7. Reprinted in *C&H* I.

Hischboeck, K. K.: 1980, 'A New Worldwide Chronology of Volcanic Eruptions (with a Summary of Historical Ash-Producing Activity and Some Implications for Climatic Trends of the Last 100 Years)', *PPP* **29**, 223ff.

Ingram, Martin *et al.*: 1978, 'Historical Climatology', *Nature* **276**, 329–34.

Ingram, Martin, Underhill, David J., and Farmer, G.: 1979, 'The Use of Documentary Sources for the Study of Past Climates', Paper presented at the ICCH, University of East Anglia. Includes bibliography of documentary sources. Published in *C&H* II,

Kington, John A.: 1974, 'An Application of Phenological Data to Historical Climatology', *Weather* **29**, 320–8.

Kington, John A.: 1977, 'Fluctuations Climatiques: Une etude synoptique du climat, fin XVIIIe-debut XIXe siecle', *Annales* **32**, 2, 227–36.

Kington, John A.: 1975, 'An Analysis of Seasonal Characteristics of 1781–1784 and 1968–71 Using PSCM Indices', *Weather* **30**, 109–114.

Kington, John A.: 1976, 'An Examination of Monthly and Seasonal Extremes Using Historical Weather Maps from 1781: October 1981', *Weather* **31**, 151–8.

Kington, John A.: 1978, 'Historical Daily Synoptic Weather Maps from the 1780's', *JM* **3**, 65–71.

Kraus, E. B.: 1955, 'Secular Changes of East Coast Rainfall Regimes', *QJRMS* **81**, 430–9.

Kraus, E. B.: 1956, 'Secular Changes of the Standing Circulation', *QJRMS* **82**, 289–300.

Kukla, G. L. and Mathews, R. K. *et al.*: 1977, 'New Data on Climatic Trends', *Nature* **270**, 573–80.

Kutzbach, John and Bryson, R. A.: 1974, 'Variance Spectrum of Holocene Climatic Fluctuations in the North Atlantic Sector', *Journal of the Atmospheric Sciences* **31**, 1958–1963,

Kutzbach, J. E. and Guettner, R. J.: 1980, 'On the Design of Paleoenvironmental Data Networks for Estimating Large-Scale Patterns of Climate', *QR* **14**, 169ff.

LaMarche, V. C.: 1974, 'Paleoclimatic Inferences from Long Tree-Ring Records', *Science* **183**, 1043–8.

LaMarche, V. C.: 1978, 'Tree-Ring Evidence of Past Climatic Variability', *Nature* **376**, 334–8.

Lamb, H. H. and Johnson, A. I.: 1966, 'Secular Variations of the Atmospheric Circulation since

1750', *Geophysical Memoirs, 110*. See also *idem*. 'Climatic Variation and Observed Changes in the General Wind Circulation, Parts I–III', *Geografiska Annaller* **41**, 1959, 94–134 and 43, 1961, 363–400.

Lamb, H. H.: 1966, *The Changing Climate*, London.

Lamb, H. H.: 1974, 'Climate, Vegetation and Forest Limits in Early Civilized Times', *Philosophical Transactions of the Royal Society*, A, **276**, 193–230.

Lamb, H. H.: 1977, *Climate: Present, Past and Future; Vol. 2: Climatic History and the Future*, London. *The* standard work on the chronology of past climates, with an invaluable bibliography.

Lamb, H. H. and North, H. T.: 1978, 'Arctic Ice, Atmospheric Circulation and World Climate', *Geographical Journal* **144**, 1–22.

Lamb, H. H.: 1979, 'Climatic Variation and Changes in the Wind and Ocean Circulation: The Little Ice Age in the Northeast Atlantic', *QR* **11**, 1–20.

Lamb, H. H.: 1980, 'Weather and Climate Patterns of The Little Ice Age', in Oeschger, H., Messerli, B. and Suilar, M. (eds.), *Das Klima – Analysen und Modelle, Geschichte und Zukunft*, Berlin.

Landsberg, Helmut: 1981, 'Using Early Weather Records', *Weatherwise* **34**, 197ff.

Landsberg, H. E.: 1980, 'Past Climates from Unexploited Written Sources', *JIH* **10**, 4, 631–42. Includes footnote citations of medieval and early modern sources, and extant secondary literature on them. Reprinted *C&H* I.

Landsberg, H. E. and Albert, J. M. 1974, 'Summer of 1816 and Volcanism', *Weatherwise* **27**, 63–6.

Mercer, J. H.: 1976, 'Glacial History of Southernmost South America', *QR* **6**, 125–66.

Mitchell, J. M., Jr.: 1963, 'On the World-Wide Pattern of Secular Temperature Change', in *Changes of Climate: Proceedings of the UNESCO/WMO Rome 1961 Symposium*, Paris, pp. 161–81.

Moodie, D. W. and Catchpole, A. J. W.: 1975, 'Environmental Data from Historical Documents by Content Analysis: Freeze-up and Break-up of Estuaries of Hudson Bay 1714–1871', *Manitoba Geographical Studies* **5**, Winnipeg.

Oeschger, H., Messerli, B., and Svilar, M. (eds.): 1980, *Das Klima – Analysen and Modelle, Geschichte und Zukunft*, Berlin.

Oliver, J. and Kington, J. A.: 1970, 'The Usefulness of Ships' Log-Books in the Synoptic Analysis of Past Climates', *Weather* **25**, 520–8.

Pattersson, O.: 1914, *Climatic Variations in Historic and Prehistoric Times*. Berlin; reprint of *Svenska Hydrografisk-Biologiska Skriften*, Vol. 5.

Pierce, J.: 1974, 'Cultural Sensitivity to Environmental Change: II, 1816, The Year Without a Summer', Center for Climate Research, Institute for Environmental Studies, IES Report No. 15. University of Wisconsin, Madison,

Porter, S. C.: 'Glaciological Evidence of Holocene Climatic Change'. Paper presented to the ICCH, University of East Anglia, 1979. Published in *C&H* II.

Potter, H. R.: 1978, 'The Use of Historic Records for the Augmentation of Hydrological Data', Institute of Hydrology, Rep. No. 46, Wallingford, Oxfordshire. Includes extensive bibliography of sources.

Proceedings of the Conference on the Climate of the Eleventh and Sixteenth Centuries. Aspen, Colorado, 1962. National Center for Atmospheric Research, NCAR Technical Note, 63–1, 1963. See also Ladurie's note on the conference: *Annales*, 1963.

Ratcliffe, R. A. S. *et al.*: 1978, 'Variability in the Frequency of Unusual Weather Over Approximately the Last Century', *QJRMS* **104**, 243–55.

Reed, Arden: 'The Climates of Coleridge and Baudelaire: A Meteorology', (unpublished dissertation), Johns Hopkins University.

Reimherr, George: 1976, *Paleoclimatology: A Bibliography*. National Technical Information Service.

Robock, Alan: 1979, 'The "Little Ice Age": Northern Hemisphere Average Observations and Model Calculations', *Science* **206**, 1402–4.

Salinger, M. J.: 1976, 'New Zealand Temperatures Since 1300 A.D.', *Nature* 260, 310–1.

Schwartzbach, M.: 1963, *Climates of the Past*, American Meteorological Society, reprinted 1977.

Sorenson, Curtis J. *et al.*: 1971, 'Paleosols and the Forest Border in Keewatin, N.W.T.', *QR* **1**, 468–73.

Stommel, Henry and Elizabeth: 1979, 'The Year Without A Summer', *Scientific American*, **240**, 176–86.

Stow, C. D.: 1978, 'The Voyage of the Beagle – Meteorological Notes and Curiosities', *Weather* **33**, 340–50.

'Tales The Ice Can Tell', *Mosaic* 9, 1978, 15–21.

Thompson, L. *et al.*: 1979, 'Climatic Ice Core Records from the Tropical Quelccaya Ice Cap', *Science* **203**, 1240–3.

Turner, J.: 1965, 'A Contribution to the History of Forest Clearance', *Proceedings of the Royal Society* **B161**, 343–54.

Van der Zwan, C. J.: 1981, 'Palynology, Phytogeography, and Climate of the Long Carboniferous', *PPP* **33**, 279ff.

Wardle, R.: 1973, 'Variations of the Glaciers of Westland National Park and the Hooker Range, New Zealand', *New Zealand Journal of Botany* **11**, 349–88.

Webb, Thompson: 1980, 'The Reconstruction of Climatic Sequences from Botanical Data', *JIH* **10**, 4, 749–72. Reprinted in *C&H* I.

Wendland, William and Bryson, Reid: 1974, 'Dating Climatic Episodes of the Holocene', *QR* **4**, 9–24.

Wigley, T. M. L.: 1977, 'Geographical Patterns of Climatic Change: 1000 B.C. – 1700 A.D.', Interim Final Report to the National Oceanic and Atmospheric Administration.

Wilburn, J. C.: 1976, *Estimates of Climatic Changes Following Volcanic Eruptions from Tree Growth Records*. Fort Huachuca, Arizona.

Willett, H. C.: 1950, 'Temperature Trends of the Past Century', *Centenary Proceedings*. Royal Meteorological Society, London, pp. 195–206.

Williams, I. M.: 1978, 'Inland Ice Sheet Thinning Due to Holocene Warmth', *Science* **201**, 1014–16.

Wilson, A. T.: 1980, 'Isotope Evidence from Past Climatic and Environmental Change', *JIH* **10**, 4, 795–812. Reprinted in *C&H* I.

C. Reconstructions of the British Climate

Baker, J. N. L.: 1932, 'The Climate of England in the Seventeenth Century', *QJRMS* 58, 421–39.

Burroughs, W. J.: 1979, 'An Analysis of Winter Temperatures in Central England and Newfoundland', *Weather* **34**, 19–23.

Craddock, J. M.: 1976, 'Annual Rainfall in England Since 1725', *QJRMS* **102**, 823–40.

Craddock, J. M.: 1977, 'A Homogeneous Record of Monthly Rainfall Totals for Norwich for the Years 1836 to 1976', *MM* **106**, 267–278.

Craddock, J. M.: 1977, 'Rainfall at Oxford from 1767 to 1814', *MM* **109**, 361–72.

Craddock, J. M. and Wales-Smith, B. G.: 1977, 'Monthly Rainfall Totals Representing the East Midlands for the Years 1726 to 1975', *MM* **106**, 97–111.

Davis, N. E.: 1972, 'The Variability of the Onset of Spring in Great Britain', *QJRMS* **98**, 763–777.

Douglas, K. *et al.*: 1978, 'A Meteorological Study of July to October 1588: The Spanish Armada Storms', *University of East Anglia, Climatic Research Unit, Research Publication* **6**.

Dyer, T. G. J.: 1976, 'An Analysis of Manley's Central England Temperature Data', *QJRMS* **102**, 871–88.

Glasspoole, J.: 1933, 'The Rainfall Over the British Isles of Each of the Eleven Decades During the Period 1820 to 1929', *QJRMS* **59**, 253–60.

Gray, B. M.: 1976, 'Medium Term Fluctuations of Rainfall in Southeastern England', *QJRMS* **102**, 627–38.

Hallam, H. E.: 1979, 'The Climate of East Anglia, 1250–1350', Paper delivered at the ICCH, University of East Anglia, Norwich.

Hughes, M. K. *et al.*: 1978, 'Climatic Signals in British Isles Tree-Ring Chronologies', *Nature* **272**, 605–6.

Jackson, M. C.: 1978, 'Snow Cover in Great Britain', *Weather* **33**, 298–309.

Kemp, D. D.: 1976, 'Winter Weather in West Fife in the 18th Century', *Weather* **31**, 400–4.

Kington, John A.: 1975, 'An Analysis of Seasonal Characteristics of 1781–1784 and 1968–71 Using PSCM Indices', *Weather* **30**, 109–114.

Kington, John A.: 1976, 'An Examination of Monthly and Seasonal Extremes Using Historical Weather Maps from 1781: October 1781', *Weather* **31**, 151–8.

Kington, John A.: 1978, 'Historical Daily Synoptic Weather Maps from the 1780's', *JM* **3**, 65–71.

Kington, John A.: 1975, 'A Comparison of British Isles Weather Type Frequencies in the Climatic Record from 1781 to 1971', *Weather* **30**, 21–4.

Kington, John A.: 1976, 'An Introduction to an Examination of Monthly and Seasonal Extremes in the Climatic Record of England and Wales . . . from 1781 Onward', *Weather* **31**, 72–8 and 151–8.

Lamb, H. H.: 1967, 'Britain's Changing Climate', *Geographical Journ.* **133**, 4, 445–68.

Ledger, D. C. and Thom, A. S.: 1977, '200 Years of Potential Moisture Deficit in Southeast Scotland', *Weather* **32**, 342–9.

Manley, G.: 1940, 'Snowfall in the British Isles', *MM* **75**, 41–8.

Manley, G.: 1947, 'Snow Cover in the British Isles', *MM* **76**, 28–36.

Manley, G.: 1949, 'The Snowline in Britain', *Geografiska Annaler* **36**, 179–93.

Manley, G.: 1953, 'The Mean Temperature of Central England, 1698–1952', *QJRMS* **79**, 242–61.

Manley, G.: 1951, 'The Range of Variation of the British Climate', *Geographical Journ.* **117**, 43–68.

Manley, G.: 1952, 'Thomas Barker's Meteorological Journals, 1748–1763 and 1777–1789', *QJRMS* **78**, 255–9.

Manley, G.: 1959, 'Temperature Trends in England 1698–1957', *Archiv für Meteorologische Geophysik und Bioklimat* B, **9**, 413–33. A seminal article.

Manley, G.: 1961, 'A Preliminary Note on Early Meteorological Observations in the London Region with Estimates of Monthly Mean Temperature 1680–1706', *MM* **90**, 303–10.

Manley, G.: 1962, 'Early Meteorological Observations and the Study of Climatic Fluctuations', *Endeavour* **21**, 43–50.

Manley, G.: 1969, 'Snowfall in Britain Over the Past 300 Years', *Weather* **24**, 428–37.

Manley, G.: 1971, 'The Mountain Snows of Britain', *Weather* **26**, 192–300.

Manley, G.: 1974, 'Central England Temperatures: Monthly Means 1659–1973', *QJRMS* **100**, 389–405.

Manley, G.: 1975, '1684: The Coldest Winter in the English Instrumental Record', *Weather* **30**, 383–8.

Manley, G.: 1978, 'Variation in the Frequency of Snowfall in East-Central Scotland, 1708–1975', *MM* **107**, 1–16.

Meyer, G. M.: 1927, 'Early Water Mills in Relation to Changes in the Rainfall of East Kent', *QJRMS* **53**.

Nicholas, F. J. and Glasspoole, J.: 1931, 'General Monthly Rainfall Over England and Wales, 1727–1931', *British Rainfall* 299–306.

Norgate, T. B.: 1977, 'Early 17th Century Weather in Norfolk', *Weather* **32**, 305–6.

Pearson, M. G.: 1973, 'Snowstorms in Scotland, 1782–86', *Weather* **28**, 195–201.

Pearson, M. G.: 1973, 'The Winter of 1739–40 in Scotland', *Weather* **28**, 20–4.

Pearson, M. G.: 1975, 'Never Had it So Bad', *Weather* **30**, 14–21.

Pearson, M. G.: 1975, '1779–A Year of Remarkable Weather in Eastern Scoland', *Weather* **30**, 34–42.

Pearson, M. G.: 1976, 'Snowstorms in Scotland, 1729–1830', *Weather* **31**, 390–2.

Pearson, M. G.: 1978, 'Snowstorms in Scotland, 1831–1861', *Weather* **33**, 392–9.

Pennington, W.: 1969, *The History of British Vegetation*, London.

Pilcher, J. J. and Baillie, M. G. L.: 1979, 'Climate Reconstructions of the 15th Century Using British Isles Tree-Rings', Paper presented at the ICCH, University of East Anglia.

Polland, C. K. and Wales-Smith, B. G.: 1977, 'Richard Towneley and 300 Years of Regular Rainfall Measurement', *Weather* **32**, 438–45.

Schove, D. J.: 1973, 'Weather in Scotland, 1659–1660: The Diary of Andrew Hay', *Annals of Science* **30**, 165–77.

Schove, D. J.: 1974, 'Dendrochronological Dating of Oak from Old Windsor, Berkshire, A.D. 650–906', *Medieval Archaeology* **18**, 165–72.

Schweingruber, F. H., Braeker, O. U., and Schaer, E.: 1978, 'Dendroclimatic Studies in Great Britain and in the Alps', in *Evolution of Atmospheres and Climatology of the Earth*, Toulouse.

Schweingruber, F. H., Braeker, O. U., and Schaer, E.: 'Dendroclimatic Studies from Central Europe and England', *Boreas* (forthcoming).
Sissons, J. B.: 1979, 'Palaeoclimatic Inferences from Former Glaciers in Scotland and the Lake District', *Nature* **278**, 518–21.
Smith, C. G.: 1974, 'Monthly, Seasonal and Annual Flucturations of Rainfall at Oxford Since 1815', *Weather* **29**, 2–17.
Smithson, P. A.: 1976, 'Precipitation Fluctuations and the Incidence of Heavy Fall of Rain in South Yorkshire', *Weather* **31**, 246–55.
Titow, J. Z.: 1960, 'Evidence of Weather in the Account Rolls of the Bishopric of Winchester, 1209–1350', *EHR*, Second Series, **12**, 3, 360–407.
Titow, J. Z.: 1970, 'Le Climat a Travers les Roles de Comptabilite de l'eveche de Winchester (1350–1450)', *Annales* **2**.
Unsworth, M. *et al.*: 1979, 'The Frequency of Fog in the Midlands of England', *Weather* **34**, 72–7.
Varley, G. *et al.*: 1979, 'Climatic Signals in Tree-Ring Chronologies for British Oaks (and reply)', *Nature* **278**, 282–3.
Wales-Smith, B. G.: 1971, 'Monthly and Annual Totals of Rainfall Representative of Kew, Surrey, from 1697 to 1970', *MM* **100**, 345–62.
Wales-Smith, B. G.: 1977, 'An Analysis of Monthly Potential Evaporation Totals Representative of Kew from 1698 to 1976', *MM* **106**, 297–313.
Wigley, T. and Atkinson, T.: 1977, 'Dry Years in South-East England Since 1698', *Nature* **265**, 431–4.

D. Reconstructions of the European Climate

Alexandre, P.: 1974, 'Histoire du Climat et Sources Narratives du Moyen Age', *Le Moyen Age* **1**, 101–16.
Alexandre, P.: 1976, *Le Climat au Moyen Age en Belgique et Dans les Regions Voisins*, Louvain.
Alexandre, P.: 1977, 'Les Variations Climatiques au Moyen Age (Belgique, Rhenanie, Nord de la France)', *Annales* **32**, 183–97.
Amberg, B.: 'Beitrage zur Chronik der Witterung und verwandter Naturerscheinungen mit besonder Rucksicht auf das Gebiet der Reuss und der angrenzenden Gebiete der Aare und des Rheines', 3 Vols., *Beilage zu den Jahresbericht über die Hoehere Lehrant*. Luzern.
Angot, Alfred: 1885, 'Etude sur les vendages en France', *Annales du Bureau central meterologique de France*.
Baulant, M. and Ladurie, E. LeRoy: 1973, 'Les dates de vendages au XVIe siecle, elaboration d'une serie septentrionale', in *Melanges en l'honneur de Fernand Braudel, Methodologie de l'Histoire et des Sciences humaines*. Toulouse.
Bell, W. T. and Ogilvie, A. E. J.: 1978, 'Weather Complications as a Source of Data for the Reconstruction of European Climate During the Medieval Period', *Climatic Change* **1**, 331–48.
Bergthorssen, P.: 1969, 'An Estimate of Drift Ice and Temperature in Iceland for 1000 Years', *Jokull* **19**, 94–101.
Betin, Vasily V. and Preobrajenskii, Yu. V.: 1962, *Surovost zim v Evrope i ledovitost Baltiki* (Severe Winters in Europe and Ice Cover of the Baltic). Leningrad.
Bider, Max: 1960, 'Untersuchung an einer 67–jaehrigen Reihe von Beobachtungen der Kirschblute bei Liestal (Basel-Landschaft)', *Wetter und Leben* **10**, 36–50.
Bider, Max, Schuepp, M. and von Rudloff, H.: 1958, 'Die Reduktion der 200 jaehrigen Basler Temperaturreihe', *Archiv für Meteorologische Geophysik und Bioklimat*. B, **9**, 360–412.
Dansgaard, W., Johnsen, S. J., Moller, J., and Langway, C.: 1979, 'One Thousand Centuries of Climatic Record from Camp Century on the Greenland Ice Sheet', *Science* **166**, 215–23.
Dettwiller, Jacques: 1978, 'L'evolution seculaire de la temperature a Paris', *La Meteorologie* **13**, 95–130.
deVries, Jan: 1977, 'Histoire du climat et economie: des faits nouveaux, une interpretation differente', *Annales* **XXXII**, 202–207.
Die Schweiz und Ihre Gletscher, von der Eiszeit bis zur Gegenwart: 1979, Switzerland.
Duchaussoy, H.: 1934, 'Les bans de vendages de la region parisienne', *La Meteorologie*, 111–188.

Easton, C.: 1928, *Les hivers dans l'Europe occidentale*, Leyden.

8000 Jahre Walliser Gletschergeschichte: Ein Beitrag zur Erforschung des Klimaverlaufs in der Nacheiszeit: 1976, *Die Alpen* **3/4**. Includes articles on fluctuations of glacier extent by Walter Schneebeli and Friedrich Rothlisberger. Includes extensive bibliography.

Emery, F. V. and Smith, C. G.: 1976, 'A Weather Record from Snowdonia, 1697–8', *Weather* **31**, 142–50.

Fischer, Rudolf: 'Die Kaltesten Wintermonate in Berlin 1719–1941', *Zeitschrift für Angewandte Meteorologie* **110**.

Fletcher, J.: 1978, *Dendrochronology in Europe: Principles, Interpretations and Applications to Archaeology and History*. British Archeological Reports, International Series 51.

Flohn, H.: 1949, 'Klima und Witterungs-Ablauf in Zuerich in 16. Jahrhundert', *Vierteljahrsschrift der Naturforschenden Gesellschaft Zürich* **95**, 28–41.

Flohn, H.: 1972, 'Klimaschwankung im Mittelalter und ihre historisch-geographische Bedeutung', *Berichte zur Deutsche Landeskunde* **7**, 347–57.

Frenzel, B.: 1977, *Dendrochronologie und Klimaschwankungen in Europa*, Erdwissenschaftliche Forschung, Vol. 13.

Garnier, M.: 1955, 'Contribution de la phenologie a l'etude des variations climatiques', *La Meteorologie*, **40**, 4, 291–300.

Gottschalk, M. K. E.: 1971, *Stormvloeden en rivieroverstromingen in Nederland. Deel I: De Periode voor 1400*, Assen.

Gottschalk, M. K. E.: 1975, *Stormvloeden en rivieroverstromingen in Nederland. Deel II: De Periode 1400–1600*, Assen.

Gottschalk, M. K. E.: 1977, *Stormvloeden en rivieroverstromingen in Nederland. Deel III: De Periode 1600–1700*, Assen.

Haeberli, W. and Schweingruber, F. H.: 1979, 'Klima seit der Eiszeit', in *Die Schweiz und ihre Gletscher. Von der Eiszeit bis zur Gegenwart*, pp. 26–45.

Hamberg, H. E.: 1906, 'Moyennes mensuelles et annuelles de la temperature et extremes de temperatures mensuels pendant les 150 annees 1756–1903 a l'observatoire de Stockholm', *Kunglis Svenska Vetenskapakademiens handlingar* **40**, 1–59.

Hughes, Malcolm: 1978, 'European Tree-Rings and Climate', *New Scientist* **77**, 500–2.

Huttunen, P. and Stober, Julie: 1980, 'Dating of Palaeomagnetic Records from Finnish Lake Sediment Cores Using Pollen Analysis', *Boreas* **9**, 193ff.

Jeanneret, Francois and Vautier, Philippe: 1977, *Kartierung der Klimaeignung für die Landwirtschaft in der Schweiz*, Bern.

Jensen, F.: 1976, (Title translated from the Dutch) *Summers in the Netherlands Since 1706 from a Thermal Point of View*. Royal Netherlands Meteorological Institute.

Kington, J. A.: 1980, 'Daily Weather Mapping from 1781: A Detailed Synoptic Examination of Weather and Climate During the Decade Leading up to the French Revolution', *Climatic Change* **3**, 7ff.

Labrijn, A.: 1945, 'Het klimaat van Nederland gedurende de laatse twee en een halve eeuw', *Koninklijk Nederlandsch Meteorologisch Instituut Mededelingen en Verhandelingen* **102**.

Ladurie, E. LeRoy: 'Fluctuations meteorologiques et bans de vendages au XVIIIe siecle', *Federation historique du Languedoc mediterraneen et du Roussillon*, 30th and 31st Congresses. Sete-Beaucaire, 1956–7, Montpellier, N.D.

Ladurie, E. LeRoy: 1960, 'Climat et recoltes aux XVIIe et XVIIIe siecles', *Annales*.

Ladurie, E. LeRoy: 1965, 'Le climat des XIe et XVIe siecles, series comparees', *Annales*.

Ladurie, E. LeRoy: 1971, *Times of Feast, Times of Famine*. New York. Enlarged translation of *Histoire du climat depuis l'an mil*. Paris, 1967. An important synthetic work with a copious bibliography.

Ladurie, E. LeRoy: 1978, 'Une synthese provisoire; les vendages du XVe au XIXe siecle', *Annales* **33**, 763–71.

Ladurie, E. LeRoy and Baulant, Micheline: 1980, 'Grape Harvests from the Fifteenth Through the Nineteenth Centuries', *JIH* **10**, 4, 839–49. Reprinted in *C&H* I.

LaMarche, V. C. and Fritts, H.: 1971, 'Tree Rings, Glacial Advance, and Climate in the Alps', *Zeitschrift für Gletscherkunde und Glazialgeologie* 7, 1–2, 125–31.

Lamb, H. H.: 1965, 'The Early Medieval Warm Epoch and Its Sequel', *PPP* **1**, 13–37.

Lamb. H. H.: 1974, 'Contributions to Historical Climatology: the Middle Ages and After; Christmas Weather and Other Aspects', in *Klimatologische Forschung/Climatological Research (The Hermann Flohn 60th Anniversary Volume)*. Bonn, 549–67.

Lenke, Walter: 1960, *Klimadaten von 1621–1651 nach Beobachtungen des Landgrafen Hermann IV. von Hessen*, Offenbach.

Lenke, Walter: 1961, *Bestimmung der alten Temperaturwerte von Tübingen und Ulm mit Hilfe von Haufigkeitsverteilunge*. Offenbach.

Lenke, Walter: 1964, *Untersuchung der altesten Temperaturmessungen mit Hilfe des strengen Winters 1708–1709*, Berichte des Deutschen Wetterdienstes, Nr. 92.

Lenke, Walter, 1968, *Das Klima des 16. und Anfang des 17. Jahrhunderts nach Beobachtungen von Tycho de Brahe auf Hven, Leonard III Treutwein in Furstenfeld und David Fabricius in Ostenfriesland*. Offenbach.

Liljequist, G. H.: 1943, 'The Severity of Winters at Stockholm, 1757–1942', *Geografiska Annaler* 81–97.

Lindgren, S. and Neumann, J.: 1981, 'The Cold and Wet Year 1695 – A Contemporary German Account', *Climatic Change* **3**, 173ff.

Long, C.: 1974, 'The Oldest European Weather Diary', *Weather* **29**, 233–7.

Matthews, J. A.: 1977, 'Glacier and Climatic Fluctuations Inferred from Tree-Growth Variations Over the Last 250 Years, Central Southern Norway', *Boreas* **6**, 1–24.

Messerli, Bruno *et al.*: 1975, 'Die Schwankungen des Unteren Grindelwaldgletschers seit dem Mittelalter. Ein interdisziplinarer Beitrag zur Klimageschichte', *Zeitschrift für Gletscherkunde und Glazialgeologie* **11**, 1, 3–110.

Messerli, Bruno *et al.*: 1978, 'Fluctuations of Climate and Glaciers in the Bernese Oberland, Switzerland, and their Geo-Ecological Significance, 1600–1975', *Arctic and Alpine Research* **10**, 247–60.

Morner, Nils-Axel and Wallin, Bill: 1977, 'A 10,000-Year Temperature Record from Gotland, Sweden', *PPP* **21**, 113–138.

Mueller, K.: 1953, *Geschichte des Badischen Weinbaus*, 2nd edition.

Patzelt, G. and Bortenschlager, S.: 1973, 'Die Post-glazialen Gletscher und Klimaschwankungen in der Venedigergruppe (Hohe Tauern, Ostalpen)', *Zeitschrift für Geomorphologie*, N. F., Supp. **16**, 25–72.

Pfister, C.: 1975, 'Die Schwankungen des Untern Grindelwaldgletschers im Vergleich mit historischen Witterungsbeobachtungen und Messungen', *Zeitschrift für Gletscherkunde und Glazialgeologie,* **11**, 1, 74–90.

Pfister, C.: 1977, 'Zum Klima des Raumes Zürich im spaten 17. und fruhen 18. Jahrhundert', *Vierteljahrsschrift der Naturforschenden Gesellschaft Zürich.* **CXXII**, 447–471.

Pfister, C.: 1972, 'Phanologische Beobachtungen in der Schweiz der Aufklarung', *Inf. und Beitr. z. Klimaforschung*, Edited by the Geographical Institute of the University of Bern. Vol. 9, 15–30.

Pfister, C.: 1978, 'Fluctuations in the Duration of Snow-Cover in Switzerland Since the Late Seventeenth Century', *Proceedings of the Nordic Symposium on Climatic Changes and Related Problems*. Copenhagen, 1–6.

Pfister, C.: 1978, 'Die Alteste Niederschlagsreihe Mitteleuropas: Zürich 1708–1754', *Meteorologische Rundschau* **31**, 56–62.

Pfister, C.: 'Getreide-Erntebeginn und Frühsommertemperaturen im Schweizerischen Mittelland seit dem Frühen 17. Jahrhundert', *Geographica Helvetica* **XXXIV**, 23–35.

Pfister, C.: 1979, 'The Reconstruction of Past Climate: The Example of the Swiss Historical Weather Documentation'. Paper delivered at the ICCH, University of East Anglia.

Pfister, C.: 1980, 'The Little Ice Age: Thermal and Wetness Indices for Central Europe', *JIH* **X**, 4, 665–96. Reprinted in *C&H* I.

Pfister, C.: 'Local Phenological Time Series from the Canton of Schaffhausen (Switzerland) and Their Application for the Interpretation of Historical Records'. Unpub. ms.

Portmann, J. P.: 1977, 'Variations Glaciares, Historiques et Prehistoriques dans les Alpes Suisses', *Les Alpes* **4**, 145–72.

Renou, E.: 1889, 'Etudes sur le Climat de Paris, 3e partie. Temperature', *Annales du Bureau central meteorologique de France*.

Riggenbach, A.: 1891, 'Die Niederschlags-Verhaltnisse von Basel', *Denkschriften der Schweizerischen Naturforschenden Gesellschaft* **32**, 2, Zürich.

Rima, Alessandro: 1963, 'Considerazioni su una serie agraria bisecolare; la produzione di vino nel Rheingau (1719–1950)', *Geofisica e Meteorologia* **XII**, 25–31.

Rothlisberger, F.: 'Late Pleistocene and Holocene Glacier and Climatic Fluctuations in the Valaisian Alps, Switzerland: Radiocarbon Dating of Fossil Soils and Woods from Moraines and Glaciers'. Forthcoming in *Geographica Helvetica*.

Ruddiman, W. F. and Cline, R. M.: 1981, 'Deglacial Warming of the Northeastern Atlantic Ocean: Connection with the Paleoclimatic Evolution of the European Continent', *PPP* **35**, 111ff.

Sandon, F.: 1974, 'A Millenium of West European Climate', *Weather* **29**, 162–6.

Schaller, A.: 1937, *Chronik der Naturereignisse im Urnerland 1000–1800*, Altdorf.

Schneebeli, W., Leuzinger, H., Rothlisberger, F., and Muller, H. N.: 1974, 'Les Variations Climatiques Postglaciaries dans les Alpes – Quatre Communications'. *Congress of the International Glacier Society*, Courmayeur. Zürich.

Schnelle, Fritz: 1950, 'Hundert Jahre Phanologische Beobachtungen im Rhein-Main Gebiet 1841–1939, 1867–1947. Ein Beitrag zur Klimageschichte des Rhein-Main Gebiets', *Meteorologische Rundschau* **II**, 150–6.

Schnelle, Fritz: 1959, 'Temperaturverhaltnisse und Pflanzenentwicklung in der Zeit von 1731 bis 1740 in Mittel-und Westeuropa', *Meteorologische Rundschau* **XI**, 58–63.

Schuepp. M.: 1960, 'Lufttemperatur 2. Teil, Klimatologie der Schweiz', *Beiheft zu den Annalen der Schweizerische Meteorologischen Zentralanstalt*, Zürich.

Schweingruber, F. H. *et al.*: 1979, 'Stand und Anwedung der Dendrochronologie in der Schweiz', *Zeitschrift für Schweizerische Archaologie und Kunstgeschichte* **36**, 2, 69–90.

Schweingruber, F. H., Braeker, O. U., and Schaer, E.: 1978, 'Dendroclimatic Studies in Great Britain and in the Alps', *Evolution of Atmospheres and Climatology of the Earth*, Toulouse.

Schweingruber, F. H., Braeker, O. U. and Schaer, E.: "Dendroclimatic Studies from Central Europe and England", *Boreas*. Forthcoming.

Schweitzer, H. J.: 1980, 'Environment and Climate in The Early Tertiary at Spitsbergen', *PPP* **30**, 297ff.

Stauffer, Bernhard and Luthi, Alfred: 1975, 'Wirtschaftsgeschichtliche Quellen im Dienste der Klimaforschung', *Geographica Helvetica* **XXX**, 49–56.

Steensberg, A.: 1951, 'Archaeological Dating of the Climatic Change in North Europe About A.D. 1300', *Nature* **168**, 672–4.

Thornes, J. B.: 1974, 'The Rain in Spain', *Geographical Magazine* **46**, 337–43.

Van den Drool, H. *et al.*: 1978, 'Average Winter Temperatures at De Bilt (The Netherlands), 1634–1977', *Climatic Change* **1**, 319–330.

Vanderlinden, E.: 1924, 'Chronique des evenements meteorologiques en Belgique jusqu'en 1834', *Memoires de l'Academie Royale de la Belge*, 2nd series, Vol. 5.

Von Rudloff, Hans: 1967, *Die Schwankungen und Pendelungen des Klimas in Europa seit dem Beginn der regelmassigen Instrumenten-Beobachtungen (1670)*, Braunschweig.

Wallen, C. C.: 1953, 'The Variability of Summer Temperatures in Sweden and Its Connection with Changes in the General Circulation', *Tellus* **5**, 2, 157–78.

Weger, Nikolaus: 1952, 'Weinernten und Sonnenflecken', *Berichte des Deutschen Wetterdienstes in der US-Zone*, 229–37.

Weikinn, C.: 1958, *Quellentexte zur Witterungsgeschichte Europas von der Zeitwende bis zum Jahre 1850, I: Hydrographie (Zeitwende–1500)*, Berlin.

Winistorfer, Jorg: 1980, 'Late Pleistocene and Holocene Glacier Extents in the Central Valaisian Alps'. Will be published in *Geographica Helvetica* for the IGU Congress, Tokyo. Manuscript copy includes extensive and up-to-date bibliography on the glacier history of Europe, Alaska, and New Zealand.

Wright, Peter: 1968, 'Wine Harvests in Luxembourg and the Biennial Oscillations in European Summers', *Weather* **XXIII**, 300–4.

Zumbuehl, H. J.: 1979, 'Die Schwankungen der Grindelwaldgletscher in den historischen Bild- und

Schriftquellen des 12. bis 19. Jahrhunderts. Ein Beitrag zur Gletschergeschichte und Erforschung des Alpenraumes', *Denkschriften der Schweizerischen Naturforschenden Gesellschaft*, Vol. 92. Zürich.

E. Reconstructions of the Chinese and Japanese Climate

Arakawa, H.: 1954, 'Fujiwhara on Five Centuries of Freezing Dates of Lake Suwa in Central Japan', *Archiv für Meteorologie, Geophysik und Bioklimatologie*, **B, 6**.

Arakawa, H.: 1955, 'Remarkable Winters in Japan from the Seventh Century', *Geofisica Pura E Applicata* **30**, 144–6.

Arakawa, H.: 1956, 'On the Secular Variation of Annual Totals of Rainfall at Seoul from 1770 to 1944', *Archiv für Meteorologie, Geophysik und Bioklimatologie,* **VII**, 406–12.

Arakawa, H.: 1956, 'Dates of First or Earliest Snow Covering for Tokyo Since 1632', *QJRMS* **82**.

Arakawa, H.: 1956, 'Climatic Change as Revealed by the Blooming Data of the Cherry Blossoms at Kyoto', *JM* **13**.

Arakawa, H.: 1957, 'Climatic Change as Revealed by the Data from the Far East', *Weather* **XII**.

Central Meteorological Research Bureau of China: 1977, 'A Study of Occurrences of Flood and Drought in the Last 500 Years in Northern and Northeastern China', (translated title), *Collected Papers on Climatic Change and Long Range Weather Forecasts*, Peking.

Chu, K.: 1973, 'A Preliminary Study on the Climatic Fluctuations During the Last 5000 Years in China', *Scientia Sinica* **16**, 226–56.

Co-Ching, Chu: 1954, 'Climatic Change During Historic Times in China', *Collected Scientific Papers: Meteorology, 1919–1949*, Peking.

Gray, Barbara: 1974, 'Early Japanese Winter Temperatures', *Weather* **29**, 103–7.

Gray, Barbara: 1975, 'Japanese and European Winter Temperatures', *Weather* **30**, 359–68.

Gribbin, J. R.: 1973, 'Climatic Change in China Over the Past 5000 Years', *Nature* **246**, 375–7.

Hsieh, C. M.: 1976, 'Chu K'o-chen and China's Climatic Change', *Geographical Journ.* **142**, 248–56.

Landsberg, H. E. and Kaylor, R. E.: 1977, 'Statistical Analysis of Tokyo Winter Temperature Approximations, 1443–1970', *Geophysical Research Letters* **4**, 105–7.

Maejima, I. and Koika, Y.: 1976, 'An Attempt at Reconstructing the Historical Weather Situations in Japan', *Geographical Report of Tokyo Metropolitan University* **11**, 1–12.

Sakaguchi, Yutaka: 1978, 'Climatic Changes in Central Japan Since 38,400 BP', *Bulletin of the University of Tokyo Geography Department* **10**, 1–10.

Schove, D. J.: 1949, 'Chinese 'Raininess' Through the Centuries', *MM* 78, 11–6.

Schove, D. J.: 1975, 'World Climatic Chronology and Lake Biwa', in S. Horie (ed.), *Paleoclimatology of Lake Biwa and the Japanese Pleistocene*, Otsu, Japan.

Wang, Pao-Kuan: 1980, 'On The Relationship Between Winter Thunder and the Climatic Change in China in the Past 2200 Years', *Climatic Change* **3**, 37ff.

Yamamoto, T.: 1971, 'On The Nature of the Climatic Change in Japan Since the 'Little Ice Age' Around A.D. 1800', *Journal of the Meteorological Society of Japan* **49**, 75–89.

F. Reconstructions of the North American Climate

Baron, William R., Smith, David C., Barns, Harold W., Fastook, James, and Bridges, Anne E.: 1980, *Long Time Series Temperature and Precipitation Records for Maine, 1808–1978*, Orono.

Berks, A. J. B.: 1981, 'Late Wisconsin Vegetational and Climatic History at Kylen Lake, Northeastern Minnesota', *QR* **16**, 322ff.

Bernabo, Christopher, and Webb, T.: 1977, 'Changing Patterns in the Holocene Pollen Record of Northeastern North America: A Mapped Summary', *QR* **8**, 70–1.

Blasing, T. J.: 1975, 'Methods for Analyzing Climatic Variations in the North Pacific Sector and Western North America for the Last Few Centuries', Unpub. dissertation, University of Wisconsin.

Blasing, T. J. and Fritts, H. C.: 1975, 'Past Climate of Alaska and Northwestern Canada as Reconstructed from Tree Rings', in G. Weller and S. A. Bowling (eds.), *Climate of the Arctic* (Proceedings of the 24th Alaskan Science Conference), 48–58.

Blasing, T. J. and Fritts, H. C.: 1976, 'Reconstructing Past Climatic Anomalies in the North Pacific and Western North America from Tree-Ring Data', *QR* **6**, 563–580.

Borchert, J. R.: 1950, 'Climate of the Central North American Grassland', *Annals of the Association of American Geographers* **40**, 1–39.

Bradley, R. and England, J.: 1979, 'A Synoptic Climatology of the Canadian High Arctic', *Geografiska Annaler* **61**, 187ff.

Bradley, R. S.: 1976, *Precipitation History of the Rocky Mountain States*.

Bradley, R. S.: 1980, 'Climatic Fluctuations in the Rocky Mountains During the Period of Instrumental Records'. Paper delivered at the ICCH, University of East Anglia, July 1979. Expanded version of paper published in *Monthly Weather Review*, Boston.

Bray, J. R.: 1965, 'Forest Growth in North West North America', *Nature* **205**.

Bryson, R. A., Baerreis, D. A., and Wendland, W. M.: 1970, 'The Character of Late Glacial and Postglacial Climatic Changes', in *Pleistocene and Recent Environments of the Central Great Plains*, Department of Geology Special Publication No. 3, Lawrence University. Lawrence, pp. 53–74.

Bryson, R. A. and Hare, F. K.: 1974, *Climates of North America*, New York, Volume 2 of H. E. Landsberg (ed.), *World Survey of Climatology*.

Bryson, R. A., Irving, M. W., and Larsen, J. A.: 1965, 'Radiocarbon and Soils Evidence of Former Forest in the Southern Canadian Tundra', *Science* **147**, 46–8.

Catchpole, A. J. W. *et al.*: 1976, 'Freeze-up and Break-up of Estuaries in Hudson Bay in the Eighteenth and Nineteenth Centuries', *Canadian Geographer* **20**, 279–97.

Catchpole, A. J. W.: 1977–78, 'Historical Evidence of Climatic Change in Western and Northern Canada', in C. R. Harrington (ed.), 'Climatic Change in Canada', *Syllogeus*. No. 26. National Museum of Natural Sciences Project on Climatic Change in Canada During the Past 20 000 Years. Excellent bibliography.

Cook, Edward R. and Jacoby, Gordon C.: 1977, 'Tree-Ring – Drought Relationships in the Hudson Valley, New York', *Science* **198**, 399–401.

Cropper, John P. and Fritts, Harold, C.: 1981, 'Tree-Ring Width Chronologies From the North American Arctic', *Arctic and Alpine Research* **13**, 245ff.

Currie, B. W.: 1956, 'Climatic Trends on the Canadian Prairies', *Meteorological Abstracts and Bibliography* **14**, 712, 551–583.

Davis, M. B.: 1969, 'Climatic Changes in Southern Connecticut Recorded by Pollen Deposition at Rogers Lake', *Ecology* **50** 409–22.

Davis, M. B., Spear, R. W., and Shane, L. C. K.: 1980, 'Holocene Climate of New England', *QR* **14**, 240ff.

Dean, Jeffrey and Robinson, William: 1977, *Dendroclimatic Variability in the American Southwest, A.D. 680 to 1970*, Tucson.

Denton, G. H. and Karlen, W.: 1977, 'Holocene Glacial and Tree-Line Variations in White River Valley and Skolai Pass, Alaska and Yukon Territory', *QR* **7**, 63–111.

Dey, Balarm: 1973, 'Synoptic Climatological Aspects of Summer Dry Spells in the Canadian Prairies', University of Saskatchewan, Unpub. dissertation.

Diaz, H. F. and Quayle, R. G.: 1979, 'The Climate of the United States since 1895: Spatial and Temporal Changes', Paper given at the ICCH, to be published in *Monthly Weather Rev*.

Elchenlaub, V. L.: 1976, 'Climatic Change in the Southern Great Lakes – Eastern Corn Belt Areas', *East Lakes Geographer* **10**, 124–34.

Flora, S. D.: 1948, *Climate of Kansas*, Topeka.

Fritts, H. C.: 1956, 'Tree Ring Evidence for Climatic Change in Western North America', *Monthly Weather Rev.* **93**, 421–43.

Fritts, H. C. *et al.*: 1971, 'Multivariate Techniques for Specifying Tree-Growth and Climate Relationships for Reconstructing Anomalies in Paleoclimate', *Journ. of Applied Meteorology* **10**, 845–64.

Fritts, H. C.: 1976, *Reconstruction of Past Climatic Variability: Final Report*.

Fritts, H. C., Lofgren, G. Robert, and Gordon, Geoffrey A.: 1979, 'Reconstructing Seasonal to Centenary Variations in Climate from Tree-Ring Evidence', Paper presented at the ICCH, University of East Anglia.

Fritts, Harold C., Lofgren, G. Robert, and Gordon, Geoffrey A.: 1979, 'Variations in Climate Since 1602 as Reconstructed from Tree Rings', *QR* **XII**, 18–46.

Fritts, Harold C., Lofgren, G. Robert, and Gorden, Geoffrey A.: 1980, 'Past Climate Reconstructed from Tree Rings', *JIH* **X**, 4, 773–94, reprinted in *C&H* I.

Georgiades, A. P.: 1977, 'Trends and Cycles of Temperatures in the Prairies', *Weather* **32**, 99–101.

Havens, James: 1955, 'The "First" Systematic American Weather Observations', *Weatherwise* **8**.

Havens, James: 1958, *Annotated Bibliography of Meteorological Observations in the United States, 1715–1818*, Washington.

Heusser, C. J.: 1956, 'Postglacial Environments in the Canadian Rocky Mountains', *Ecological Monographs* **26**, 263–302.

Hughes, Patrick: 1979, '1816 – The Year Without a Summer', *Weatherwise* **32**, 108 ff.

Hughes, Patrick: 1981, 'The Blizzard of "88"', *Weatherwise* **34**, 250 ff.

Ives, R. L.: 1953, 'Climatic Studies in Western North America', *Proceedings of the Toronto Meteorological Conference*, 218–22.

Jacoby, G. C. and Cook, E. R.: 1981, 'Past Temperature Variations Inferred from a 400-Year Tree Ring Chronology from Yukon Territory, Canada', *Arctic and Alpine Res.* **13**, 409 ff.

Landsberg, H. E.: 1967, 'Two Centuries of New England Climate', *Weatherwise* **30**, 52–7.

Landsberg, H. E. *et al.*: 1968, *Preliminary Reconstruction of a Long Time Series of Climatic Data for the Eastern United States*.

Lawson, Merlin: 1974, *The Climate of the Great American Desert: Reconstruction of the Climate of Western Interior United States, 1800–1850*. Lincoln. Includes extensive bibliography.

Ludlum, D. L.: 1963, *Early American Hurricanes*.

Ludlum, D. L.: 1966, *Early American Winters, 1604–1820*.

Ludlum, D. L.: 1968, *Early American Winters, 1821–1870*.

Ludlum, D. L.: 1970, 'A Century of American Weather: Decade 1881–1890', *Weatherwise* **23**, 3, 131–5.

Ludlum, David M.: 1971, *Weather Record Book*, Princeton.

Ludlum, David M.: 1973, 'The Weather of American Independence', *Weatherwise* **26**, 152–9.

Ludlum, David M.: 1974, 'The Weather of American Independence', *Weatherwise* **27**, 162–8.

Ludlum, David M.: 1975, 'The Weather of American Independence', *Weatherwise* **28**, 118–21, 147, 172–6.

Ludlum, David M.: 1976, 'The Weather of American Independence', *Weatherwise* **29**, 236–40, 288–90.

Ludlum, David M.: 1981, 'The Weather of Independence–9:1781 The Surrender of Yorktown', *Weatherwise* **34**, 208 ff.

Lynch, H.: 1931, *Rainfall and Stream Run-Off in Southern California Since 1769*, Los Angeles.

McKay, G.: 1978, 'Climatic Variability and Its Impact on the Canadian Prairies', *Workshop on Research in Great Plains Drought Management Strategies*, University of Nebraska, Lincoln.

McKenna, M. C.: 1980, 'Eocene Paleolatitude, Climate and Mammals of Ellesmere Island', *PPP* **30**, 349 ff.

Mock, S. J. and Hibler, W. D.: 1976, 'The 20-Year Oscillation in Eastern North American Temperature Records', *Nature* **261**, 484–6.

Passow, Michael J.: 1977, 'A New Jersey Doctor's Weather', *New Jersey History*.

Pfaller, Louis (ed.): 1967, 'North Dakota Weather; An Analytic Summary of Conditions Over the Last Century', *North Dakota History* **34**.

Porry, Samuel: 1977, *Climate of the United States and Its Endemic Influences*, American Meteorological Society reprint of a volume first published in 1842.

Roden, G.: 1966, 'A Modern Statistical Analysis and Document of Historical Temperature Records in California, Oregon, and Washington, 1821–1964', *J. of Applied Meteorology* **5**, 3–24.

Smith, H. G. and Worrall, J. (eds.): 1970, *Tree-Ring Analysis with Special Reference to Northwest America*.

Swain, A.: 1978, 'Environmental Change During the Past 2000 Years in North-Central Wisconsin', *QR* **10**, 55–68.

Swan, Susan: 1981, 'Mexico in the Little Ice Age', *JIH* **11**, 633–48.

Thompson, Kenneth: 1980, 'Forests and Climatic Change in America: Some Early Views', *Climatic Change* **3**, 47 ff.

Thompson, Kenneth: 1981, 'The Question of Climatic Stability in America Before 1900', *Climatic Change* **3**, 227.

Thornthwaite, C. W.: 1941, *Atlas of Climatic Types in the United States, 1900–1939*, U.S. Conservation Service Miscellaneous Publication No. 421. Washington.
Travis, Paul D.: 1978, 'Changing Climate in Kansas: A Late 19th Century Myth', *Kansas History* **1**, 48–58.
U.S. Department of Agriculture Yearbook: 1941, *Climate and Man*, Washington. Reprinted in 1974.
Van Devender, Thomas: 1979, 'Development of Vegetation and Climate in the Southwestern United States', *Science* **204**, 701–10.
Wahl, E. W.: 1968, 'A Comparison of the Climate of the Eastern United States During the 1830's with the Current Normals', *Monthly Weather Rev.* **96**, 2, 73–82.
Wahl, E. W. and Lawson, T. L.: 1970, 'The Climate of the Mid-Nineteenth Century United States Compared to the Current Normals', *Monthly Weather Rev.* **98**, 4, 259–65.
Wax, Charles L.: 1977, 'An Analysis of the Relationships between Water Level Fluctuations and Climate, Coastal Louisiana'. Unpublished dissertation, Louisiana State University.
Webb, Thompson and Bryson, Reid: 1974, 'The Late- and Post-Glacial Sequence of Climatic Events Midwest, U.S.A.', *QR* **2**, 70–115.
Webb, Thompson and Bryson, Reid: 1974, 'The Late- and Post-Glacial Sequence of Climatic Events in Wisconsin and East-Central Minnesota: Quantitative Estimates Derived from Fossil Pollen Spectra by Multivariate Statistical Analysis', *QR* **II**, 70–115.
Zubrow, Ezra: 1974, *Population, Contact, and Climate in the New Mexican Pueblos*, Tucson. Mainly a reconstruction of past climatological conditions.

G. Reconstructions of the Climate of Other Regions

Brice, W. C. (ed.): 1978, *The Environmental History of the Near and Middle East Since the Last Ice Age*, London.
Burrows, C. J. and Greenland, D. E.: 1979, 'An Analysis of the Evidence for Climatic Change in New Zealand for the Last Thousand Years. Evidence from Diverse Natural Phenomena and from Instrumental Records', *Journal of the Royal Society of New Zealand* **9**, 3, 321–73.
Diester-Haass, L.: 1973, 'Holocene Climate in the Persian Gulf as Deduced from Grain-Size and Pteropod Distribution', *Marine Geology* **14**, 207–23.
Dumont, Henri: 1978, 'Neolithic Hyperarid Period Preceded the Present Climate of the Central Sahel', *Nature* **274**, 356–8.
Gabriel, B.: 1977, 'Early and Mid-Holocene Climate in the Eastern Central Sahara', *African Environment, Special Report* **6**, 65–7.
Issar, A.: 1980, 'Stratigraphy and Paleoclimates of the Pleistocene of Central and Northern Israel', *PPP* **29**, 261 ff.
Kutzbach, J. H.: 1980, 'Estimates of Past Climate at Paleolake Chad, N. Africa Based on a Hydrological and Energy Balance Model', *QR* **14**, 210 ff.
Neumann, J. and Sigrist, R. M.: 1978, 'Harvest Dates in Ancient Mesopotamia as Possible Indicators of Climatic Variations', *Climatic Change* **1**, 239–52; addendum, 253–6.
Nicholson, Sharon: 1976, 'A Climatic Chronology for Africa: Synthesis of Geological, Historical, and Meteorological Information and Data'. Univ. of Wisc., Unpub. diss.
Nicholson, Sharon E.: 1978, 'Climatic Variations in the Sahel and Other African Regions during the Past Five Centuries', *Journal of Arid Environments* **I**, 3–24. Includes bibliography of recent work in African climate reconstruction.
Nicholson, S. E.: 1981, 'The Historical Climatology of Africa', *C&H* II.
Schove, D. J.: 1975, 'Chronology of Pluvials in Northern Africa', *Geographical Journ.* **141**, 195–6.
Seth, S. K.: 1963, 'A Review of Evidence Concerning Changes of Climate in India During the Proto-Historical and Historical Periods', in *Changes of Climate, Proceedings UNESCO/WMO Rome 1961 Symposium*, Paris, pp. 443–54.
Shao-Wu, Wang and Zong-Ci, Zhao: 1981, 'Drought and Floods in China, 1470–1979', *C&H* II.
Talbot, M. and Delibrias, G.: 1977, 'Holocene Variations in the Level of Lake Bosumtwi, Ghana', *Nature* **268**, 722–4.
van Zeist, W., Timmers, R., and Bottema S.: 1972, 'Studies of Modern and Holocene Pollen Precipitation in Southeastern Turkey', *Palaeohistoria* **14**, 19–38.

H. Drought as a Meteorological Phenomenon

Alves, Jaoquim: 1953, *History of Droughts, 17th–19th Centuries* (Title translated from the Portuguese).

Bark, L. Dean: 1978, 'History of American Droughts', in Norman Rosenberg (ed.), *North American Droughts*, Boulder, pp. 9–21.

Birks, J.: 1977, 'The Reaction of Rural Populations to Drought: A Case Study from South-East Arabia', *Erdkunde* **31**, 299–305.

Borkar, V. V. and Nadkarni, M. V.: 1975, *Impact of Drought on Rural Life*, Bombay.

Brooks, Reuben: 'Drought Perception as a Force in Migration from Ceara'. Unpublished dissertation; order from University Microfilms No. 72–25, 146.

Bryson, R. A.: 1967, 'Possibilities of Major Climatic Modification and Their Implications: Northwest India, A Case for Study', *Bulletin of the American Meteorological Society* **48**, 3, 136–42.

Center for Technology, Environment, and Development: Climate and Society Research Group of Clark University: 1979, "The Effect of Climate Fluctuations on Human Populations: A Progress Report, No. 2'. Studies of the changing impact of drought in the U.S. Great Plains, the Sahel, and the Tigris-Euphrates basin (during the ancient period). See also for impact of irrigation technology.

Claxton, Robert H.: 1978, 'How Guatemala Responded to Drought: 1563–1922', Aspen Institute *Food and Climate Review*, 40–5.

Cunniff, Roger: 'The Great Drought: Northeastern Brazil, 1877–1880', unpublished Univ. of Texas dissertation.

Dando, William A.: 1977, 'Droughts in North Dakota', *Great Plains-Rocky Mountains Geographical Journ.* **6**, 197–203.

Dick, Everett: 1973, 'The Great Nebraska Drought of 1894: the Exodus', *Arizona and the West* **15**, 333–334.

Glantz, Michael and Katz, Richard: 1977, 'When is a Drought a Drought?' *Nature* **267**, 192–3.

Hall, Anthony: 1978, *Drought and Irrigation in North-East Brazil*, Cambridge.

Heathcote, R. L: 1969, 'Drought in Australia; a Problem in Perception', *Geographical Review* **59**, 175–94.

Heathcote, R. L.: 1973, 'Drought Perception', in J. V. Lovett (ed.), *The Environmental, Economic and Social Significance of Drought*, Melbourne.

Hughes, P.: 1976, 'Drought: The Land Killer', *American Weather Stories*, Washington, D.C.

Kraenzel, Carl Frederick: 1955, *The Great Plains in Transition*, Norman, Oklahoma.

Kraenzel, Carl Frederick: 1963, 'Great Plains: A Region Basically Vulnerable', in Carle Hodge (ed.), *Aridity and Man*, Washington, pp. 539–48.

Krishnan, A.: 1977, 'Climatic Changes Relating to Desertification in the Arid Zone of Northwest India', *Annals of the Arid Zone* **16**, 302–9.

Landsberg, H. E.: 1975, 'Drought, a Recurrent Element of Climate', *Special Environmental Report No. 5*, World Meteorological Organization. Geneva.

Lawson, M. P. *et al.*: 1971, *Nebraska Droughts: A Study of Their Past Chronological and Spatial Extent with Implications for the Future*, Occasional paper No. 1, Department of Geography, University of Nebraska.

Lovett, J. V. (ed.): 1973, *The Environmental, Economic and Social Significance of Drought*, Melbourne.

Malin, J. C.: 1946, 'Dust Storms, 1850–1900', *Kansas Historical Quarterly*. **14**.

Markham, C. G.: 1975, 'Twenty-Six-Year Cyclical Distribution of Drought and Flood in Ceara, Brazil', *Professional Geographer* **27**, 454–6.

Mooley, D. A. and Pant, G. B.: 1979, 'Droughts in India Over the Last 200 Years, Their Socio-Economic Impact and Remedial Measures for Them', Unpublished paper presented at the ICCH, University of East Anglia.

Nace, R. L. and Pluhowski, E. J.: 1965, *Drought of the 1950's with Special Reference to the Mid-continent*, Geological Survey Water-Supply Paper No. 1804. Washington.

Namias, Jerome: 'Severe Drought and Recent History', *JIH* **X**, 4, 697–712. Reprinted in *C&H* I.

Osborn, Ben: 1950, 'Some Effects of the 1946–1948 Drought on Ranges in Southwestern Texas', *Your Range Management* **3**, 1–15.

Palmer, W. C.: 1965, *Meteorological Drought*, U.S. Department of Commerce, Weather Bureau Research Paper No. 45. Washington, D.C.
Palmer, W. C. and Denny, L. M.: 1971, *Drought Bibliography*, NOAA Technical Memo EDS 20, U.S. Department of Commerce, Washington, D.C.
Rosenberg, Norman J. (ed.): 1978, *North American Droughts*, Boulder, Colorado.
Saarinen, T. F.: 1966, *Perception of the Drought Hazard on the Great Plains*, University of Chicago, Geography Department, Research Paper No. 106.
Special Assistant to the President for Public Works Planning: 1958, *Drought: A Report*, Washington.
Stockton, C. W. and Mekom D. M.: 1975, 'Long-Term History of Drought Occurrence in the Western United States as Inferred from Tree-Rings', *Weatherwise* **28**, 244–9.
Tannehill, Ivan Ray: 1947, *Drought: Its Causes and Effects*, Princeton.
Thomas, H. E.: 1962, *The Meteorologic Phenomenon of Drought in the Southwest*, U.S. Geological Survey, Professional Paper No. 372-H. Washington, D.C.
Thomas, H. E.: 1963, *General Summary of Effects of the Drought in the Southwest*, U.S. Geological Survey, Professional Paper, Washington, D.C.
Tomanek, G. W. and Hulett, G. K.: 1970, 'Effects of Historical Droughts on Grassland Vegetation in the Central Great Plains', in Dort, W. and Jones, J. K., (eds.), *Pleistocene and Recent Environments of the Central Great Plains*, Lawrence, pp. 203–10.
Warrick, Richard A.: 1975, *Drought Hazard in the United States: A Research Assessment*, Boulder, Colorado.
Wedel, W. R.: 1937, 'Dust Bowls of the Past', *Science* **86**, 2232, Supplement, 8–9.
Young, Vernon: 1956, 'The Effect of the 1949–1954 Drought on the Ranges of Texas', *J. of Range Management*.

I. The Socio-Economic Impact of Drought

Barnhart, J. D.: 1925, 'Rainfall and the Populist Party in Nebraska', *American Political Sci. Rev.* **19**, 527–40.
Battacharya, M.: 1975, 'Emergency Administration: A Study of Drought-Relief Operations in an Indian State', *J. of Administration Overseas* **14**, 259–65.
Brooks, R. H.: 1975, 'Drought and Public Policy in Northeastern Brazil', *Ekistics* **39**, 30–5.
Harschbarger, C. E. and Duncan, M.: 1977, 'The Economic Realities of Drought', *Monthly Review of the Federal Reserve Bank of Kansas City*, 3–13.
Mann, D. E.: 1969, 'The Political Implications of Migration to the Arid Lands of the U.S.', *Natural Resources Journ.* **9**, 212–27.
Miewald, Robert D.: 1978, 'Social and Political Impacts of Drought', in N. H. Rosenberg (ed.), *North American Droughts*, Boulder, Colorado, pp. 79–102.
Mooley, D. A. and Pant, G. B.: 'Droughts in India Over the Last 200 Years, Their Socio-Economic Impacts, and Remedial Measures for Them'. *C&H* II.
Mukerjee, Tapan: 1972, 'Economic Analysis of the Alternative Adjustments to Drought in the Great Plains', *Contract Report for the National Water Commission*. Stockton, California.
Riefler, R. F.: 1978, 'Drought: An Economic Perspective', in N. J. Rosenberg (ed.), *North American Droughts*, Boulder, Colorado, pp. 63–78.
Riefler, R. F.: 'Economic Impact of Prolonged Drought on the State of Nebraska'. Unpublished ms. University of Nebraska. Undated.

J. The American Dust Bowls

Bonnifield, Paul: 1979, *The Dust Bowl: Men, Dirt and Depression*.
Coffey, Marilyn: 1978, 'The Dust Storms', *Natural History* **87**, 72–82.
Cronin, Francis and Beers, Howard W.: *Areas of Intense Drought Distress, 1930–1936*, WPA, Division of Social Research Bulletin, Series 7, No. 1. Washington, D.C.
Fossey, W. Richard: 1977, 'Talkin' Dust Bowl Blues: A Study of Oklahoma's Cultural Identity During the Great Depression', *Chronicles of Oklahoma*, 12–33.

Grant, H. Roger and Purcell, L. Edward (eds.): 1976, *Years of Struggle: The Farm Diary of Elmer G. Powers, 1931–1936*, Ames, Iowa.
Hargreaves, Mary: 1976, 'Land Use Planning in Response to Drought: The Experience of the Thirties', *Agricultural History* **50**.
Hearst, James: 1978, 'We All Worked Together: A Memory of Drought and Depression', *Palimpsest* **59**.
Herman, Alan: 1977, 'Dust, Depression and Demagogues: Political Radicals of the Great Plains, 1930–36', *J. of the West* **16**.
Hoyt, J. C.: 1938, *Drought of 1936 with Discussion on the Significance of Drought in Relation to Climate*, United States Geological Survey, Water Supply Paper 820. Washington, D.C.
Hurt, R. Douglas: 1977, 'Dust Bowl: Drought, Erosion, and Despair on the Southern Great Plains', *American West* **14**, 22–27, 56.
Idso, Sherwood: 1976, 'Dust Storms', *Scientific American* **235**, 20, 108–11, 144.
Johnson, Vance: 1947, *Heaven's Tableland: The Dust Bowl Story*.
Kifer, R. S. and Stewart, H. L.: 1938, *Farming Hazard in the Drought Area*, WPA Division of Social Research Monograph No. 16. Washington, D.C.
Lambert, C. Roger: 1971, 'The Drought Cattle Purchase, 1934–1935', *Agricultural History* **45**.
Lambert, C. Roger: 1972, 'Drought Relief for Cattlemen: The Emergency Purchase Program of 1934–35', *Panhandle-Plains Historical Review* **45**.
Link, Irene: 1937, *Relief and Rehabilitation in the Drought Area*, WPA, Division of Social Research Bulletin, Series 5, No. 3. Washington, D.C.
Mangus, A. R.: 1938, *Changing Aspects of Rural Relief*, WPA Division of Social Research Monograph No. 14. Washington, D.C.
Schuyler, Michael W.: 1975, 'Drought and Politics, 1936: Kansas as a Test Case', *Great Plains J.* **15**, 3–27.
Schuyler, Michael W.: 1976, 'Federal Drought Relief Activities in Kansas, 1934', *Kansas Historical Quarterly* **42**, 403–26.
Skaggs, R. H.: 1975, 'Drought in the United States, 1931–40', *Annals of the Association of American Geographers,* **65**, 391–402.
Stein, Walter J.: 1973, *California and the Dust Bowl Migration*, Westport, Connecticut.
Svobida, Lawrence: 1940. *An Empire of Dust*, Caldwell, Idaho.
Taeuber, Conrad and Taylor, Carl C.: 1937, *The People of the Drought States*, WPA, Division of Social Research Bulletin, Series 5, No. 2. Washington, D.C.
Thornthwaite, C. Warren: 1936, 'The Great Plains', in Carter Goodrich *et al., Migration and Economic Opportunity, The Report of the Study of Population Redistribution*, Philadelphia, Pennsylvania.
Thornthwaite, C. Warren: 1941, 'Climate and Settlement in the Great Plains', in *Climate and Man* (The Yearbook of Agriculture for 1941). Washington, D.C. pp. 177–87.
United States Great Plains Committee: 1936, *The Future of the Great Plains*, Washington, D.C.
Ware, James Wesley: 'Black Blizzard: The Dust Bowl of the 1930's'. Unpublished dissertation; order from University Microfilms, No. 78–11077.
Woodruff, Nan Elizabeth: 1977, 'The Great Southern Drought of 1930–1: A Study in Rural Relief'. Unpublished dissertation; order from University Microfilms, No. 78–07734.
Worcester, Donald: 1977, 'Grass to Dust: Ecology and the Great Plains in the 1930's', *Environmental Rev.* **3**, 2–11.
Worcester, Donald: 1979, *Dust Bowl: The Southern Plains in the 1930's,* Oxford.

K. The Sahel Drought

Baker, S.: 1977, 'Background to the Study of Drought in East Africa', *African Environment Special Reports* **6**, 74–82.
Bein, F.: 1977, 'Peasant Response to Drought in the Sudanese Sahel', *Great Plains – Rocky Mountain Geographical J.* **6**, 137–46.
Berg, E.: 1975, *The Recent Economic Evolution of the Sahel*, Ann Arbor.
Bonte, P.: 'Drought in the Sahel: Transformation of the Sahelian Pastoral and Agricultural Systems', in R. Garcia (ed.), *The Roots of Catastrophe,* **3** (an IFIAS Report on "Drought and Man: the 1972 Case History"), Chapter 2. (Draff).

Bryson, Reid A.: 1973, 'Drought in Sahelia: Who or What is to Blame?', *Ecologist* 366–71.

Caldwell, J. C.: 1975, *The Sahelian Drought and Its Demographic Implications*, Overseas Liaison Committee Paper No. 8, American Council on Education. Washington.

Campbell, David: 1977, 'Strategies for Coping with Drought in the Sahel.' Unpublished dissertation, Clark University.

Campbell, Ian: 1975, 'Diagnosis of a Famine: Human Mismanagement as a Major Factor in the Sahelian Drought', *Ekistics* **39**, 26–9.

Center for Disease Control: 1973, 'Nutritional Surveillance in Drought Affected Areas of West Africa. (Mali, Mauritania, Niger, Upper Volta) August-September 1973'. Unpublished report.

Central for Disease Control: 1975, 'Protein/Energy Undernutritional Surveys in the Sahel, 1974 and 1975'. Unpublished report.

Charney, J. *et al.*: 1975, 'Drought in the Sahara: A Bio-Geographical Feedback Mechanism', *Science* **187**, 434–5.

Comite d'Information Sahel: 1974, *Qui Se Nourrit de la Famine en Afrique*? Paris.

Dalby, D. and Church, R. J. Harrison: 1973, *Drought in Africa*, University of London, School of Oriental and African Studies. London.

Dalby, D., Church, R. J. Harrison, and Bezzaz, F. (eds.): 1977, *Drought in Africa 2*, African Environment Special Report No. 6, International African Institute. London.

deGoyet, E. J. and deVille, C.: 1976, 'Preliminary Report of an Anthropometric Survey During a Nutritional Relief Program in Niger', *Social and Occupational Medicine* **4**, 70.

Diggs, Charles C.: 1974, 'The Drought in the Sahel', *Black Scholar* **5**, 37–42.

DuBois, V. D.: 1974, 'The Drought in Niger, Part 1, The Physical and Economic Consequences'. American Universities Field Staff Reports, West Africa Series 15(4). New York.

DuBois, V. D.: 1975, 'A Note on the Sahel'. American Universities Field Staff Reports, West Africa Series 16(4). New York.

DuBois, V. D.: 1975, 'The Drought in West Africa', *Development Digest* **13**, 49–61.

Faulkingham, R. H.: 1977, 'Ecologic Constraints and Subsistence Strategies: The Impact of Drought in a Housa Village, A Case Study from Niger', in D. Dalby, R. J. Harrison Church, and F. Bezzaz (eds.), *Drought in Africa 2*. London, pp. 148–58.

Glantz, M. (ed.): 1976, *The Politics of Natural Disaster: The Case of the Sahel Drought*, New York.

Glantz, M.: 1977, 'Nine Fallacies of a Natural Disaster: Case of the Sahel', *Climatic Change* **1**, 69–84.

Joyce, S. and Beudot, F.: 1976, *Elements for a Bibliography of the Sahel Drought*, OECD.

Kamrany, Nake M.: 1978, 'The Sahel Drought: Major Development Issues', Aspen Institute *Food and Climate Review* 46–52.

Kates, Robert W.: 1981, *Drought Impact in the Sahelian Sudanic Zone of West Africa: A Comparative Analysis of 1910–15 and 1968–74*, Clark University, Worcester, Mass.

Kloth, T. I.: 1974, 'Sahel Nutrition Survey, 1974'. Center for Disease Control. Unpublished report. Atlanta.

Lamb, H. H.: 'Some Comments on the Drought in Recent Years in the Sahel-Ethiopian Zone of North Africa', *African Environment Special Report* **6**, 33–7.

Lovejoy, Paul and Baier, Stephen: 1975, 'The Desert-Side Economy of the Central Sudan', *International Journal of African Historical Studies* **8**, 551–81.

Newman, James L. (ed.): 1975, *Drought, Famine and Population Movements in Africa*.

Oguntoyinbo, J. and Richards, P.: 1977, 'The Extent and Intensity of the 1969–1973 Drought in Nigeria', *African Environment Special Report* **6**, 114–26.

Pourafzal, H.: 1979, 'Comparative Vulnerability to Drought in the Sahel, 1910–1914 and 1968–1974', Unpublished masters thesis, Clark University.

Rapp, A.: 1974, 'A Review of Desertization in Africa: Water, Vegetation, and Man', Secretariat for International Ecology, Stockholm, Sweden, Report 1.

Schove, D. J.: 1977, 'African Drought and Weather History', in D. Dalby and R. Harrison Church (eds.), *Drought in Africa*, London, pp. 29–30.

Schove, D. J.: 1977, 'African Droughts and the Spectrum of Time', in *ibid.* London, pp. 38–53.

Seamon, J. *et al.*: 1973, 'An Inquiry into the Drought Situation in Upper Volta', *The Lancet* **2**, 774–8.

Sheets, H. and Morris, R.: 1974, *Disaster in the Desert: Failures of International Relief in the West African Drought*, Washington, D.C.
Sircoulon, J.: 1976, 'The Recent Drought in the Sahel Regions of West Africa' (title translated from the French). *Houille Blanche* **6–7**, 537ff.
Sircoulon, J.: 1976, 'Les donnees hydropluviometriques de la secheresse recente in Afrique inter-tropicale: comparaison avec les secheresses "1913" et "1940",' *Chaiers de l'Office de la Recherche scientifique des Territoiries d'Outre-Mer*, Series Hydrologiques **13**, 2, 75–174.
Swift, J.: 1977, 'Sahelian Pastoralists: Underdevelopment, Desertification and Famine', *Annual Review of Anthropology* **6**, 457–78.
Ware, H.: 'The Sahelian Drought: Some Thoughts on the Future'. United Nations, Special Sahelian Office. New York (undated).
Wisner, Benjamin: 1978, 'The Human Ecology of Drought in Eastern Kenya', Unpublished dissertation.
Wisner, Benjamin: 'An Overview of Drought in Kenya', *Natural Hazard Research Working Paper, No. 30.*

L. The Effect of Climatic Fluctuations on Agriculture and Economy

Anderson, J. L.: 1981, 'History and Climate: Some Economic Models', *C&H* II.
Arakawa, H.: 1955, 'Meteorological Conditions of the Great Famines in the Last Half of the Tokugawa Period, Japan', *Papers in Meteorology and Geophysics* **6**, 101–115.
Beveridge, W. H.: 1920, 'British Exports and the Barometer, Parts I and II', *Economic Journ.* **30**, 13–25 and 209–13.
Beveridge, W. H.: 1921, 'Weather and Harvest Cycles', *Economic Journ.* **31**, 421–53.
Beveridge, W. H.: 1922, 'Wheat Prices and Rainfall', *J. of the Royal Statistical Society* **85**, 418–54.
Brandon, R. F.: 1971, 'Late Medieval Weather in Sussex and Its Agricultural Significance', *Transactions of the Institute of British Geographers* **54**, 1–17.
deVries, Jan: 1980, 'Measuring the Impact of Climate on the Economy: Separating Real from False Assumptions', *JIH* **X**, 4, 599–630. Reprinted in *C&H* I.
Flohn, H.: 'Short-Term Climatic Fluctuations and Their Economic Role', *C&H* II.
Florescano, Enrique: 1968, 'Meterologia y ciclos agricolas en las antiquas economias. El case de Mexico', *Historia Mexicana* **17**, 516–35.
Gourou, P.: 'Man and Monsoon in Southern Asia', *Natural Resources of Humid Tropical Asia*, UNESCO, 448–56.
Jones, Eric L.: 1964, *Seasons and Prices: The Role of Weather in English Agricultural History*, London.
Libby, L.: 'Correlation of Historic Climate with Historic Prices and Wages', *Indian Journal of Meteorology and Geophysics* (forthcoming).
Martell, G.: 1976, 'A Climatic Analysis of English Wheat Prices 1200 to the Present', Unpublished Research Paper, Department of Meteorology, University of Wisconsin at Madison.
Meteorologie: 1954 (Special issue on agricultural meteorology). Series 4, No. 9.
Oliver, J.: 1965, 'Problems in Agro-Climatic Relationships in Wales in the Eighteenth Century', in Taylor, J. A. (ed.), *Climatic Change with Special Reference to the Highland Zone of Britain*, Aberystwyth.
Parry, M. L.: 1975, 'Secular Climatic Change and Marginal Agriculture', *Transactions of the Institute of British Geographers*, Publication No. 64, 1–13.
Parry, M. L.: 1976, 'The Significance of the Variability of Summer Warmth in Upland Britain', *Weather* **31**, 212–17.
Parry, M. L.: 1978, *Climate Change: Agriculture and Settlement*, Folkestone, England.
Parry, M. L.: 1979, 'Climatic Change and the Agricultural Frontier'. Unpublished paper presented at the ICCH, University of East Anglia, Published in *C&H* II.
Pfister, Christian: 1975, *Agrarkonjunktur und Witterungsverlauf im westlichen Schweizer Mittelland 1755–1797*, Bern.
Pfister, Christian: 1978, 'Climate and Economy in Eighteenth-Century Switzerland', *JIH* **IX**, 223–43.
Pfister, Christian: 'The Impact of the 'Little Ice Age' Upon the Agricultural Production of Switzerland,

1525–1825'. Paper presented at the ICCH Published in *C&H* II., as 'An Analysis of the Little Ice Age in Switerland and Its Consequences for Agricultural Production'.
Pichard, G.: 1979, 'La Part du Climat dans le Marasme Economique et Social en Provence de 1680 a 1718'. Unpublished paper presented at the ICCH, University of East Anglia.
Post, John: 1974, 'A Study in Meteorological and Trade Cycle History: The Economic Crisis Following the Napoleonic Wars', *J. Ec. Hist.* **24**, 315–49.
Post, John. *The Last Great Subsistence Crisis in the Western World*. Baltimore, Maryland, 1977.
Post, John: 1980, 'The Impact of Climate on Political, Social, and Economic Change', *JIH* **X**, 4, 719–24. Reprinted in *C&H* I.
Puiz, Anne-Marie: 1974, 'Climat, recoltes et vie des hommes a Geneve, XVIe-XVIIIe siecle', *Annales* **XXIX**, 608.
Salinger, M. J.: 1979, 'Agricultural Implications of Climate Change in New Zealand'. Paper presented at the ICCH, University of East Anglia.
Skaggs, Richard: 1978, 'Climatic Change and Persistence in Western Kansas', *Annals of the American Association of Geographers* **68**, 73–80.
Smith, David C., Borns, Harold, Baron, W. R., and Bridges, Anna: 1981, 'Climatic Stress and Maine Agriculture 1785–1885'. Paper presented at the ICCH. Published in *C&H* II.
Spufford, Margaret: 1974, *Contrasting Communities: English Villagers in the Sixteenth and Seventeenth Centuries*, Cambridge.
Starr, Thomas: 1977, 'The Role of Climate in American Agriculture: Past, Present and Future', *Ecologist* 262–7.
Sutherland, Donald: 'Weather and the Peasantry of Upper Brittany, 1780–1789'. Paper presented at the ICCH. Published in *C&H* II.
Swan, Susan Linda: 'Climate, Crops, and Livestock: Some Aspects of Colonial Mexican Agriculture', Order from University Microfilms No. 77–20, 120.
Tucker, William F.: 'The Effects of Famine in The Medieval Islamic World', Unpub. Ms.
Tucker, William F.: 1981, 'Natural Disasters and the Peasantry in Mamluk, Egypt', *J. of The Economic and Social History of the Orient* **24**, 215ff.
Utterstrom, G.: 1955, 'Climatic Fluctuations and Population Problems in Early Modern History', *Scandinavian Economic History Review* **3**, 1, 3–47.
Utterstrom, G.: 1961, 'Population and Agriculture in Sweden', *Scandinavian Economic History Review* **9**, 176–94.
Walton, K.: 1952, 'Climate and Famines in North-East Scotland', *Scottish Geographical Magazine* **68**, 13–22.
Wood, D. J.: 1965, 'The Complicity of Climate in the 1816 Depression in Dumfriesshire', *Scottish Geographical Magazine* **81**, 5–17.

M. The Role of Climate in Cultural Decline

Baerreis, D. A. and Bryson, R; A.: 1965, 'Climatic Episodes and the Dating of the Mississippian Cultures', *The Wisconsin Archaeologist* **46**, 203–220.
Bell, Barbara: 1970, 'The Oldest Records of the Nile Floods', *Geographical Journ.* **136**, 569–73.
Bell, Barbara: 1971, 'The Dark Ages in Ancient History, I: The First Dark Age in Egypt', *American Journal of Archaeology* **75**, 1–26.
Bell, Barbara: 1975, 'Climate and the History of Egypt: The Middle Kingdom', *American Journal of Archaeology* **79**, 223–69.
Bridgman, Howard: 1979, 'Climatic Change Influences and the Polynesian Migrations'. Paper presented at the ICCH, University of East Anglia.
Bryson, R. A. and Baerreis, D. A.: 1968, 'Climatic Change and the Mill Creek Culture of Iowa', *Journal of the Iowa Archaeological Society* **15–6**, 1–358.
Bryson, R. A., Lamb, H. H., and Donley, D. L.: 1974, 'Drought and the Decline of Mycenae', *Antiquity* **48**, 46–50.
Carpenter, Rhys: 1968, *Discontinuity in Greek Civilization*, New York.

Das, M. N.: 1979, 'Climatic Factors Responsible for the Decline of the Indian Kalingad Race', Paper presented at the ICCH, University of East Anglia.
Donley, D. L.: 1971, 'Analysis of the Winter Climatic Pattern at the Time of the Mycenaean Decline'. Unpublished dissertation, University of Wisconsin.
Gunn, Joel *et al.*: 1979, 'The Impact of Climatic Change on Modern and Prehistoric Cultures in Texas: A Progress Report'. Paper presented at the ICCH, University of East Anglia.
Lehmer, D. J.: 1970, 'Climatic and Culture in the Middle Missouri Valley', in Dort, W. and Jones, J. K. (eds.), *Pleistocene and Recent Environments of the Central Great Plains*, Lawrence, pp. 117–29.
Madsen, David B. and Berry, Michael S.: 1975, 'A Reassessment of Northeastern Great Basin Prehistory', *American Antiquity* **40**, 391–405.
Manley, G.: 1965, 'Possible Climatic Agencies in the Development of Postglacial Habitats', *Proceedings of the Royal Society*, **B, 161**, 363–75.
McGhee, Robert: 1970, 'Speculations on Climatic Change and Thule Culture Development', *Folk* **11–12**, 173–84.
McGhee, Robert: 1979, 'Archaeological Evidence for Climatic Change During the Past 5000 Years'. Unpublished paper presented at the ICCH, University of East Anglia. A review essay of recent archaeological work attempting to relate cultural change and decline to climatic change; attempts to find valid correspondences between culture and climate in order to allow dating of past climatic change. Published in *C&H* II.
McGovern, Thomas: 1979, 'The Economics of Extinction in Norse Greenland', Unpublished paper presented at the ICCH, University of East Anglia. Published in *C&H* II.
Paulsen, Allison C.: 1976, 'Environment and Empire: Climatic Factors in Prehistoric Andean Culture Change', *World Archaeology* **8**, 121–32.
Raikes, Robert: 1965, 'The Mohenjo-Daro Floods', *Antiquity* **39**, 196–203.
Raikes, Robert: 1967, *Water, Weather and Prehistory*, London.
Sears, Paul B.: 1958, 'Environment and Culture in Retrospect', in Terah L. Smiley (ed.), *Climate and Man in the Southwest*, Univ. of Arizona Bulletin, **28**, 4.
Shaw, Brent D.: 1976, 'Climate, Environment and Prehistory in the Sahara', *World Archaeology* **8**, 133–49.
Shaw, Brent D.: 1979, 'Climate, Environment and History: The Case of Roman North Africa'. Unpublished paper presented at the ICCH, University of East Anglia. Includes extensive bibliography. Published in *C&H* II.
Singh, G.: 1971, 'The Indus Valley Culture', *Archaeology and Physical Anthropology in Oceania* **6**, 2, 177–89.
Wedel, W. R.: 1941, 'Environment and Native Subsistence Economies in the Central Great Plains', *Smithsonian Miscellaneous Collections* **101**, 1–29.
Wedel, W. R.: 1953, 'Some Aspects of Human Ecology on the Great Plains', *American Anthropology* **55**, 499–514.
Weiss, Barry: 1979, 'The Decline of Late Bronze Age Civilization as a Possible Response to Climatic Change'. Unpublished manuscript, National Center for Atmospheric Research, Boulder.
Wright, H. E. Jr.: 1968, 'Climatic Change in Mycenaean Greece', *Antiquity* **42**, 123–7.
Wright, S.: 1976, 'Barton Blount: Climate or Economic Change?', *Medieval Archaeology* **20**, 148–52.

N. Other Socio-Cultural Effects of Climate

Arakawa, H.: 1960, 'The Weather and Great Historic Events in Japan', *Weather* **15**, 152–164.
Bastiampillai, B.: 'Climate and Man in Ceylon (Sri Lanka): An Historical Perspective'. Unpublished paper presented at the ICCH, University of East Anglia.
Carpanetto, Secondo.: 1974, 'History of Medicine and Social History' (title translated from the Italian), *Rivista Storica Italiana* **86**, 123–35.
Cassedy, J. H.: 1969, 'Meteorology and Medicine in Colonial America', *Journal of the History of Medicine*.

Emmons, David: 1971, 'Theories of Increased Rainfall and the Timber Culture Act of 1873', *Forest History* **15**.
Goldenberg, S.: 1974, 'Climate and History: Contributions to a History of Climate in the Rumanian Provinces, 16th–17th Centuries' (title translated from the Rumanian), *Revue Roumaine d'Histoire* **13**, 305–21.
Jenks, S.: 1979, 'Attitudes Toward Weather As Shown by Astrological Tracts on Weather Prediction'. Unpublished paper presented at the ICCH, University of East Anglia.
Kellogg, W. W. and Schware, Robert: 1981, *Climate Change and Society: Consequences of Increasing Atmospheric Carbon Dioxide*. Westview Press, Boulder, Colo.
Lal, K. S.: 1979, 'Influence of Rainfall on the Course of Medieval Indian History (13th–15th Centuries)'. Unpublished paper presented at the ICCH, University of East Anglia.
Loftness, Vivian: 1978, 'Climate and Architecture', *Weatherwise* **31**, 212–7.
Lindgren, S. and Neumann, J.: 1979, 1980, 'Great Historical Events That Were Significantly Affected by the Weather, 4 and 5', *Bulletin of the American Meteorological Society* **60**, **61**, 770ff., 1570ff.
Mackay, A.: 'Climate and Popular Unrest in Late Medieval Castile'. *C&H* II.
Manley, G.: 1957, 'Climate Fluctuations and Fuel Requirements', *Scottish Geographical Magazine* **73**, 1, 19–28.
Neuberger, Hans: 1970, 'Climate in Art', *Weather* **25**, 46–56. Reprinted in *Climate in Review*, Geoffrey McBoyle (ed.), 308–13. Boston.
Wahl, Eduard: 1973, 'The Influence of Climate on the Development of the Law in Europe and in Asia: A Contribution to Regionalism in Comparative Law' (title translated from the French), *Revue International de Droit Compare* **2**, 261–76.
Yamamoto, Takeo: 1979, 'Climatic Fluctuation in Prehistoric Japan and Its Influence on the Change of Burial Styles'. Unpublished paper presented at the ICCH, University of East Anglia.

O. Disasters Other Than Drought

Baker, George W. and Chapman, Dwight W. (eds.): 1962, *Man and Society in Disasters*. New York.
Boyer, Richard: 1973, 'Mexico City and the Great Flood, 1629–1635'. Unpublished dissertation. Order University Microfilms 73–28, 511.
Daniel, Pete: 1977, *Deep'n as it Comes: The 1927 Mississippi River Flood*, Oxford.
Daniels, Robert V. (ed.): 1975, 'Military Reminiscences of the Flood of 1927', *Vermont History* **43**.
Davis, R. Bruce: 1973, 'The Tornado of 1840 Hits Mississippi', *J. of Mississippi History* **35**, 43–51.
Ericksen, Neil J.: 1975, 'A Tale of Two Cities: Flood History and the Prophetic Past of Rapid City, South Dakota', *Economic Geography* **51**, 305–20.
Flora, S. D.: 1954, *Tornadoes of the U.S.*, Norman.
Galway, Joseph G.: 1981, 'Ten Famous Tornado Outbreaks', *Weatherwise* **34**, 100 ff.
Gusewelle, C. W.: 1978, 'The Winds of Ruin: Tornadoes on the American Land', *American Heritage* **29**, 90–7.
Heidorn, Keith: 1978, '1829 Twister Damaged Guelph, Ontario', *Early Canadian Life* **2**.
Heidorn, Keith: 1978, 'Guelph Tornado of 1829', *Weatherwise* **31**, 226–7.
Heidorn, Keith: 1978, 'The Great Hurricane of 1635 and the Legend of Thacker's Island', *New England Galaxy* **19**, 32–8.
Heidorn, Keith: 1974, 'The Weather Luck of Christopher Columbus', *Sea Frontiers* **20**, 302–11.
Heidorn, Keith: 1978, 'The Winds of November', *Early Canadian Life* **2**.
Hoberman, Louisa: 1974, 'Bureaucracy and Disaster: Mexico City and the Flood of 1629', *J. of Latin American Studies* **6**, 211–30.
Hoberman, Louisa: 1976, 'The Colonial Urban Legacy Re-examined: The Politics of Flood Control in 17th Century Mexico City', *Proceedings of the Pacific Coast Council on Latin American Studies* **5**, 69–75.
Hofsommer, Donovan L.: 1976, 'Steel Plows and Iron Men: The Illinois Central Railroad and Iowa's Winter of 1936', *Annals of Iowa* **43**, 292–8.
Knauth, Otto W.: 1960, 'The Winter of 1935–36', *Annals of Iowa* **35**, 288–93.
Mattison, Ray H.: 1967, 'The Flood of 1881', *North Dakota History* **34**.

Nurnberger, Ralph: 1974, 'The Great Baltimore Deluge of 1817', *Maryland Historical Magazine* **29**, 405–8.
Rackley, Barbara: 1971, 'The Hard Winter, 1886–7', *Montana: The Magazine of Western History* **21**.
Ravensdale, J. R.: 1974, *Liable to Floods: Village Landscape on the Edge of the Fens, A. D. 450–1850*. Cambridge.
Salivia, Luis: 1972, *History of Windstorms in Puerto Rico and the Antilles, 1492–1970*. (Title translated from the Spanish) San Juan.
Smith, Roland: 1975, 'The Politics of Pittsburgh Flood Control, 1908–1936', *Pennsylvania History* **42**, 5–24.
Smith, Roland: 1977, 'The Politics of Pittsburgh Flood Control, 1936–1960', *Pennsylvania History* **44**, 3–24.
Steele, Maryland S.: 1970, 'Water, Water Everywhere: The 1913 Fort Wayne, Indiana Flood', *Old Fort News*.
U.S. Weather Service: 1971, *Memorable Hurricanes in the United States Since 1873*.
Waite, Paul J.: 1970, 'Outstanding Iowa Storms', *Annals of Iowa* **40**.
Waite, Paul J.: 1968, 'Iowa Blizzards', *Iowa Farm Science* **22**, 22–3.
Wilson, William E.: 1969, 'Blizzards and Buffalo – 1880', *Montana: The Magazine of Western History* **29**.

P. Demography and Subsistence

Abel, Wilhelm: 1966, *Agrarkrisen und Agrarkonjonktur in Mitteleuropa vom 13. bis zum 19. Jahrhundert*, 2nd Edition. Berlin.
Abel, Wilhelm: 1974, *Massenarmut und Hungerkrisen im vorindustriellen Europa*, Hamburg.
Aleati, Giuseppe: 1957, *La Popolazione di Pavia Durante il Dominio Spagnolo*, Milan.
Alvarez, Gonzolo Anes: 1970, *Las Crisis Agrarias en la Espana Moderna*, Madrid.
Appleby, Andrew B.: 1978, *Famine in Tudor and Stuart England*, Stanford.
Appleby, Andrew B.: 1979, 'Grain Prices and Subsistence Crises in England and France, 1590–1740', *J. Ec. Hist.* **XXXIX**, 4, 865–887.
Armengaud, Andre: 1961, *Les Populations de l'Est-Aquitain au Debut de l'Epoque Contemporaine*, Paris.
Blayo, Yves: 1975, 'La Mortalite en France de 1740 a 1829', *Population*, Numero special, 123–37.
Bondois, P.-M.: 1924, 'La Disette de 1662', *Revue d'Histoire Economique et Sociale*, **12**.
Chambers, J. D.: 1957, *The Vale of Trent, 1670–1800. EHR* Supplement. III. London.
Chambers, J. D.: 1972, *Population, Economy and Society in Preindustrial England*, Oxford.
Cobb, Richard: 1965, *Terreur et Subsistances, 1793–1795*, Paris.
Colloque de Nice: 1969, *Villes de l'Europe Mediterraneenne et de l'Europe Occidentale du Moyen Age au XIXe Siecle*, Annales de la Faculte des Lettres Humaines de Nice, Nos. 9 and 10. Nice.
Connell, K. H.: 1950, *The Population of Ireland 1750–1845*, Oxford.
Del Panta, L. and Bacci, M. Livi: 1975, 'Cronologia, Intensita e Diffusione delle Crisi de Mortalita in Italia, 1600–1850'. Unpublished paper. Presented to the International Colloquium of Historical Demography, Montreal.
Deyon, Pierre: 1967, *Amiens, Capitale Provinciale*, Paris.
Drake, Michael: 1961–62, 'An Elementary Exercise in Parish Register Demography', *EHR* 2nd ser., **14**, 432–6.
Drake, Michael: 1969, *Population and Society in Norway 1735–1865*, Cambridge.
Drake, Michael: 1974, *Historical Demography: Problems and Projects*, Milton Keynes, Bucks.
Dupaquier, Jacques: 1968, 'Sur la Population Francaise au XVIIe et au XVIIIe Siecle', *Revue Historique* **239**, 43–79.
Finlay, Roger A. P.: 1977, 'The Population of London, 1580–1650'. Unpublished dissertation, Cambridge University.
Flinn, M. W.: 1974, 'The Stabilisation of Mortality in Pre-industrial Western Europe', *J. Eur. Ec. Hist.* **3**, 285–318.
Gautier, Etienne, and Henry, Louis: 1958, *La Population de Crulai*, Paris.

Flinn, M. W. (ed.): 1977, *Scottish Population History from the Seventeenth Century to the 1930's*, Cambridge.
Glass, D. V. and Eversley, D. E. C. (eds.): 1965, *Population in History*, London. Includes articles on historical population trends in the major European states.
Godechot, Jacques, and Moncassin, Suzanne: 1964, 'Demographie et Subsistances en Languedoc (du XVIIIe au debut du XIXe siecle)', *Bulletin d'Histoire Economique et Sociale de la Revolution Francaise* 19–60.
Gooder, A.: 1974, 'The Population Crisis of 1727–30 in Warwickshire', *Midland History* **1**, 4, 1–22.
Goubert, Pierre: 1954, 'Une Richesse Historique: les Registres Paroissiaux', *Annales* **9**.
Goubert, Pierre: 1960, *Beauvais et la Beauvaisis de 1600 a 1730*, Paris.
Goubert, Pierre: 1966, *Louis XIV et Vingt Millions de Francais*. Paris.
Goubert, Pierre: 1970, 'Historical Demography and the Reinterpretation of Early Modern French History: A Research Review', *JIH* **I**, 37–48.
Goubert, Pierre: 1970, 'Le "Tragique" XVIIe Siecle', in Ernest Labrousse *et al., Histoire Economique et Sociale de la France*. Vol. 2. Paris.
Goubert, Pierre: 1970, 'Le Regime Demographique Francais au Temps de Louis XIV', in Ernest Labrousse *et al., Histoire Economique et Sociale de la France*. Vol. **2**, Paris.
Guillaume, P. and Poussou, P.-J.: 1970, *Demographie Historique*, Paris.
Harrison, R. J.: 1973, 'The Spanish Famine of 1904–6', *Agricultural History* **47**, 300–7.
Helin, Etienne: 1963, *La Demographie de Liege au XVIIe et XVIIIe Siecles*, Memoire de l'Academie Royale de Belgique. Brussels.
Helleiner, Karl F.: 1967, 'The Population of Europe from the Black Death to the Eve of the Vital Revolution', in E. E. Rich and C. H. Wilson (eds.), *The Cambridge Economic History of Europe*. Vol. IV, 1–95. Cambridge.
Higounet, Charles (ed.): 1971, *Histoire de l'Aquitaine*, Toulouse.
Hofstee, E. W.: 1977, *De Demografische Ontwikkeling van Nederland in de Eerste Helft van de 19e Eeuw*, Wageningen.
Ibarra y Rodriguez, Eduardo: 1926, 'El Problema de las Subsistencias en Espana al Comenzar la Edad Moderna. La Carne', *Nuestro Tiempo* **25**, 5–29; **26**, 206–250.
Ibarra y Rodriguez, Eduardo: 1941, 1942, 'El Problema Cerealista en Espana Durante el Reinade de los Reyes Catolicos (1475–1516)', *Anales de Economia* **1**, 163–217, 299–300; **2**, 3–30, 119–136.
Imhof, A. E. and Lindskog, B. J.: 1974, 'Les Causes de Mortalite en Suede et en Finlande entre 1749 et 1773', *Annales* **29**.
Jutikkala, E.: 1955, 'The Great Finnish Famine in 1696–97', *Scandinavian Economic History Review* **3**, 48–63.
Kershaw, I.: 1973, 'The Great Famine and Agrarian Crisis in England, 1315–1322', *P&P* **59**, 1–50.
Lachiver, Marcel: 1969, *La Population de Meulan du XVIIe au XIXe Siecle*, Paris.
Langer, William: 1963, 'Europe's Initial Population Explosion', *American Historical Review* **69**, 1–17.
Laslett, P.: 1971, *The World We Have Lost*, Second edition, London.
Lebrun, Francois: 1971, *Les Hommes et la Mort en Anjou aux 17e et 18e Siecles*, Paris.
Lee, Ronald: 1973, 'Population in Preindustrial England: An Econometric Analysis', *Quart. J. of Economics* **87**, 581–607.
Lee, Ronald: 1980, 'Shortrun Fluctuations in Vital Rates, Prices and Weather', forthcoming as Chapter 9 of E. A. Wrigley and Roger Schofield (eds.), *Population Trends in Early Modern England*.
Lemarchand, Guy: 1963, 'Les Troubles de Subsistances dans la Generalite de Rouen (second Moitie du XVIII Siede)', *Annales Historiques de la Revolution Francoise* **174**, 401–427.
Levine, David and Wrightson, Keith: 'The Social Context of Illegitimacy in Early Modern England', in P. Laslett (ed.), *Comparative Studies in the History of Bastardy*, (forthcoming).
Lucas, H. S.: 1962, 'The Great European Famine of 1315, 1316, and 1317', in Carus-Wilson, E. D. (ed.), *Essays in Economic History* **2**, 49–72, London.
Manry, Andre-Georges (ed.): 1974, *Histoire de l'Auvergne*, Toulouse.
McAlpin, Michelle B.: 1979, 'Dearth, Famine, and Risk: The Changing Impact of Crop Failures in Western India, 1870–1920', *J. Ec. H.* **XXXIX**, 143–57.

Mendels, Franklin: 1917, 'Industrialization and Population Pressure in 18th Century Flanders'. Unpublished doctoral dissertation, University of Wisconsin.

Mendels, Franklin: 1970, 'Recent Research in European Historical Demography', *American Historical Review* **75**, 1065–1073.

Meuvret, Jean: 1946, 'Les Crises de Subsistances et la Demographie de la France d'Ancien Regime', *Population* **1**, 643–50.

Meuvret, Jean: 1977, *Le Probleme des Subsistances a l'Epoque Louis XIV*, Paris and The Hague, Two parts.

Mitchell, J. Clyde: 1961, 'Wage Labour and African Population Movements in Central Africa', in K. M. Barvour and R. M. Prothero (eds.), *Essays on African Population*, London.

Mols, Roger: 1954–6, *Introduction a la Demographie Historique des Villes d'Europe du XIVe au XVIe Siecle*, Louvain. Three volumes.

Morineau, Michel: 1971, *Les Faux-Semblants d'un Demarrage Economique: Agriculture et Demographie en France au XVIIIe Siecle*, Cahiers des Annales, Paris.

Morineau, Michel: 1974, 'Revolution Agricole, Revolution Alimentaire, Revolution Demographique', *Annales de Demographie Historique*.

Nadal, Jorge: 1966, *La Poblacion Espanola (Siglos XVI a XX)*, Barcelona.

Oeppen, James: 1979, 'Short-Run Fluctuations in English Vital Rates, 1541–1871'. Unpublished ms., SSRC Cambridge Group for the History of Population and Social Structure. (An important article).

Poitrineau, Abel: 1965, *Le Vie Rurale en Basse-Auvergne au XVIIIe Siecle*, Paris.

Queruel, Julian: 1974, 'La Crise de Subsistance des Annees 1740 dans le Ressort du Parlement de Paris', *Annales de Demographie Historique*, 281–333.

Razzell, P. E.: 1965, 'Population Change in Eighteenth-Century England: A Reinterpretation', *EHR* **18**, 312–332.

Reinhard, Marcel, Armengaud, Andre and Dupaquier, Jacques: 1968, *Histoire Generale de la Population Mondiale*, Paris.

Rogers, Colin D.: 1975, *The Lancashire Population Crisis of 1623*, Manchester.

Russell, J. C.: 1948, *British Medieval Population*, Albuquerque.

Salvesen, H.: 1975–76, 'The Agrarian Crisis in Norway in the Late Middle Ages', in Dyer, C. C. (ed.), *Medieval Village Research Group Report*.

Sanchez-Albornoz, Nicolas: 1963, *Las Crisis de Subsistencias en el Siglo XIX*. Rosario, Argentina.

Schofield, R. S.: 1970, 'Perinatal Mortality in Hawkshead, Lancashire, 1581–1710', *Local Population Studies*.

Schofield, R. S.: 1972, ' "Crisis" Mortality', *Local Population Studies*.

Schofield, R. S. and Wrigley, E. A.: 1981, *The Population History of England 1541–1871*.

Shea, William: 'Famines in the Early History of Egypt and Syro-Palestine'. Order from University Microfilms No. 77–8035.

1973, *Sur la Population Francaise au XVIIIe et au XIXe Siecles: Hommage a Marcel Reinhard*, Paris.

Toutain, Jean-Claude: 1965, *La Population de la France de 1700 a 1959*, Paris.

Utterstrom, G.: 1954, 'Some Population Problems in Pre-industrial Sweden', *Scandinavian Economic History Review* **2**, 103–65.

Vandier, Jacques: 1936, *La Famine dans l'Egypte Ancienne*, Cairo.

Wrigley, E. A.: 1968, 'Mortality in Pre-industrial England. The Example of Colyton, Devon, Over Three Centuries', *Daedalus*.

Wrigley, E. A.: 1969, *Population and History*, New York. Includes extensive bibliography.

Wrigley, E. A.: 1978, 'English Mortality in the Industrial Revolution Period', in E. G. Forbes (ed.), *Proceedings of XIth International Congress of the History of Science*, Edinburgh.

Q. Disease and Health

Appleby, Andrew: 1973, 'Disease or Famine? Mortality in Cumberland and Westmoreland, 1580–1640', *EHR* **26**, 3, 403–32.

Appleby, Andrew: 1975, 'Nutrition and Disease: The Case of London, 1550–1750', *JIH* **6**.

Appleby, Andrew: 1980, 'Epidemics and Famine in the Little Ice Age', *JIH* **10**, 4. Reprinted in *C&H* I.

Appleby, Andrew: 'The Disappearance of the Plague: A Continuing Puzzle', *EHR* (forthcoming).

Bennassar, B.: 1969, *Recherches sur les grandes epidemies dans le nord de l'Espagne a la fin du XVIe Siecle*, Paris.

Bergman, M.: 1967, 'The Potato Blight in the Netherlands and Its Social Consequences, 1845–47', *International Review of Social History* **XII**, 390–431.

Beveridge, W. I. B.: 1977, *Influenza: The Last Great Plague*, London.

Biraben, J. N.: 1968, 'Certain Demographic Characteristics of the Plague Epidemic in France 1720–1722', *Daedalus* **97**, 536–45.

Biraben, J. N.: 1975, 1976, *Les Hommes et la Peste en France et dans les Pays Europeens et Mediterraneens*. Tome I: *La Peste dans l'Histoire*, Paris and The Hague. Tome **II**: *Les Hommes Face a la Peste*, Paris.

Brownlee, John: 1925, 'The Health of London in the Eighteenth Century', *Proceedings of the Royal Society of Medicine* **XVII**, 73–85.

Caporael, L. R.: 'Ergotism: The Satan Loosed in Salem', *Science* **192**, 21–26.

Carefoot, G. L. and Sprott, E. R.: 1974, *Famine on the Wind: Plant Diseases and Human History*, Sydney.

Carriere, Charles *et al.*: 1968, *Marseille, Ville Morte. La Peste de 1720*, Marseilles.

Cipolla, Carlo: 1973, *Cristofano and the Plague*, London.

Cousens, S. H.: 1963, 'The Regional Variation in Mortality During the Great Irish Famine', *Proceedings of the Royal Irish Academy* **LXIII**.

Creighton, Charles: 1894, *A History of Epidemics in Britain*, Cambridge. (Two volumes).

Dorwart, R. A.: 1959, 'Cattle Disease (Rinderpest?) – Prevention and Cure in Brandenburg, 1665–1732', *Agricultural History* **33**, 79–85.

Edwards, R. Dudley and Williams, T. Desmond (eds.): 1957, *The Great Famine: Studies in Irish History, 1845–52*, New York.

Faber, J. A.: 1962, 'Cattle-Plague in the Netherlands During the Eighteenth Century', *Mededelingen van de Landbouwhogeschool te Wageningen* **62**, 1–7.

Fisher, F. J.: 1965, 'Influenza and Inflation in Tudor England', *EHR* 2nd ser. **18**, 120–9.

Flinn, M.: 'The Disappearance of the Plague', *J. Eur. Ec. Hist.* (forthcoming).

Goubert, J. P.: 1969, 'Le Phenomene Epidemique en Bretagne a la fin du XVIIIe Siecle', *Annales* **XXIV**, 1562–88.

Goubert, J. P.: 1974, *Malades et Medecins en Bretagne, 1770–90*, Paris.

Gray, Malcolm.: 1955, 'The Highland Potato Famine of the 1840's', *EHR* **VII**.

Hackett, C. J.: 1963, 'On the Origin of the Human Treponematoses (Pinta, Yaws, Endemic Syphilis, and Venereal Syphilis)', *Bulletin of the World Health Organization* **XXIX**, 7–41.

Hackett, L. W.: 1949, 'Conspectus of Malaria Incidence in Northern Europe, the Mediterranean Region and the Near East', in Mary Boyd (ed.), *Malariology*, Philadelphia.

Hannaway, Caroline: 1972, 'The Societe Royale de Medecine and Epidemics in the Ancien Regime', *Bulletin of the History of Medicine* **46**.

Hirst, L. Fabian: 1953, *The Conquest of Plague*, Oxford.

Hodge, William: 1978, 'Weather and Mortality', *Environmental Data and Information Service*, 12–4.

Hollingsworth, M. F. and Hollingsworth, T. H.: 1971, 'Plague Mortality Rates by Age and Sex in the Parish of St. Botolph's Without Bishopsgate, London, 1603', *Population Studies* **25**.

Hopkinson, G. G.: 1973, 'Cattle Plague in South Yorkshire, 1746–57', *Transactions of the Hunter Archaeological Society* **X**.

Howe, Melvyn: 1972, *Man, Environment and Disease in Britain*, New York.

Howson, W. G.: 1961, 'Plague, Poverty and Population in Parts of North-West England, 1580–1720', *Transactions of the Historical Society of Lancashire and Cheshire* **112**, 29–55.

Hughes, J.: 1971, 'The Plague at Carlisle, 1597–98', *Transactions of the Cumberland and Westmorland Antiquarian and Archaeological Society* **71**.

Ladurie, Emmanuel Le Roy: 1975, 'Famine Amenorrhoea (Seventeenth-Twentieth Centuries)', in Robert Forster and Orest Ranum (eds.), *Biology of Man in History*, Baltimore, pp. 163–78.

Landsberg, H. E.: 1969, *Weather and Health*.
Lebrun, Francois: 1977, 'Les Epidemies en Haute-Bretagne a la Fin de l'Ancien Regime (1770–1789)', *Annales de Demographie Historique*.
McNeill, William: 1976, *Plagues and Peoples*, New York.
Menken, Jane, Trussel, James, and Watkins, Susan: 1981, 'The Nutrition Fertility Link: An Examination of the Evidence'. *JIH* **11**, 425–441. A major review essay with an up-to-date bibliography.
Mirowski, Philip: 'The Plague and the Penny-Loaf: The Disease-Dearth Nexus in Stuart and Hanoverian London'. Unpublished manuscript. Department of Economics, University of Michigan.
Mokyr, Joel: 1980, 'Industrialization and Poverty in Ireland and the Netherlands', *JIH* **X**, 3, 429–58.
Mokyr, Joel: 'The Deadly Fungus: An Economic Investigation into the Short-Term Demographic Impact of the Irish Famine, 1846–1851', in J. Simon (ed.), *Research in Population Economics* **II**.
Mullett, C. F.: 1946, 'The Cattle Distemper in Mid-Eighteenth Century England', *Agricultural History* **20**.
Mullett, C. F.: 1949, 'A Century of English Quarantine (1709–1825)', *Bulletin of the History of Medicine* **XXIII**, 527–45.
Newman, Marshall: 1976, 'Aboriginal New World Epidemiology and Medical Care, and the Impact of Old World Disease Imports', *American Journal of Physical Anthropology* **XLV**, 667–72.
Palliser, David: 1974, 'Dearth and Disease in Staffordshire, 1504–1670', in C. W. Chalklin and M. A. Havinden (eds.), *Rural Change and Urban Growth, 1500–1800*, London.
Panzac, Daniel: 1973, 'La Peste a Smyrne au XVIIIe Siecle', *Annales* **28**.
Pickard, Ranson: 1947, *The Population and Epidemics of Exeter in Pre-Census Times*, Exeter.
Post, John: 1976, 'Famine, Mortality, and Epidemic Disease in the Process of Modernization', *EHR* 2nd ser., **29**, 14–37.
Razzell, Peter: 1977, *The Conquest of Smallpox*, Firle, Sussex.
Revel, J.: 1970, 'Autour d'une Epidemie Ancienne: la Peste de 1666–1670', *Revue d'Histoire Moderne et Contemporaine* **17**.
Rothenberg, G. E.: 1973, 'The Austrian Sanitary Cordon and the Control of the Bubonic Plague, 1710–1871', *J. of the History of Medicine and Allied Sciences* **XXVIII**.
Schofield, Roger: 1977, 'An Anatomy of an Epidemic: Colyton, November 1645 to November 1646', in *The Plague Reconsidered* (*Local Population Studies* Supplement, 1977), 95–126.
Schove, David J.: 1972, 'Chronology and Historical Geography of Famine, Plague and other Pandemics', *Proceedings of the XXIII Congress of the History of Medicine*, 1265–72, London.
Shrewsbury, J. F. D.: 1970, *A History of Bubonic Plague in the British Isles*, Cambridge.
States, Jerome S.: 'Weather and Death in Three American Cities'. Unpublished dissertation, University of Pittsburgh. Order from University Microfilms No. 77–23, 577.
Terlouw, Frida: 1971, 'De Aardappelziekte in Nederland in 1845 en volgende Jaren', *Economisch en Sociaal-Historish Jaarboek* **XXXIV**, 261–308.
The Plague Reconsidered: 1977, *Local Population Studies* Supplement.
Woehlkens, Eric: 1954, *Pest und Ruhr im 16. und 17. Jahrhundert*, Hamburg.
Woodham-Smith, Cecil: 1961, *The Great Hunger: Ireland 1845–1849*, New York.

R. Human Migration

Abel, Wilhelm: 1965, 'Desertions Rurales: Bilan de la Recherche Allemande', in *Villages Desertes et Histoire Economique, XIe–XVIIIe Siecle*, 515–532, Paris.
Beresford, Guy: 1979, 'The Long-Term Variation of the Climate in Britain and Its Effect Upon Rural Ecology and Settlement'. Paper presented at the ICCH, University of East Anglia.
Beresford, Maurice: 1954, *The Lost Villages of England*, London.
Beresford, Maurice: 1965, 'Villages Desertes: Bilan de la Recherche Anglaise', in *Villages Desertes et Histoire Economique, XIe–XVIIIe Siecle*, 533–580, Paris.
Bremer, R. G.: 1974, 'Patterns of Spatial Mobility: A Case Study of Nebraska Farmers, 1890–1970', *Agricultural History* **48**, 529–42.
Clark, P.: 1972, 'The Migrant in Kentish Towns, 1580–1640', in Clark and P. Slack, *Crisis and Order in English Towns 1500–1700*, London.

Connell, K. H.: 1950, 'The Colonization of Waste Land in Ireland, 1780–1845', *EHR* **III**.

Greenwood, M. J.: 1975, 'Research on Internal Migration in the United States: A Survey', *Journal of Economic Literature* **13**, 397–433.

Pesez, J. M. and Ladurie, E. Le Roy: 1965, 'Le Cas Francais: vue d'Ensemble', in *Villages Desertes et Histoire Economique, XIe-XVIIIe Siecle*, pp. 127–252.

S. Agriculture

Abel, Wilhelm: 1967, *Geschichte der Deutschen Landwirtschaft vom Fruher Mittelalter bis zum 19. Jahrhundert*, Volume II of Gunther Franz (ed.), *Deutsche Agrargeschichte* 2nd edition. Stuttgart.

Alvarez, Gonzalo Anes: 1968, 'Production et Productivite Agricoles dans les deux Castiles de la fin du XVIIe Siecle a 1836', in *Third International Conference of Economic History, Munich 1965*, Vol. II, 85–91. Paris.

Baker, D: 1970, 'The Marketing of Corn in the First Half of the 18th Century: Northeast Kent', *Agricultural History Review* **XVIII**.

Bernard, R. J.: 'L'Alimentation Paysanne en Gevaudan au XVIIIe Siecle' *Annales* **24**, 1449–1467.

Bielmann, J.: 1972, *Die Lebensverhaltnisse im Urnerland wahrend des 18. und zu Beginn des 19. Jahrhundert*, Basel.

Bourde, Andre: 1967, *Agronomes et Agronomie au XVIIIe Siecle*, Paris, 3 vols.

Bourke, P. M. Austin: 1968, 'The Use of the Potato Crop in Prefamine Ireland', *Journal of the Statistical and Social Inquiry Society of Ireland* **XXI**.

Braudel, F.: 1972, *The Mediterranean and the Mediterranean World in the Age of Philip II*. New York. Original French version 1949, Revised 1966.

Braudel, F.: 1973, *Capitalism and Material Life, 1400–1800*, New York. Original French version, 1967.

Breen, Timothy: 'The Culture of Agriculture: Tobacco and Society in Pre-Revolutionary Virginia'. Unpublished manuscript.

Chambers, J. D. and Mingay, G. E.: 1966, *The Agricultural Revolution, 1750–1880*, London.

Collins, E. J. T.: 1975, 'Dietary Change and Cereal Consumption in Britain in the Nineteenth Century', *Agricultural History Review* **23**, 97–115.

de Silva, Jose Gentil: 1963, 'Villages Castillans et Types de Production au XVIe Siecle', *Annales* **18**, 729–744.

de Silva, Jose Gentil: 1965, *En Espagne, Developpement Economique, Subsistances, Declin*, Paris.

deVries, Jan: 1976, *The Economy of Europe in an Age of Crisis, 1600–1750*. Cambridge.

Drummond, J. C. and Wilbraham, Anne: 1939, *The Englishman's Food: A History of Five Centuries of English Diet*, London.

Everitt, Alan: 1967, 'The Marketing of Agricultural Produce', in Joan Thirsk' (ed.), *The Agrarian History of England and Wales, IV, 1540–1640*, Cambridge, pp. 466–592.

Everitt, Alan: 1968, 'The Food Market of the English Town, 1660–1760', in *Third International Conference of Economic History, Munich, 1965,* **I**, 57–72. Paris.

Fay, C. R.: 1932, *The Corn Laws and Social England*, Cambridge.

Ferry, Guy and Mulliez, Jacques: 1970, *L'Etat et la Renovation de l'Agriculture au XVIIe Siecle*, Paris.

Festy, Ocatave: 1947, *L'Agriculture Pendant le Revolution Francaise; Les Conditions de Production et de Recolte des Cereales*, Paris.

Fisher, F. J.: 1954, 'The Development of the London Food Market, 1540–1640'. Reprinted in E. M. Carus-Wilson (ed.), *Essays in Economic History* **I**, 135–151. London.

Florescano, Enrique: 1976, *The Origin and Development of the Agrarian Problems of Mexico, 1500–1821*, Title from the Spanish. Mexico.

Forster, Elborg and Robert (eds.): 1975, *European Diet from Pre-Industrial to Modern Times*, New York.

Fourastie, Jean and Grandamy, Rene: 1968, 'Remarques sur les Prix Salariaux des Cereales et al Productivite du Travail Agricole en Europe du XVe au XXe Siecles', in *Third International Conference of Economic History, Munich 1965,* **I**, 647–58. Paris.

Fourquin, Guy: 1964, *Les Campagnes de La Region Parisienne a la Fin du Moyen Age*. Publications de la Faculte des Lettres et Sciences Humaines de Paris, 'Recherches', 10. Paris.

Franz, Gunther: 1970, *Geschichte des Deutschen Bauernstandes vom frühen Mittelalter bis zum 19. Jahrhundert*. Volume IV of Gunther Franz (ed.), *Deutsche Agrargeschichte*, Stuttgart.

Goubert, Pierre. "Les Techniques Agricoles dans les Pays Picard aux XVIIe et XVIIIe Siecles", *Revue d'Histoire Economique et Sociale*. 35, 1957.

Goubert, Pierre: 1974, 'Societes Rurales Francaises du XVIIIe Siecle', in Fernand Braudel *et al.*, *Conjuncture Economique, Structures Sociales: Hommage a Ernest Labrousse*, Paris.

Gould, J. D.: 1962, 'Agricultural Fluctuations and the English Economy in the Eighteenth Century', *J. Ec. Hist.* **22**, 313–333.

Gras, N. S. B.: 1915, *The Evolution of the English Corn Market from the Twelfth to the Eighteenth Century*, Cambridge, Massachusetts.

Heitz, G.: 1968, 'Produktion und Produktivitat in der Landwirtschaft unter Gutsherrschaftlichen Bedingungen im 17. und 18. Jahrhundert', in *Third International Conference of Economic History, Munich 1965*, Vol. **II**, 197–204. Paris.

Hoffman, Richard C.: 1975, 'Medieval Origins of the Common Fields', in William Parker and E. L. Jones, (eds.), *European Peasants and Their Markets: Essays in Agrarian Economic History*, Princeton.

Hohenberg, Paul H.: 1977, 'Maize in French Agriculture', *J. Eur. Ec. Hist.* **6**.

Hoskins. W. G.: 1964, 'Harvest Fluctuations and English Economic History, 1480–1619', *Agricultural History Review* **12**, 1, 28–46.

Hoskins, W. G.: 1968, 'Harvest Fluctuations and English Economic History, 1620–1759', *Agricultural History Review* **16**, 1, 15–31.

Houtte, J. A. and Verhulst, A.: 1968, 'L'Approvisionnement des Villes dans les Pays-Bas (Moyen Age et Temps Modernes)', in *Third International Conference of Economic History, Munich 1965*, Vol. **I**. 73–8. Paris.

Jacquart, J.: 1966, 'La Production Agricole dans la France du XVIIe Siecle', *XVIIe Siecle*, 70–1.

Jacquart, J.: 1968, 'La Productivite Agricole dans la France du Nord aux XVIe et XVIIe Siecles', in *Third International Conference of Economic History, Munich 1965*, Vol. **II**, 64–74. Paris.

Jacquart, J.: 1974, *La Crise Rurale en Ile-de-France 1550–1670*, Paris.

John, A. H.: 1965, 'Agricultural Productivity and Economic Growth in England, 1700–1760', *J. Ec. Hist.* **25**, 19–34.

John, A. H.: 1976, 'English Agricultural Improvements and Grain Exports, 1660–1765', in D.C. Coleman and A. H. John (eds.), *Trade, Government and Economy in Pre-Industrial England*, London.

Jones, E. L.: 1965, 'Agriculture and Economic Growth in England, 1660–1750: Agricultural Change', *J. Ec. Hist.* **25**, 1–18.

Jones, E. L.: 1967, *Agriculture and Economic Growth in England, 1650–1815*, London.

Jones, E. L.: 1975, 'Afterword' in W. N. Parker and E. L. Jones (eds.), *European Peasants and Their Markets: Essays in Agrarian Economic History*, Princeton.

Jones, E. L. and Woolff, S. J. (eds.): 1969, *Agrarian Change and Economic Development: The Historical Problems*, London.

Jones, G. R. J.: 1965, 'Agriculture in North-West Wales During the Later Middle Ages', in Taylor, J. A. (ed.), *Climatic Change with Special Reference to Wales and Its Agriculture*, Aberystwyth.

Kerridge, Eric: 1968, *The Agricultural Revolution*, New York.

Labrousse, Ernest: 1944, *La Crise de l'Economie Francaise a la Fin de l'Ancien Regime et au Debut de la Revolution Francaise*, Paris.

Labrousse, Ernest *et al.*: 1970, *Histoire Economique et Sociale de la France*, Vol. 2. Paris.

Ladurie, E. LeRoy: 1966, *Les Paysans de Languedoc*, Paris.

Ladurie, E. LeRoy and Joseph Goy (eds.): 1972, *Les Fluctuations du Produit de la Dime*, Paris.

Ladurie, E. LeRoy and Joseph Goy (eds.): 1978, 'La Dime et La Peste XIVe-XVIIIe Siecle', *Revue Historique* **CCLX**, 123–42.

Langer, William: 1975, 'American Foods and Europe's Population Growth, 1750–1850', *Journal of Social History*, 51–66.

Leon, Pierre, *et al.*: 1976, *Histoire Economique et Sociale de la France*, Vol. 3, Paris.

Leskiewicz, Janina: 1965, 'Les Entraves Socials au Developpement de la "Nouvelle Agriculture" en Pologne', in *Third International Conference of Economic History, Aix-en-Provence 1962*, Vol. II, 237–47, Paris.

Lutge, Friedrich: 1967, *Geschichte der Deutschen Agrarverfassung vom frühen Mittelalter bis zum 19. Jahrhundert*. Vol. III of Gunther Franz (ed.), *Deutsche Agrargeschichte* Stuttgart, 2nd ed.

Makkai, Laszlo: 1968, 'Production et Productivite Agricole en Hongrie a l'Ere du Feodalisme Tardif (1550–1850)', in *Third International Conference of Economic History, Munich, 1966*, Vol. **II**, 171–180, Paris.

Masefield, G. B.: 1967, 'Crops and Livestock', in E. E. Rich and C. H. Wilson (eds.), *The Economy of Expanding Europe in the Sixteenth and Seventeenth Centuries*, Cambridge Economic History of Europe, IV, Cambridge, pp. 276–307.

Matejek, Francois: 1968, 'La Production Agricole dans les Pays Tschecoslovaques à Partir de XVIe Siecle Jusqu' à la Premiere Guerre Mondiale', in *Third International Conference of Economic History, Munich, 1965*, Vol. **II**, 203–220, Paris.

McCloskey, Donald N.: 1972, 'The Enclosure of Open Fields: Preface to a Study of Its Impact on the Efficiency of English Agriculture in the Eighteenth Century', *J. Ec. Hist.* 15–35.

McCloskey, Donald N.: 1976, 'English Open Fields as Behavior Toward Risk', in Paul Uselding (ed.), *Research in Economic History* **I**, 124–170.

Meuvret, Jean: 1968, 'Production et Productivite Agricoles', in *Third International Conference of Economic History, Munich 1965*, Vol. **II**, 11–22, Paris.

Morineau, Michel: 1970, 'La Pomme de Terre au XVIIIe Siecle', *Annales* **25**, 1767–84.

Neveux, Hughes: 1973, 'L'Alimentation du XIVe au XVIIIe Siecle: Essai de Mise au Point', *Revue d'Histoire Economique et Sociale* **51**, 336–79.

Pach, S. P.: 1968, 'Die Getreideversorgung der Ungarischen Stadte vom XV. bis XVII. Jahrhundert', in *Third International Conference of Economic History, Munich, 1965*, Vol. **I**, 97–108, Paris.

Parain, Charles: 1966, 'The Evolution of Agricultural Techniques', in the *Cambridge Economic History of Europe*, 2nd ed., Vol. **I**, Cambridge.

Philippe, R.: 1970, 'Une Operation Pilote: L'Etude du Ravitaillement de Paris au Temps de Lavoisier', in J. J. Hemardinquer (ed.), *Pour une Histoire de l'Alimentation*, Paris.

Roessingh, H. K.: 1979, 'Tobacco Growing in Holland in the Seventeenth and Eighteenth Centuries: A Case Study of the Innovative Spirit of Dutch Peasants', *The Low Countries History Yearbook; Acta Historiae Neerlandica*, Vol. **XI**, 1978, The Hague.

Schuermann, M.: 1974, *Bevolkerung, Wirtschaft und Gesellschaft in Appenzell-Innerhoden im 18. und frühen 19. Jahrhundert*. Geschichtsfreund, Appenzell.

Skipp, Victor: 1978, *Crisis and Development. An Ecological Case Study of the Forest of Arden*, Cambridge.

Thirsk, Joan: 1964, 'The Common Fields', *P&P* **29**.

Thirsk, Joan: 1966, 'The Origin of the Common Fields', *P&P* **33**.

Thirsk, Joan (ed.): 1967, *The Agrarian History of England and Wales, IV, 1540–1640*. Indispensable, with copious bibliography.

Thompson, F. M. L.: 1968, 'The Second Agricultural Revolution, 1815–1880', *EHR* **21**, 62–77.

Titow, J. Z.: 1965, 'Medieval England and the Open-Field System', *P&P* **32**.

Titow, J. Z.: 1972, *Winchester Yields: A Study in Medieval Agricultural Productivity*, Cambridge.

Usher, S. P.: 1913, *The History of the Grain Trade in France, 1400–1710*, Cambridge, Mass.

Van Bath, B. H. Slicher: 1963, *The Agrarian History of Western Europe, A.D. 500–1800* (Trans. O. Ordish), London.

Van Bath, B. H. Slicher: 1967, 'The Yields of Different Crops (mainly cereals) in Relation to the Seed, ca. 810–1820', *Acta Historiae Neerlandica* **2**, 26–106.

Van Bath, B. H. Slicher: 1968, 'La Productivite Agricole. Les problemes fondamentaux de la societe pre-industrielle en Europe occidentale,' in *Third International Conference of Economic History, Munich 1965*, Vol. **II**, 23–30, Paris.

Van Bath, B. H. Slicher: 1977, 'Agriculture and the Vital Revolution', in Rich and Wilson (eds.), *The Cambridge Economic History of Europe*, Vol. 5, Cambridge. Includes extensive bibliography.

Vicens-Vives, Jaime: 1969, *An Economic History of Spain*, Princeton.

Wallerstein, Immanuel: 1974, *The Modern World-System, Capitalist Agriculture and the Origins of the European World Economy in the Sixteenth Century*, New York. Includes extensive bibliography.

Yatkunsky, V. K.: 1968, 'Principaux moments de l'histoire de la production agricole en Russie du XVIe siecle a 1917', in *Third International Conference of Economic History, Munich 1965*, Vol. **II**, 221–237, Paris.

Yelling, J. A.: 1977, *Common Field and Enclosure in England, 1450–1850*, London.

Zytkowicz, Leonid: 1968, 'Production et productivite de l'economic agricole en Pologne aux XVIe–XVIIIe siecles', in *Third International Conference of Economic History, Munich 1965*, Vol. **II**, 149–170, Paris.

T. Grain Price Fluctuations

Appleby, Andrew B.: 1979, 'Grain Prices and Subsistence Crises in England and France, 1590–1740', *J. Ec. Hist.* **39**, 865 ff.

Baulant, Micheline and Meuvret, Jean: 1962, *Prix des cereales extraits de la mercuriale de Paris* (*1520–1698*), Paris, 2 vols.

Beveridge Sir William H. *et al.*: 1939, *Prices and Wages in England*, Vol. 1, London.

Bowden, Peter: 1967, 'Agricultural Prices, Farm Profits and Rents', in Joan Thirsk (ed,), *The Agrarian History of England and Wales, IV: 1500–1640*, Cambridge.

Braudel, Fernand and Spooner, Frank: 1967, 'Prices in Europe from 1450 to 1750', in E. E. Rich and C. H. Wilson, *The Cambridge Economic History of Europe*, Vol. 4, Cambridge.

Dupaquier, Jacques, Lachiver, Marcel and Meuvret, Jean: 1968, *Mercuriales du pays de France et du Vexin francais 1640–1792*, Paris.

Farmer, D. L.: 1957, 'Some Grain Price Movements in Thirteenth-Century England', *EHR* **X**.

Florescano, Enrique: 1969, *Precios del maiz y crises agricolas en Mexico* (*1708–1810*).

Freche, Georges and Genevieve: 1967, *Les prix des grains, des vins, et des legumes a Toulouse* (*1486–1868*), Paris.

Granger, C. W. J. and Hughes, A. O.: 1971, 'A New Look at Some Old Data: The Beveridge Wheat Price Series', *Journ. of the Royal Statistical Society*, Series V, **134**, 3, 413–28.

Hauser, Henri (ed.): 1936, *Recherches et Documents sur l'Histoire des Prix en France de 1500 a 1800*, Paris.

Lee, Ronald D.: 1975, 'Comment on "A New Look at Some Old Data: The Beveridge Wheat Price Series" ', *Journ. of the Royal Statistical Society*, Series V. **138**, 2, 295.

Mestayer, Monique: 1963, 'Les prix du ble et de l'avoine a Douai de 1329 a 1793', *Revue du Nord* **45**, 167–76.

Meuvret, Jean: 1951, 'La geographie des prix des cereales et les anciennes economies europeennes: Prix mediterraneens, prix continentaux, prix atlantiques a la fin du XVII siecle', *Revista de Economia* **4**, 63–69.

Meuvret, Jean: 1969, 'Les oscillations des prix des cereals aux XVIIe et XVIIIe siecles en Angleterre et dans les pays du bassin Parisien', *Revue d'Histoire Moderne et Contemporaine* **16**.

Meuvret, Jean: 1970, 'Prices, Population and Economic Activities in Europe, 1688–1715: A Note', in J. S. Bromley (ed.), *The New Cambridge Modern History, VI: The Rise of Great Britain and Russia, 1688–1715/25*, 874–897, Cambridge.

Phelps Brown, E. H. and Hopkins, S. V.: 1956, 'Seven Centuries of the Price of Consumables, Compared with Builders' Wage Rates', *Economica* **33**, 92, 311–4.

Rogers, J. E. T.: 1866–1902, *A History of Agriculture and Prices in England, From the Year After the Oxford Parliament* (*1259*) *to the Commencement of the Continental War* (*1793*), Oxford, 7 vols.

Tits-Dieuaide, M. J.: 1975, *La Formation des Prix Ceraliers en Brabant et en Flandre au XVe Siecle*, Brussels.

U. Administrative Studies

Barnes, Thomas: 1961, *Somerset 1625–1640: A County's Government During the 'Personal Rule'*, Cambridge, Massachusetts.

Beier, A. L.: 1974, 'Vagrants and the Social Order in Elizabethan England', *P&P* **64**, 3–29. See also *P&P* **71**, 1976, 130–134.

Beloft, M.: 1938, *Public Order and Popular Disturbances, 1660–1714*, Oxford, 2nd ed., London 1963.

Booth, A.: 1977, 'Food Riots in North-West England 1790–1801', *P&P* **77**, 84–107.

Coats, A. W.: 1972, 'Contrary Moralities: Plebs, Paternalists, and Political Economists', *P&P* **54**, 130–3.

Cromack, A.: 1923, *Poor Relief in Scotland*, Aberdeen.

Darivas, B.: 1952, 'Etude sur la Crise Economique de 1593–1597 en Angleterre et la loi des Pauvres', *Revue d'Histoire Economique et Sociale* **XXX**, 382–98.

Davies, C. S. L.: 1969, 'Les Revoltes Popularies en Angleterre, 1500–1700', *Annales* **24**, 24–60.

Dawson, William H.: 1914, *Municipal Life and Government in Germany*, London.

Delumeau, Jean: 1957, *Vie Economique et Sociale de Rome dans le Seconde Moitie du XVIe Siecle*, Paris, 2 vols.

Diaz de Moral, Juan: 1969, *Historia de las Agitaciones Campesinas Andaluzas-Cordoba*, 2nd ed., Madrid.

Fairchilds, Cissie: 1976, *Poverty and Charity in Aix-en-Provence*, Baltimore.

Genovese, Elizabeth Fox: 1973, 'The Many Faces of Moral Economy: A Contribution to a Debate', *P&P* **58**, 161–8.

Gutton, J.-P.: 1971, *La Societe et les Pauvres*, Paris.

Hampson, E. M.: 1934, *The Treatment of Poverty in Cambridgeshire 1597–1834*, Cambridge.

Hobsbawm, E. J.: 1967, 'Le Agitazioni Rurali in Inghilterra nel Primo Ottocento', *Studi Storici* **8**, 257–281.

Hufton, O.: 1972, 'Begging, Vagrancy, Vagabondage and the Law: An Aspect of the Problem of Poverty in 18th Century France', *European Studies Review* **ii**.

James, M.: 1930, *Social Problems and Policy During the Puritan Revolution*, London.

Jones, Eric L.: 1978, 'Disaster Management and Resource Saving in Europe, 1400–1800', in *Natural Resources in European History*. Antoni Maczak and William Parker (eds.), Washington, D.C. pp. 114–38.

Jordan, W. K.: 1959, *Philanthropy in England 1480–1660*, London.

Kamen, Henry: 1971, *The Iron Century: Social Change in Europe 1550–1660*, New York.

Lallemand, L.: 1910, *Histoire de la Charite*, Paris, 4 vols.

Leonard, E. M.: 1900, *The Early History of English Poor Relief*, Cambridge.

Meyer, Jean: 1966, *La Noblesse Bretonne au XVIIIe Siecle*, Paris, 2 vols.

Outhwaite, R. B.: 1978, 'Food Crises in Early Modern England: Patterns of Public Response'. Unpublished paper delivered at the International Economic History Conference, Edinburgh, Scotland.

Paultre, C.: 1906, *La Repression de la Mendicite et du Vagabondage en France sous l'Ancien Regime*, Paris.

Ponko, Vincent, Jr.: 1968, *The Privy Council and the Spirit of Elizabethan Economic Management, 1558–1603*, Transactions of the American Philosophical Society. Philadelphia.

Pound, J. F.: 1971, *Poverty and Vagrancy in Tudor England*, London. See also *P&P* **71**, 1976, 126–9.

Pullan, Brain: 1971, *Rich and Poor in Renaissance Venice: The Social Institutions of the Catholic State to 1620*, Cambridge, Massachusetts.

Rose, R. B.: 1959, 'Eighteenth-Century Price-Riots, the French Revolution and the Jacobin Maximum', *International Review of Social History* **4**, 432–45.

Rose, R. B.: 1961, 'Eighteenth Century Price Riots and Public Policy in England', *International Review of Social History* **6**, 2, 277–92.

Rumeu de Armas, A.: 1944, *Historia de la Prevision Social en Espana*, Madrid.

Salas, M. Jimenz: 1968, *Historia de la Assistencia Social en Espana en la Edad Moderna*, Madrid.

Shelton, W. J.: 1973, *English Hunger and Industrial Disorders*, London.

Slack, Paul: 1972, 'Poverty and Politics in Salisbury 1597–1666', in Peter Clark and Paul Slack (eds.), *Crisis and Order in English Towns, 1500–1700*, London.
Slack, P. A.: 1974, 'Vagrants and Vagrancy in England, 1598–1664', *EHR* 2nd ser., **XXVII**.
Stevenson, J. and Quinault, R.: 1974, *Popular Protest and Public Order*, London.
Thompson, E. P.: 1971, 'The Moral Economy of the English Crowd in the Eighteenth Century', *P&P* **50**, 76–136.
Tilly, Charles: 1975, 'Food Supply and Public Order in Modern Europe', in Charles Tilly (ed.), *The Formation of National States in Western Europe*, Princeton. A major review essay.
Tilly, Louise: 1971, 'The Food Riot as a Form of Political Conflict in France', *JIH* **II**, 1, 23–57.
Van den Eerenbeemt, H. F. J. M.: 1972, 'Het Huwelijk tussen Filantropie en Economie: een Patriotse Illusie', *Economisch-en Sociaal-Historisch Jaarboek* **XXXV**, 28–64.
Walter, John and Wrightson, Keith: 1976, 'Dearth and the Social Order in Early Modern England', *P&P* **71**, 22–42.
Zanetti, Dante: 1963, 'L'Approvisionnement de Pavie au XVIe Siecle', *Annales* **18**, 44–62.

V. War

Benedict, Philip: 1975, 'Catholics and Huguenots in Sixteenth-Century Rouen: The Demographic Effects of the Religious Wars', *French Historical Studies* **9**, 209–34.
Clark, George: 1958, *War and Society in the 17th Century*, Cambridge.
Franz, Gunther: 1961, *Der Dreizigjahrige Krieg und das Deutsche Volk. Unterzuchungen zur Bevolkerungs und Agrargeschichte*, 3rd ed. Stuttgart.
Gutmann, Myron: 1976, 'War and Rural Life in the Seventeenth Century: The Case of the Basse-Meuse'. Unpublished dissertation, Princeton University.
Gutmann, Myron: 1980, *War and Rural Life in the Early Modern Low Countries*, Princeton.
Jacquart, Jean: 1960, 'La Fronde des Princes dans le Region Parisienne et ses Consequences Materielles', *Revue d'Histoire Moderne et Contemporaine* **VII**, 257–290.
Kamen, Henry: 1968, 'The Social and Economic Consequences of the Thirty Years' War', *P&P* **39**, 44–61.
Rabb, Theodore: 1962, 'The Effects of the Thirty Years' War on the German Economy', *Journ. of Modern History* **34**, 40–51.
Roupnel, Gaston: 1922, *La Ville et la Campagne au XVIIe Siecle: Etude sur les Populations de Pays Dijonais*, Paris.

W. The Environment and Man

Albion, R. G.: 1926, *Forests and Sea Power: The Timber Problem of the Royal Navy 1652–1862*, Cambridge, Massachusetts.
Ashton, T. S.: 1951, *Iron and Steel in the Industrial Revolution*, 2nd ed. Manchester.
Bagley, William: 1942, *Soil Exhaustion and the Civil War*, Washington.
Bamford, Paul: 1956, *Forests and French Sea-Power 1660–1789*, Toronto.
Bennett, John: 1969, *Northern Plainsmen*, Chicago.
Blanchard, Ian: 1978, 'Resource Depletion in European Mining and Metallurgical Industries, 1400–1800', in Antoni Maszak and William N. Parker (eds.), *Natural Resources in European History*, 85–113, Washington.
Blouet, Brian and Lawson, Merlin (eds.): 1975, *Images of the Plains: The Role of Human Nature in Settlement*, Lincoln.
Bogucka, Maria: 1978, 'North European Commerce as a Solution to Resource Shortage in the Sixteenth to Eighteenth Centuries', in Antoni Maszak and William N. Parker (eds.), *Natural Resources In European History*, 9–42, Washington.
Bogue, A. G.: 1963, *From Prarie to Corn Belt*, Chicago.
Bowden, Martyn J.: 1975, 'Desert Wheat Belt, Plains Corn Belt: Environmental Cognition and Behavior of Settlers in the Plains Margin, 1850–99', in Brian Blouet and Merlin Lawson (eds.), *Images of the Plains*, Lincoln.

Brenner, Robert: 1976, 'Agrarian Class Structure and Economic Development in Pre-Industrial Europe', *P&P* **70**, 30–75. A major critical review of the neo-Malthusian (ecological) model of socio-economic change. See also *P&P* **79**, 1978, and **97**, 1982.

Brimblecombe, P.: 1975, 'Industrial Air Pollution in Thirteenth Century Britain', *Weather* **30**, 388–96.

Brimblecombe, P.: 1977, 'London Air Pollution, 1500–1900', *Atmospheric Environment* **11**, 1157–62.

Brimblecombe, P.: 1978, 'Interest in Air Pollution Among Early Fellows of the Royal Society', *Notes and Records of the Royal Society of London* **32**, 2, 123–129.

Brimblecombe, P. and Ogden, C.: 1977, 'Air Pollution in Art and Literature', *Weather* **32**, 285–91.

Brimblecombe, P. and Wigley, T. M.: 1978, 'Early Observations of London's Urban Plume', *Weather* **33**, 215–20.

Claw, A. and N. L.: 1956, 'Timber and the Advance of Technology', *Annals of Science* **12**.

Craven, Avery Odette: 1926, *Soil Exhaustion as a Factor in the Agricultural History of Virginia and Maryland, 1606–1860*, University of Illinois Studies in the Social Sciences, Vol. XIII, No. 1. Urbana, Illinois, reprinted 1965.

Danhof, Clarence H.: 1969, *Change in Agriculture: The Northern United States, 1820–1870*, (especially Chapter 10, 'Utilization of the Soil'), Cambridge, Massachusetts.

Darby, H. C.: 1956, 'The Clearing of the Woodland in Europe', in W. L. Thomas (ed.), *Man's Role in Changing the Face of the Earth*, 183–216 Chicago. Extensive bibliography through 1955.

De Zeeuw, J. W.: 1978, 'Peat and the Dutch Golden Age: The Historical Meaning of Energy Attainability', *A.A.G. Bijdragen* **XXI**, 3–31.

Earle, Carville: 1975, *The Evolution of a Tidewater Settlement System: All Hallow's Parish, Maryland, 1650–1783*, Chicago (particularly Chapter 2, 'Parameters of the Settlement System').

Erkirch, Arthur: 1963, *Man and Nature in America*, New York.

Flinn, M. W.: 1978, 'Technical Change as an Escape from Resource Scarcity: England in the Seventeenth and Eighteenth Centuries', in *Natural Resources in European History*, Antoni Maszak and William N. Parker (eds.), 139–59. Washington. A major critique of the scarcity-induced technological change model.

Greven, Philip J.: 1970, *Four Generations: Population, Land and Family in Colonial Andover, Massachusetts*, Ithaca.

Gross, Robert: 1976, *The Minutemen and Their World*, New York (especially Chapter 4, 'A World of Scarcity').

Hargreaves, M. W, M.: 1957, *Dry Farming in the Northern Great Plains, 1900–1925*, Cambridge, Massachusetts.

Henretta, James A.: 1978, 'Farmers and Farms: Mentalite in Pre-Industrial America', *William and Mary Quarterly* **35**, 3–32. A major review essay on American local studies.

Hewes, Leslie: 1958, 'Wheat Failure in Western Nebraska 1931–1954', *Annals of the Association of American Geographers* **48**, 375–97.

Hewes, Leslie: 1967, 'The Conservation Reserve of the American Soil Bank as an Indicator of Regions of Maladjustment in Agriculture, with Particular Reference to the Great Plains'. Paper presented to the *Festschrift Leopold G. Schiedl zum 60. Geburtstag, Vienna*, Lincoln, Nebraska.

Hewes, Leslie: 1973, 'The Great Plains One Hundred Years After Major John Powell', in *Images of the Plains: Conference Proceedings*, Lindoln, Nebraska.

Hewes, Leslie: 1973, *The Suitcase Farming Frontier*, Lincoln, Nebraska.

Hillard, Sam B.: 1975, 'The Tidewater Rice Plantation: An Ingenious Adaptation to Nature', *Geoscience and Man* **12**, 57–66.

Jacks, G. U. and Whyte, R. O.: 1939, *Vanishing Lands: A World Survey of Soil Erosion*, New York. Reprinted 1972.

Jones, Eric L.: 1979, 'The Environment and the Economy', *The New Cambridge Modern History*, XIII (Companion Volume), Peter Burke (ed.), 15–42.

Jordan, Terry G.: 1964, 'Between the Forest and the Prairie', *Agricultural History* **38**, 205–16.

Jordan, Terry G.: 1966, *German Seed in Texas Soil: Immigrant Farmers in Nineteenth-Century Texas*, Austin.

Jordan, Terry G.: 1967, 'The Imprint of the Upper and Lower South on Mid-Nineteenth-Century Texas', *Annals of the Association of American Geographers* **57**, 667–90.

Kawamoto, Thomas, M.: 1981, 'Via U.S. Mail – Early Weather Forcasts', *Weatherwise* **34**, 110ff.

Ladurie, E. Le Roy: 1966, *The Peasants of Languedoc*, Paris. English translation 1974.

Lemon, James T.: 1966, 'The Agricultural Practices of National Groups in Eighteenth-Century Southeastern Pennsylvania', *Geographical Review* **56**, 467–96.

Lemon, James T.: 1972, *The Best Poor Man's Country: A Geographical Study of Early Southeastern Pennsylvania*, Baltimore.

Lindgren, S. and Neumann, J.: 1981, 'Benvenuto Cellini (1500–71) and 'Rainstopping', *Bulletin of the American Meteorological Society* **62**, 145 ff.

Lockridge, Kenneth: 1968, 'Land, Population and the Evolution of New England Society, 1630–1790', *P&P* **39**, 62–80.

Lockridge, Kenneth: 1970, *A New England Town: The First Hundred Years*, New York.

Ludlum, David M.: 1977, 'A Balloon to Cross the Ocean in 1859: *The Atlantic*', *Weatherwise* **30**, 154 ff.

Malin, James: 1944, *Winter Wheat in the Golden Belt of Kansas: A Study in Adaptation to Subhumid Geographical Environment*, Lawrence, Kansas.

Malin, James: 1950, *Grassland Historical Studies: Natural Resources Utilization in a Background of Science and Technology*, Lawrence, Kansas.

Malin, James: 1956, 'The Grassland of North America: Its Occupance and the Challenge of Continuous Reappraisals', in W. L. Thomas (ed.), *Man's Role in Changing the Face of the Earth*, 350–66. Chicago.

Malin, James: 1956, *The Grasslands of North America*.

McCraken, Eileen: 1971, *The Irish Woods Since Tudor Times: Distribution and Exploitation*, Newton Abbot.

McManis, Douglas R.: 1964, *The Initial Evaluation and Utilization of the Illinois Prairies, 1815–1840*, Chicago: University of Chicago, Department of Geography, Research Paper No. 94.

Morris, Brian: 1981, 'Changing Conceptions of Nature', *The Ecologist* **11**, 30 ff.

Ottoson, Howard *et al.*: 1966, *Land and People in the Northern Plains Transition Area*, Lincoln, Nebraska.

Pipes, Richard: 1974, *Russia Under the Old Regime* (Chapter I: 'The Environment and its Consequences'). New York.

Rice, John G.: 1977, 'The Role of Culture and Community in Frontier Prairie Farming', *Journ. of Historical Geography* **3**, 155–74.

Richardson, H. G.: 1921, 1922, 'Some Remarks on British Forest History', *Transactions of the Royal Scottish Arboricultural Society* **35**, **36**.

Rosenberg, Nathan: 1972, *Technology and American Economic Growth*, New York.

Te Brake, W. H.: 1975, 'Air Pollution and Fuel Crises in Preindustrial London, 1250–1650', *Technology and Culture* **16**.

Torry, William: 1979, 'Anthropological Studies in Hazardous Environments: Past Trends and New Horizons', *Current Anthropology* **20**, 3, 517–540. A major review article with an extensive bibliography.

Torry, William: 1979, 'Anthropology and Disaster Research', *Disasters* **3**, 1, 43–52.

Tubbs, Colin: 1968, *The New Forest: An Ecological History*, Newton Abbot.

Von Viettinghoff-Riesch, Baron: 1958, 'Outlines of the History of German Forestry', *Irish Forestry* **15**.

Webb, Walter: 1931, *The Great Plains*, New York.

Wrigley, E. A.: 1962, 'The Supply of Raw Materials in the Industrial Revolution', *EHR* 2nd ser., **XV**.

X. Principal Journals Publishing Articles on Climate in History

Agricultural History
Agricultural Meteorology
Annales Economies Societes Civilisations
Annals of the Association of American Geographers
Archiv für Meteorologische Geophysik und Bioklimat
Arctic
Arctic and Alpine Research
Atmospheric Environment

Australian Geography
Boreas
Bulletin of the American Meteorological Society
Canadian Geographer
Climatic Change
Disasters
Ecologist
Economic History Review
Environmental Conservation
Geofisica e Meteorologia
Geofisica Pura e Applicata
Geografiska Annaller
Geographica Helvetica
Geographical Bulletin
Geographical Journal
Geographical Magazine
Geographical Monthly
Geographical Review
Geophysical Memoirs
Journal of Applied Meteorology
Journal of Economic History
Journal of European Economic History
Journal of Historical Geography
Journal of Interdisciplinary History
Journal of Meteorology
Journal of the Atmospheric Sciences
Journal of the Meteorological Society of Japan
La Meteorologie
Manitoba Geographical Studies
Meteorological Magazine
Meteorological Monthly
Meteorologische Rundschau
Monthly Weather Review
Nature
Naturwissenschaften
Outlook in Agriculture
Palaeogeography, Palaeoclimatology, Palaeoecology
Past and Present
Progress in Physical Geography
Quarterly Journal of the Royal Meteorological Society
Quaternary Research
Revue d'Histoire Economique et Sociale
Revue Historique
Scandinavian Economic History Review
Science
Scientific American
Search
Weather
Weatherwise
Wetter und Leben
Zeitschrift für Angewandte Meteorologie
Zeitschrift für Gletscherkunde und Glazialgeologie

Note

The following important publications, all with extensive bibliographic references, came to the attention of the author while this manuscript was in press. The section of this Bibliography to which they belong is indicated in parentheses.

Brenner, Robert: 1982, 'The Agrarian Roots of European Capitalism', *P&P* 97, 16–113. (S)
Kupperman, Karen Ordahl: 1982, 'The Puzzle of the American Climate in the Early Colonial Period', *American Historical Review* 87, 1262–89. (F)
Lamb, H. H.: 1982, *Climate, History and the Modern World*, London. (B)
Quinn, M.-L.: 1982, 'Federal Drought Planning in the Great Plains – a First Look', *Climatic Change* 4, 273–96. (U)
Takahashi, K. and Yoshino, M. M. (eds.): 1978, *Climatic Change and Food Production*, Tokyo. (L)

Princeton University

INTRODUCTION TO: RESEARCH ON POLITICAL INSTITUTIONS AND THEIR RESPONSE TO THE PROBLEM OF INCREASING CO_2 IN THE ATMOSPHERE

National political institutions constitute perhaps the clearest link between overall societal behavior and individual perceptions and actions. On the one hand, national institutions help develop the policies relating to energy, agriculture, international trade, and so forth that generally set the stage for economic and social decisions throughout a nation. On the other hand, these institutions provide a principal mechanism by which citizens and interest groups can address issues at local, national, and international levels. This central role of national political institutions in modern-day life underpins the importance of the research discussed in the following paper by Mann.

Mann recognizes at the outset what he terms the "scepticism of existing institutions" regarding their ability to cope with long-term, global problems like the possibility of climatic change. A pessimistic view is this: might not the social costs of developing and implementing efficacious responses to either prevent climate changes or ameliorate their effects exceed the costs of, say, doing nothing? For example, a conflict over non-fossil energy resources in a world trying to abstain from fossil fuels might do just as much harm as a conflict over diminished food resources in a world of changing climatic conditions. From a less pessimistic perspective, it is still clear that society needs to have some idea of both the likelihood and the costs of successful implementation of different kinds of responses. Only then will any informed decisions about strategies for dealing with climate change (or its prospect) be possible.

Mann's agenda for research thus focuses on the role of national government itself, mechanisms for adjustment within nations, and the development or modification of national policies in light of the possibility of climate change. Although he addresses these issues mainly in the context of the political system of the United States, they are certainly relevant to other nations and political systems as well. What is inescapable of course is the critical need to understand better what the nature of political decision making is likely to be with respect to climate change. The following contribution by Mann greatly expands our foothold in this largely unexplored territory.

R. S. CHEN

R. S. Chen, E. Boulding, and S. H. Schneider (eds.), Social Science Research and Climate Change: An Interdisciplinary Appraisal, 115.

DEAN E. MANN

RESEARCH ON POLITICAL INSTITUTIONS AND THEIR RESPONSE TO THE PROBLEM OF INCREASING CO_2 IN THE ATMOSPHERE

The focus of this paper is the agenda of research on political institutions and the roles those institutions might play in meeting the challenge and resolving the issues connected with increases in the concentrations of CO_2 in the atmosphere. Institutions are conceived broadly, encompassing those structures, processes and policy approaches for making public decisions and for influencing the behavior of private individuals, groups, and firms. Institutional analysis must incorporate the private structures and processes as they relate to the public decision-making process. In the United States and other countries with substantial private economies, the interdependencies between the public and private sectors make such comprehensive analysis essential.

1. Domestic and International Institutions and Politics

This paper will deal primarily with research on the domestic institutional structures and processes and policy approaches.[1] While the emphasis will be on the United States, the same kind of analysis appropriate to the United States may be appropriate, with modifications, to other nations as well. The problem of increasing CO_2 in the atmosphere is global in character and such global problems must be translated within each nation by domestic institutions into domestic policy. In a politically decentralized world, it is necessary to examine the domestic institutions of various societies to ascertain the institutional capability to cope with the threat and the reality of changing climate. State sovereignty will remain the pattern for the foreseeable future – indeed it may become even more powerful in the future – and climate policy, like virtually every other meaningful policy, will reflect this domestic institutional capability (Morganthau, 1978). Even though agreements may be made internationally, these agreements must be translated into practical political measures through domestic structures and processes (Hanrieder, 1978).

Moreover, it is inevitable that the institutional capacity will reflect specific political conditions within the various nations and blocs of nations (Bertsch *et al.*, 1978). These political conditions at the most generalized level have to do with (1) the character of the economic system – whether essentially based on free markets or on state socialism; (2) the capability of the bureaucracy, especially if the bureaucracy has extensive authority to make decisions with respect to the allocation of scarce resources; (3) the configuration of private interest groups and their relationships to the government with respect to the latter's regulatory authority over them; (4) the condition of public opinion, whether attentive to issues of public policy or relatively uninformed and inattentive; (5) the character and quality of constitutional law, i.e., the limitations both substantive and procedural, under which authorities may act in making public policy; (6) not least of all, the specific economic and strategic interests that are damaged or benefited by the

R. S. Chen, E. Boulding, and S. H. Schneider (eds.), Social Science Research and Climate Change: An Interdisciplinary Appraisal, 116–145.

increase of CO_2 in the atmosphere. Each state may be expected to act in its own self-interest as it defines it and that interest is a summation of the economic costs and benefits perceived, military advantages and disadvantages, and the myth structure accruing to productive enterprise in each country (Kelley *et al.*, 1976).

Agreement on approaches to the solution of the problems associated with the environment and more specifically with global warming will depend on the extent to which the representatives of nations affected by this trend are willing to negotiate their differences.[2] The study of the impacts of increasing CO_2 must therefore include research on the international politics of such issues (Falk, 1971; Sprout and Sprout, 1971). Such issues include the definition of the stakes of each nation; the relationship of the global warming issue to other stakes of each individual nation; the general relationships among the major blocs of nations (the industrialized West plus Japan and the industrialized East; the Northern industrialized developed nations and the Southern less developed and less industrialized nations); the various regional systems; the resources of each nation or regional system to deal with the problems associated with global warming, either by countering the effects internally or by demanding compensation for costs imposed on them. Such studies should complement the research on the international law and institutions that might be devised or erected to make decisions in a formal way with respect to the global warming impacts.

As an international problem, the CO_2 concentration problem cannot be divorced from the more general issue of the energy future of the United States and other nations that are presently dependent on oil from the Middle East and other producer areas of the world. Coal, like nuclear power, is often proposed as the alternative to foreign oil as the principal energy source during the transition from the fossil fuel to a non-fossil fuel economy – whatever that may be. Those who argue for increased nuclear development, and particularly the breeder reactor, see nuclear power as the alternative that may make the United States and the other developed nations less dependent on the Middle East and less likely to become embroiled in a major conflict – including the possibility of a thermonuclear conflict – either with Middle Eastern nations or with the Soviet Union. Similarly, despite the problem of increased CO_2 concentrations, the burning of coal may be perceived as an alternative to major international conflicts over oil and the uncertainties associated with the supply of nuclear fuel, the safety of reactors, the possibilities of sabotage and the disposal of nuclear waste.

2. Relationship of This Agenda to Other Disciplines

This research agenda is likely to attract the skills and methodologies of political scientists. Indeed, there is hardly a facet of political science that would not be appropriate as an approach to understanding of the problem: political socialization, public administration, public law, political behavior, etc. This is part of the problem of drawing up an agenda; the CO_2 issue cuts across every subdivision of the discipline. At the same time, it should be recognized that many of the research topics of interest to political scientists are also of interest to sociologists, geographers, psychologists, economists, anthropologists, and historians. Some overlapping of research agendas is therefore inevitable, but these overlaps suggest both the recognition by representatives of several disciplines of the importance of the agenda items and perhaps the opportunity for multi-disciplinary research. Political scientists should scrutinize the research agendas of the other social sciences – psychology, anthropology, geography, and economics – for further emphasis on and elaboration of some topics on the research agenda assigned here for political scientists.

3. Scepticism of Existing Institutions

There exists a profound scepticism of existing political and economic institutions and their capability of resolving major public problems, whether concerned with the environment or other issues. Government, it is argued, tends to regulate where it should not, and spend more than it should, thus constituting a dual burden on all of society. Politicians and bureaucrats are viewed as lacking the appropriate incentives for problem solving; indeed, their incentive structure is viewed as leading them in the direction of perpetuation and aggravation of social problems. They tend to build large projects, usually in excess of need and with only casual relationship to some form of cost/benefit analysis (Niskanen, 1971). Critics often view the marketplace and its pricing system as the appropriate mode of decision-making on most resource questions (Anderson, 1977).

On the other hand, there are those who despair over the market as a device for making inter-generational decisions with regard to non-renewable resources or cumulative pollution effects that cannot easily be reversed. Reliance on the discount rate with a strong preference for current consumption and lack of concern for future values are seen as serious weaknesses in market economies. Intergenerational resource issues are found to be essentially moral issues where the role of government in establishing limits, norms and priorities is viewed as crucial (Daly, 1979). Whether government is capable of establishing standards that protect such long-term values is another matter.

Finally, there are those who see a profound need for government intervention but perceive little inclination on the part of the general public or its representatives to take anything but a short-term view of environmental/resource policy. There is evidence of persistence of a strong commitment to environmental protection among the general public (Mitchell, 1978, 1979, 1980) but it is uncertain how strong that commitment might be in the face of circumstances that portend a significant decline in real income. Moreover, there are those who believe that public support for environmental protection will wax and wane (Downs, 1972) or that the public pays serious attention to environmental issues only under conditions of crisis. The constitutional and electoral systems with relatively brief terms of office and the necessity for incumbents to satisfy pressing and current public demands lend themselves to emphasis on meeting present demands with little concern for the future environment (Cooley and Wandesforde-Smith, 1970). This same penchant for short-term problem-solving horizons leads to unspoken conspiracies between legislators and bureaucrats to build administrative structures and policy frameworks that are more to their mutual benefit and less to the benefit of the general public (Fiorina, 1977).

The appropriate role of government is controversial throughout the world, even in solidly socialist countries such as the Eastern European bloc and in the developing nations. Nevertheless, there is a clearly greater tendency to rely on public institutions for the major decisions in society in those regions of the world, in some cases because of doctrinal tenets and in others because the governmental apparatus has a virtual monopoly of capital, administrative skills and technical training. Of necessity decisions with respect to major capital investments such as the construction of centralized electrical energy facilities burning fossil fuels or the construction of nuclear plants are exclusively public decisions that require the use of public funds and public administrative machinery.

The level of support for government policy by the general public and by specific interest groups will vary from country to country. In many countries decisions will be made exclusively by government elites with little input or expectation of public participation in the political process. Where there are serious deficiencies in education and the concern of most people is to earn sufficient to feed oneself and one's family, it is unlikely that the general public will have much interest in or influence over such policy decisions. Indeed, the general public as well as major interest groups may consider an issue as problematic and uncertain as CO_2 concentrations in the atmosphere to be so trivial in comparison with other more pressing issues that it may hardly find a place on the political agenda. On the other hand, in a country like Japan, it is clear that some environmental policies have had the benefit of strong support from a relatively cohesive economic and business elite.

The scepticism and uncertainty regarding the capacities of political institutions must be challenged if the consequences of climatic change are to be dealt with effectively. In effect, is there evidence that slow, cumulative and arguably irreversible changes in the environment that threaten not only the quality of life but the future habitability of the planet can be dealt with effectively by existing institutions (Moss, 1980)? Is public sentiment incapable of perceiving a public interest beyond its own immediate gratification? Specifically, can existing institutions deal effectively with the subtle and slow changes in climate and society that arise out of the gradual increases of CO_2 in the atmosphere?

4. The Evidence of Experience and the Function of Analogy

Except as one monitors current efforts to deal with the CO_2 problem, it is necessary largely to explore analogous situations involving slow, cumulative, potentially non-reversible changes. Do analogies exist and how close are the parallels? This is a researchable question that social scientists can and should address. The following suggests only some indicators of possible parallel or analogous policy-making.

It has been argued that the recent efforts to deal with air pollution in the United States are the result of "speculative augmentation", i.e., a radically new approach to air pollution that relied on untried technology and methodology and arose out of a profound dissatisfaction with the efforts that had been tried theretofore. Neither the costs, benefits nor the economic and political consequences were clearly foreseen because of the experimental character of the new approach. Nevertheless, under appropriate public leadership, pressure from well organized environmental groups, and public sentiment that was perceived to be favorable, this legislation was passed and was implemented with some success (Jones, 1974; Jones, 1975). Recent evidence of postponements of deadlines and failure to impose strict standards suggests that this audacious policy was something less than a complete success but the experience is nevertheless instructive (Ingram and Mann, 1978; Walker and Storper, 1978).

Federal legislation dealing with DDT and other pesticides may provide additional evidence of a kind analogous to the CO_2 situation. There appears to be a consensus on the impact of DDT on the food chain and the serious implications of the accumulation of that pesticide in the tissues of specific species, including human beings. Yet the evidence is not overwhelming and there remain those in the scientific community who are unconvinced. The situation is similar for a number of other pesticides. Yet the demand for control of the use of pesticides became sufficiently great to lead to its prohibition and has led to the prohibition of several others as well.

Current concern for radiation safety and storage of nuclear wastes suggests another possible policy situation not entirely dissimilar from the CO_2 problem. Despite the fact that the record of nuclear power plant safety has been excellent, and no deaths have resulted from their operation, and despite the confidence expressed by most scientists in the energy community that storage of nuclear wastes is safe and reliable, public sentiment – and consequently the sentiments of many decision-makers – remains strongly resistant to expansion of the nuclear industry. California's action to make new nuclear plants virtually an impossibility is indicative of this mood.

A final example – the issue of the impact of super-sonic transports and fluorocarbons on the ozone layer – suggests another possible policy approach: legislation accompanied by and associated with exhortation. Recognition of the dangers created by the widespread use of fluorocarbons has led to legislation and a virtual cessation of the use of this substance as a propellant in spray cans. The problem is hardly resolved because of continued use throughout the world but the United States has virtually ceased to utilize it for this purpose, although continuing to use it in refrigerants in large quantities. A determined attack on the SST for its potential impact on the ozone layer along with a demonstration of the lack of economic viability of the program led to Congressional decisions not to continue development of the SST, with an almost one billion dollar loss in public investment (Segal, 1972). It is not beyond the realm of possibility that a strong public and private effort to alter certain practices that contribute to the CO_2 problem or that exacerbate the consequences of climate change might achieve important results on a combined legislative and voluntary basis.

Illustrations could also be drawn from foreign experience – illustrations of both successes and failures comparable to those found in the United States. Other superpowers such as the Soviet Union and Japan have faced similar environmental issues with comparable difficulties and achievements in air, and water pollution control and waste disposal (Kelley *et al.*, 1976). The efforts to deal with environmental issues in those nations reflect not only the seriousness of the problems encountered but the ideology and institutional structure for dealing with them. For example, the traditional authoritarian and paternalistic ideology, institutions and interpersonal relationships found in Japan are viewed as facilitative of environmental protection measures. A key industrial association played a major role in the success of a compensation plan for those who suffered ill health effects from pollution (Anderson *et al.*, 1977, p. 51).

The number of examples could be extended, with more or less relevance to the CO_2 situation. The purpose of analysis of analogous policy situations would be to ascertain the strategic and tactical possibilities for achieving policy goals with respect to climatic change. What was the objective condition of public sentiment? What leadership emerged to define the problem and its solutions? How was interest group mobilization accomplished? What were the problems with implementation? To what extent may the conditions at a given moment with respect to the CO_2 problem be propitious for major advances in policy?

5. Politics as Cause and Effect

Political decisions and derivative policy may be both cause and effect with respect to climatic change. Decision-makers may clearly perceive that their actions lead to significant, although uncertain, impacts on individuals and society. A deliberate policy of requiring fossil fuel burning electric utilities to convert from oil to coal may be an example of this sort. Recent steps within the United States to convert power plants presently burning oil and natural gas to coal have been taken in full recognition of the potential climatic consequences that may be encountered in the long run. At the other end of the scale are political decisions that may be taken to deal with climatic change that are less the consequences of other political decisions than individual preferences for a given style of life. The intense preference for the use of the automobile in the United States is an excellent illustration. In the developing tropical nations, the burning of trees as fuel is a continuing practice of individuals who are certainly not cognizant of the climatic consequences of the aggregate level of burning. One might argue, of course, that continued acquiescence by governments in those modes of life and utilization of given means of transportation or practices of fuel consumption for heating is a causative political decision of the first magnitude.

Political decision-making with respect to climate change may be considered a second and in some cases a third-order consequence. Direct impacts are experience by specific groups: farmers, operators of tourist and recreational facilities, cattle and sheep raisers, municipalities, an industry concerned with water supply, etc. These impacts are secondarily felt by businesses that depend on primary industry for their prosperity. These impacts are then transformed into political concerns and their expression as political demands.

These political demands are usually mediated by political interest groups such as trade associations, trade unions, local economic bodies such as chambers of commerce, professional associations, and other kinds of membership organizations. These orgnaizations perform the function of articulating the interests of the affected individuals, preparing specific remedial proposals and transmitting the collective concerns to decision-makers (Truman, 1951; Greenwald, 1977). In some instances, although probably to a lesser degree today, political parties may play an important role in this articulation process (Burnham, 1970; Sundquist, 1973).

As mediators of public policy concerns, it is important to ascertain the perceptions of interest group leaders of the issues associated with climate change and the policy options available for dealing with them. To a significant degree, these interest groups are veto groups; that is, they are able to impose a veto over specific formulations of public policy. The price of obtaining their consent is some form of compromise. Thus, their perceptions of the issues associated with climate change, the stakes they perceive – both theirs and their adversaries – the policy options they deem acceptable or anathema, and the mechanisms through which policy should be carried out are important information for those who are responsible for fashioning public policy. Information should be obtained on their level of knowledge of the issues, the intensity of concern that is exhibited, and the place climate change occupies in their scales of priorities.

The technical and unseen and unfelt nature of climate change, particularly as it is masked by natural perturbations and cyclical swings in rainfall and temperature, makes public opinion a less direct influence on the formulation of policy. Nevertheless, the attitudes and perceptions of the general public as well as specific publics can have a significant influence and impose major constraints on public policy-making. To what extent may their time preferences dictate the speed and direction of public policy? To what extent do they identify climatic change with nature or with actions of humankind? To whom do they attribute responsibility for climatic change and whom do they hold responsible for taking corrective action?

The extent to which private interest groups and public opinion play roles in policy formulation and implementation will naturally vary from nation to nation and culture to culture. In the ultimate sense, all nations must take public opinion and public support into account in making policy, but there are large differences in the discretionary authority of ruling elites and their bureaucratic subordinates. In a relatively closed society, it may be expected that an issue as subtle as climate change will hardly reach the surface of general public awareness, thus leaving the elite virtual plenary power to deal with the problem. On the other hand, it may be expected that some regimes may endeavor to incite or inflame public opinion over actual or potential adverse climate change if such stimulation seems appropriate in the circumstances.

6. Approaches to Policy Analysis and the CO_2 Problem

A fundamental assumption of policy analysis is that policies often determine politics, i.e., that the political process reconfigures with the character of the policy issue under consideration. The politics of science, for example, are different from the politics of welfare which is in turn different from the politics of defense. Each kind of policy issue evokes the participation and the concern of different sets of actors who relate themselves to each other in characteristic ways. While the formal processes through which policy are made may be similar – or may be entirely dissimilar – the pattern of interaction among the participants in the process may be highly divergent. Lowi has distinguished among four kinds of politics: distributive, regulatory, redistributive and constituent politics. Each is distinguishable by the pattern of private group interaction and the roles of the formal institutions in the process, reflecting the kinds of issues the political system is dealing with (Lowi, 1972).

Briefly, in the distributive policy arena, local interests seek benefits from the public treasury by means of projects and programs of benefit to their locality and for which they bear little or none of the cost. Coalition formation is characteristic of this area, with the familiar triangular relationship of associated local interest, congressional committees and federal bureaus playing the instrumental roles in achieving legislative goals. In the regulatory area, the politics are sectoral in character, with major economic (or other well-organized) interests lined up against each other. The treasury is usually unavailable to pay the costs, so the outcome is of advantage to some interests and entails clear-cut losses to others. The major participants are the organized interests, regulatory agencies and the Congress that determines the rules of the games by which the participants play.

In the redistributive political arena, the issues have to do with general shifts in the benefits and burdens of society: taxes, tariffs, welfare, unemployment benefits. These issues involve transferring wealth from one large sector to another large sector of society and thus correspond most closely to class politics. The major private participants are the peak economic organizations such as the trade union congresses and the large business associations on the private side, and the President and the Congress as a whole on the public side. Finally, in constituent politics, the issues have to do with the structure of government and the basic assignments of political tasks and responsibilities in society. In an ultimate sense, all four forms of politics are redistributive – transferring benefits or advantage from one group or class to another.

In examining the issues raised by the increase of CO_2 in the atmosphere, one must recognize that they will take varied forms and therefore will be dealt with through different political arenas as indicated above. Some will be dealt with through distributive politics, some through regulatory or redistributive or constituent politics, or a combination of the various modes (Mann, 1975). It will be important to examine the character of each policy issue to discern its characteristic political arena and the consequences that ensue from these two features.

The significance of such analysis lies in the strategic advantage given the analyst by an appreciation of these systematic relationships. While sounding deterministic in character, i.e., that a given policy issue determines political dynamics, in fact that designer of policy may deliberately intervene to alter the major characteristics of policy so as to alter the politics.

Another analytic approach distinguishes among three strategies for dealing with the increasing concentration of CO_2 in the atmosphere: preventive, curative and adaptive (Corbett, 1979). The preventive strategy emphasizes the restriction and reduction of activities that contribute to the concentration in the first place. Thus, there would be few or no new fossil fuel plants constructed, some existing ones might be phased out, and an emphasis would be placed on fuels that do not make contributions to CO_2 concentration. There might be restrictions on the kinds of activities that lead to demands for CO_2 production. The approach would be largely regulatory in character and involve the imposition of strong enforcement machinery for its accomplishment.

A second strategy would be curative, i.e., designed to deal with the CO_2 concentrations by effectively neutralizing it as it is produced. Stack scrubbers and afforestation projects have been suggested. The focus of such strategy would be largely technological and project-oriented. Such an approach would be both regulatory and expensive in terms of expenditures for new technology and resource programs. Those expenditures might be private or public but in either case they would increase substantially the social cost of producing energy.

The adaptive strategy assumes that CO_2 concentrations are allowed to build up without significant intervention and then society develops mechanisms for adjusting to the changes in climate. These changes would come gradually and the adaptive process would be relatively slow and incremental. Such a strategy allows experience rather than predictions to guide the specific social and technological changes and feedback may be more useful than predictions. It allows costs to be spread over time. On the other hand, it presumes that adaptation can occur without undue social costs, that the changed climate will not impose such severe disruptions that they cannot be accommodated by existing institutions.

The preventive and curative strategies clearly have the disadvantage of concentrating most efforts to deal with the climate change problem at the beginning of the learning curve with respect to the character and severity of the problem. They assume certain deleterious consequences for society and impose major costs for society generally and for certain sectors of society in quite differential ways in a relatively compressed period of time. The adaptive strategy postulates that there are no changes that cannot be dealt with by technology and/or institutions but permits the costs to be distributed over longer periods of time and allows individual and group choices to determine much of the adaptation that occurs. All will impose costs and those costs must be examined carefully.

7. Political Models and the CO_2 Issue

Political scientists and sociologists have developed various models to describe the political system of the United States and other societies. One set, chiefly used to analyze the politics of the United States, has three principal models within it: the elite model, the pluralist model, and the radical democratic model (Dye, 1976; Prewitt and Stone, 1973). The principal distinguishing characteristics of the politics in each model concern the level and quality of participation by individuals and groups within society.

In the elite model, the principal actors are the wielders of economic power in the country. They are primarily the heads of large corporations, banks, utilities, and insurance companies who are considered to have relatively homogeneous interests and stakes. Political decision-makers are secondary actors in this model, responding to the preferences of the dominant economic interests. Those interests dominate by controlling the channels of communication, by their influence over ideology and through their direct intervention in the political process through campaigns and lobbying. They are able to suppress the issues the outcome of which might affect their interests detrimentally (Domhoff, 1978).

In the pluralist models, political power is more generally distributed among a host of groups that do not necessarily have similar interests. Political decisions are made through a very real contest among coalitions of actors and groups. The various interests gain access to government through innumerable channels that are relatively open to well-organized constituencies. Governmental actors are major participants in this group struggle at both the legislative and administrative levels and their decisions record the victories and defeats.

Finally, in the radical democratic model, political decisions are made by popular majorities in open and meaningful contests on issues. The public is assumed to be well-informed and participates at high levels in terms of both numbers and intensity.

There are adherents of all of these generalized models, although those who subscribe to the radical democratic model as an *explanation* of American politics and not a normative preference must be few indeed. The most reasonable view is to conclude that on some issues, well-structured and integrated elites in the business community have powerful if not completely controlling interests. On others, however, it would appear that political decisions are clearly the result of the interplay of many powerful and conflicting groups, some of which are clearly not part of what one would call a "power elite".

In other countries facing the problems of dealing with increased concentrations of CO_2 in the atmosphere, the regimes range from the most centralized, ruthless and primitive dictatorships to enlightened, liberal and decentralized democracies with the host of nations ranging somewhere between these two polar extremes. Moreover, the character of the regimes may change radically over time as new vectors of forces come into play and new groups gain ascendancy in those societies.

Among the industrialized nations, the political structures range from thoroughly authoritarian regimes such as the Soviet Union to liberal and socialist democracies such as the United Kingdom and Sweden. Among the less developed nations, the regimes may be classified in various ways: some are mobilization regimes, with an emphasis on centralized political structures, the inculcation of mass participation in politics, and revolutionary ideologies; some are more traditionalist, with laissez-faire economics and pluralist politics characteristic of the regimes. Others emphasize decentralization and worker control as in Yugoslavia. Still others are military dictatorships with varying approaches toward the solution of social problems.

These varying regimes are likely to differ substantially in their acceptance of scientific input with respect to an ecological problem such as climate change, the extent to which they are prepared to recognize the problem as an important one, their willingness to invest or to forego economic benefits in recognition of the problem, and therefore their acceptance of the need to engage in international agreements to deal with it. An understanding of how such regimes treat highly technical issues with profound social, economic and political implications will be important in evaluating the possibilities of dealing with climate change. Indeed, it may be found that many existing political structures are unlikely to deal at all with such a subtle, cumulative problem.

It is difficult to predict the political model most applicable to the emerging issue of the impacts of climate change in the United States or elsewhere. The future may bring sweeping changes in political structures as they respond to changing economic, military and ecological challenges. Leaving aside possible political transformations that may result from war or strategic considerations, it is argued by those who predict the need for the steady-state society that future political institutions will require the steadying hand of centralized decision-makers who do not respond to the pulling and the hauling of popular and pluralist forces but rather to the ecological requirements for saving the habitability of the planet (Ophuls, 1977). The rise of the philosopher-kings of an ecological persuasion – who would surely respond to the challenge of climate change – may lead to transformations of political systems throughout the world.

The issue of climate change is clearly embedded within the more general and sweeping issue of the energy future of the developed and developing worlds. It is now and increasingly in the future will be an issue in the struggle over whether the United States and other nations will emphasize nuclear energy, coal, or more exotic fuels. And in some ways, it cuts across the environmental movement in that that movement tends to be anti-nuclear but may prefer coal. For those concerned with climate impacts, nuclear may be preferable to coal.

It seems unlikely, therefore, that the CO_2 issue will be fought out on an isolated basis. It will be used as an issue in the battle over energy and that will be hotly contested among the various groups in society. It does not appear, given the virtual cessation of nuclear development in the United States, that the dominant economic interests will dictate the terms or the results in this battle or even that they are united in the struggle. The pluralist model would appear to be the most appropriate to explain both energy politics and the narrower issue of climate impacts.

At the initial stages of the discussion of the CO_2 issue, however, an elite model may be far more useful in explaining how decisions are made. The elite involved, however, is not the concentrated economic interests of society but the elite within the government: the scientists and the administrators who make decisions about research budgets. They will determine the extent to which the society and its various groups learn more about the nature of the climatic threat (Price, 1965).

8. Approaches to Managing the Atmospheric and Climatic Commons

It is generally acknowledged that the atmosphere and the climate are common resources not subject to individual appropriation and ownership. Benefits accruing to individuals accrue to others as well. Ill effects of atmospheric or climatic change are felt by entire populations. One cannot "capture" the atmosphere or the climate in the form of property; to do so would violate the rights and interests of others.

Equally important to an understanding of the stresses that will be placed on political institutions is an appreciation of the differential effects of climate change on various sectors of society, various regions of the world, as well as various regions within the United States alone. Institutions must be capable of sorting out those differential effects, assessing public responsibility for them, and providing amelioration for the detrimental effects of those changes or reducing undeserved gains resulting from climatic changes. Moreover, institutions must be created to provide the precise formulas through which those benefits and burdens may be distributed.

The character of the atmospheric resource dictates an approach to problem-solving that necessarily involves government intervention. To prevent individuals and groups from polluting the atmosphere or altering the climate to the detriment of others, it is necessary to establish rules that govern the way all parties may use the atmosphere. It is necessary that there be "mutual coercion, mutually agreed upon', to use Garrett Hardin's phrase (Hardin, 1968).

Public intervention, however, does not necessarily dictate a given approach to managing the commons. A preventive strategy would involve the imposition of controls over fossil-fuel burning and hence the reduction of CO_2 emissions. This could be accomplished through a number of strategies that are thoroughly understood, ranging from air quality management, emissions standards and controls, and emission taxes (Stern, 1977). A strategy of adaptation, however, could utilize a much wider range of strategies, including a broad array of public policies and the possibilities of private decisions under market-like conditions to achieve societal goals. In other words, both rules and prices might be allowed to play roles in the achievement of societal adjustment to climatic change.

9. Potential Substantive Decisions to Deal with the CO_2 Problem and the Political and Institutional Research Issues They Engender

In addition to the classification of policy approaches by their tendency to be distributive, redistributive, regulatory and constituent, or their focus on prevention, cure, and adaptation, it may be convenient to classify policy in terms of three time phases: problem recognition, problem avoidance and problem adaptation.[3] In a sense, this classification assists in the setting of priorities in research in that it suggests research that can and should be done at the time that society is beginning to perceive the existence of a problem, leaving for later consideration how society will adapt to circumstances that are only dimly perceived at the present time.

It is safe to say that the scientific community, and not without controversy, is presently in the problem identification or recognition stage. Numerous scientific uncertainties exist and will probably not be resolved for several decades: the rate of production of CO_2; the extent to which CO_2 will be absorbed in natural sinks; the precise impacts the increased concentrations will have on climate globally and regionally; the impact that those global and regional changes will have on agriculture and other economic and social sectors. Scientific research goes forward on a number of fronts while decision-makers are forced to make decisions that exacerbate or diminish the potential problem and the public gradually becomes aware that such a problem exists. The impact of changed climatic regimes on society only now has become a topic of serious consideration. Basically, this is a research phase with some emphasis on education of specialized groups such as members of Congress, bureaucratic decision makers, and representatives of private interest groups such as electric utilities and farm groups.

The problem avoidance phase may correspond to the preventive and curative strategies for dealing with climate change. Once convinced that the problem is real and serious, official decision-makers, the general public and actors in specific groups potentially affected by climate change may embark on actions to prevent or cure the problem: reduce the output of CO_2 into the atmosphere by various technologies; or to change lifestyles in such a way as to reduce the demand for energy-induced CO_2 production and adopt other energy strategies that will curtail the burning of fossil fuels. Actions taken to prevent or cure the problem are likely to occasion major political battles in that such actions challenge the *status quo* or the existing pattern of behavior or development. Electric utilities may not accept the view that they must not continue to build coal-fired steam plants because of the potential CO_2 build-up (they certainly resist any implication that they are somehow responsible for the acid rain problem today). Large numbers of the general public may resist policies that curtail their present pattern of behavior that relies on ever-increasing quantities of electrical energy. Other sectors may view the *status quo* differently: they may opt for the continuation of existing climatic patterns and argue for policies that will ensure their continuation. Still others may prefer the *status quo* because of the climatic changes they may bring.

The problem adaptation phase is clearly associated with the previously discussed adaptation strategy. In this phase, incremental adjustments to climatic change are made in economic and social behavior over an extended period of time. The length of time involved and the greater knowledge of the precise nature of the impacts may make this phase less conflictive and more susceptible of solution through public policy or technology or both.

The political agenda of research necessarily focusses on an examination of the role of governmental institutions, processes and structures in achieving policy goals. But the decisions made by government may (1) prescribe a minimal role for government and a broad role for the private sector under governmental influence; (2) a narrowly defined but powerful role for government, as in the impositions of taxes for some purposes; (3) a broad role for government, cutting across many policy sectors such as regulation of production, controls of effluents, and subsidization of those adversely affected. A crucial decision made by government must be based on an examination of the capability of government to make appropriate decisions and to implement them.

Several conditions affect the possibility of achieving the goals of prevention, cure, or adaptation to climate change. One is the reality of conflicting policies. Very real and divergent interests will struggle over policies responding to climate in that those policies will conflict with other policies or changed conditions favorable to given interests. Those who benefit from climate change, those who oppose higher budgets, those who want more coal-fired steam plants, those opposed to policies that may affect their life style may all oppose given policies responding to climate change. Such conflicts may reveal the infeasibility of given programs or may suggest the need for compromises. Those compromises may or may not satisfy the requirements of a policy that intelligently and coherently deals with climatic change.

Another serious condition affecting the success of governmental programs is that of policy overload. It is argued that government has assumed so many diverse responsibilities in so many policy sectors that the chief task of future decision-makers is to sort out and remedy the errors resulting from contradictory policy (Wildavsky, 1979). One environmental responsibility overlaps another and one impact negates another (Andrews, 1979). Pollution policy overlaps with welfare policy; welfare policy has implications for manpower policy; manpower policy has an effect on educational policy; so on. All have impacts on fiscal policy. Policies are often made without recognition of and with little understanding of their implications for each other. And certainly there is no "comprehensive" plan that draws all of these disparate elements together. The critics of comprehensive planning assert that this lack of an overall plan simply reflects humankind's inability to grasp everything at once or, to put it another way, it provides decision-makers with an opportunity to learn from past mistakes unconstrained by the dogma of a foreordained and rigidly prescribed set of goals and means.

There should be explicit recognition of the need for careful examination of the conditions necessary for implementation of climate change policy. The lessons of the twentieth century, in the United States and elsewhere, reveal that much policy remains merely symbolic or falls far short of the goals established by those who fashioned the policy. In some cases, the goal was only symbolic anyway (Edelman, 1964) but in many, if not most cases, there was sincere expectation that the goals would in fact be realized. Failure to achieve goals was often blamed on lack of zeal, poor organization or some other easy scapegoat when in fact the failure lay in the choice of strategy, the lack of resources, the conflicting interests brought into play, and numerous other factors (Wildavsky, 1973; Bardach, 1977).

Recognition of the potential and very real overloads of government, the conflicts among programs supported by various interest groups and the need for continuous learning from errors leads one concerned about the research agenda to examine the nature of the public policy in terms of implications for other policy, the possibilities of policy implementation, and the kind of societal commitment involved. Such a concern may lead to a preference for sharply focused policy that deals with climate change specifically, that imposes less than total societal commitment, and that is susceptible to alteration in view of changing conditions or outcomes. On the other hand, some sweeping reform may be called for, comparable to the enactment of the National Environmental Policy Act, which might affect the warp and woof of society. Whichever is chosen, it should not be by default in considering the matter.

10. Specific Research Issues

None of the classification schemes provides a neat rubric under which to organize the research effort in political institutions. Some research topics transcend a given time frame, some cannot be neatly contained within a distributive-redistributive-regulatory-constituent framework, some may be relevant to more than one of the preventive, curative and adaptive strategies, and some may be relevant to one or all of the phases of policy-making: problem recognition, avoidance and adaptation. A mixed strategy will be followed, incorporating the above classification schemes as appropriate.

It should be recognized, moreover, that a number of the research topics indicated below are really appropriate as multi-disciplinary projects in which political analysis plays an important part along with economic, sociological and psychological analysis. The focus of political research would be on the political institutions and their incentives upon the actors within the policy system but this research would in many cases ideally form a part of a larger research effort incorporating the work of investigators of a number of disciplines.

It also seems apparent that some research on political institutions and their response to the climate change problem should be undertaken only as an element in the more general investigation of the capability of political institutions to deal with long-term, cumulative and potentially severe insults to the environment and their consequences for societies. Climate change is only one example of such cumulative and potentially threatening effects of mankind's actions and the responses to such effects through the various polities of the world may be studied as generic properties of political regimes.

10.1 Constituent Policy

This general category of research efforts has its focus on the role of government itself, its structure, priorities and the values of those holding responsible positions. They are relevant to problem recognition in particular, but also to all three of the preventive-curative-adaptive strategies and may have significant redistributional consequences in the long run.

(1) *Estimations of the Character, Extent and the Evolution of the Threat over Time.* One of the most immediate and crucial issues is the level and kind of investment to be made in the investigation of the character, extent and timing of the CO_2 threat. While CO_2 levels are rising, the temperature of the earth is not, thus making the assertions regarding the relationship arguable and controversial (Landsberg, 1975: *Science*, 1979, 1980). Moreover, there are other research needs that may seem more pressing, such as the investigation of other forms of energy that do not produce large quantities of CO_2.

Within the context of the overall budget situation – largely controlled by economic and defense considerations – this issue is likely to be fought out within the scientific community and those decision-making institutions that depend on scientific achievement for major steps forward in their missions. These would include scientists and managers in such agencies as the Department of Energy, the National Science Foundation, NOAA and the Office of Science Technology Policy as well as top advisers in the White House.

Congress and the President are likely to go along with scientific recommendations. The private sector, other than the universities and the scientific communities that they represent, is not likely to become involved.

Under these circumstances, what level of priority should the study of the CO_2 issue have on the research agenda of a scientific field that arguably has already been a neglected field (Gribbin, 1978)? This is a genus of the more general species of issues that concern the (1) perceptions of scientists as to whether research emphasis should be placed on physical phenomena that indicate potentially irreversible and devastating consequences to society and the environment or whether greater emphasis should be placed on more immediate and definable problems; (2) whether the focus should be on CO_2 impacts or more general climatic investigations in which CO_2 would be a related research question; (3) whether social impacts should receive serious attention or be considered consequences that can be investigated once (if ever) the physical uncertainties are resolved.

Research on the impact of CO_2 on the biosphere and society is pre-eminently a governmental responsibility. There is unlikely to be serious question of this subject as a proper area of activity. The only questions will relate to priorities within science itself and the appropriate institutional approach to research on this subject. Some form of Delphi research technique among the most knowledgeable people in climatology, the atmospheric sciences generally, administrators and policy-makers in the scientific community having concern for energy research, presidential and congressional officials who are involved in scientific and energy decision-making, and members of firms in the private sector who are knowledgeable about the relationship of fossil fuel combustion and atmospheric, biological and social consequences (officials of utility firms and the Electric Power Research Institute, for example) "would be appropriate".

(2) *Research on Perceptions of Needs and Values that Lead to Decisions Regarding Continuation of Large CO_2-Producing Activities*. Although there is a growing awareness of the CO_2 problem and its imminence and perhaps a consensus regarding its character, decisions must be made, particularly by administrators in the Department of Energy, that will decide the availability of energy and the production of CO_2 in the immediate future. Recent decisions to convert oil and gas-fired plants to coal is indicative of the bind decision-makers are in. The perceptions of need and values in this instance clearly place the production of energy by large centralized facilities burning fossil fuels ahead of a concern for their contribution to the CO_2 problem whose parameters can be only dimly foreseen.

In more general terms, then, the research question concerns the manner in which an elite – administrators in the various departments concerned with energy management and development, public lands management, along with White House officials – weigh competing values and scientific evidence regarding issues that involve trade-offs between present satisfaction and future potential, dramatic, devastating and irreversible detriments. Could the institutional framework be altered to change the input of information to change this decision pattern? Could the inclusion of a different set of actors – let us say, EPA administrators – make a difference in this pattern? Do their values diverge substantially from others? Indeed, is there a pattern of decisions, and if there is, what are the features and characteristics of this pattern that may be changed by scientific analysis of value input? Systematic interviews with both governmental and private officials on their perceptions of social needs and values would be valuable in assessing the possibilities for shifts in policy and administration. They would reveal the extent to which a preventive or curative stategy would prove feasible.

(3) *Research on the Relationship of Ideology to Capability of Making Necessary Adjustments.* Decisions with respect to prevention or adjustment to the threat of CO_2 impacts are inevitably filtered through a screen of political, economic and social ideology. This ideology constrains both the means of achieving social goals and the goals themselves. In the broadest sense, the ideology concerns the appropriate role of government in society, the procedures for making decisions while protecting the rights and interests of individual members in society, and the justice of the results of government action (Dolbeare and Dolbeare, 1976).

The impacts of CO_2 will be differentially experienced by various members of society. Those aggrieved by the consequences of CO_2 will inevitably seek compensation for their losses and the compensation mechanisms through which to achieve redress will inevitably involve governmental action. The decisions as to the propriety of given courses of action will reflect the political and economic ideologies of the actors. Such ideologies may favor or oppose government subsidies, regulation, taxation, insurance, or direct governmental action. Similarly, the consequences of the proposed actions may be subject to debate. Agriculture will feel the most direct impacts of changed climate and no sector has been the subject of greater ideological debate than agriculture with respect to size of operation and the social value of its activity. Will the impacts of climate change tend to strengthen the forces leading toward domination of large corporate farming? Will government programs of various kinds tend in the same direction? To what extent might government programs resist this trend?

Ideology is seldom the subject of research as part of a government program. Ideology nevertheless remains as an unspoken and implicit constraint. Research on this subject would provide insight on a powerful barrier or vehicle for effective public action. Surveys of both decision-makers and the general public would be useful in providing insights into these powerful concepts and, like the previous topic, reveal much about the feasibility of a given strategy.

(4) *Government Organization to Deal with Climatic Change.* Government organization seldom provides a mechanism by which public effort may be concentrated in achieving given policy goals. Too many agencies have a piece of the action. No single agency has sufficient authority and resources to obtain a breakthrough in achieving policy goals (Seidman, 1975). Unfortunately, when virtually all policy space is the subject of some government policy and program, it is difficult to concentrate authority in the manner that those fully committed to a particular policy may desire. Recent efforts by President Carter to create a Department of Natural Resources indicate the difficulties in accomplishing such re-organizations (Convery *et al.*, 1979).

Should climate policy be the exclusive responsibility of the Department of Energy? Might the Department of Agriculture share some authority in this area? Might not other consequences of climatic change be found in subject-matter areas under the responsibility of the Department of the Interior and the Department of the Army (water); the Department of Transportation (alternatives to automobiles); EPA (air pollution);

Commerce (economic development)? Might they not reasonably claim a piece of the action? Such shared responsibility is typical and each agency is often prepared to defend its turf out of both self interest and commitment to a given policy objective.

This does not mean, however, that attention to government organization for accomplishing objectives having to do with climatic change is of no concern. Indeed, there may be important institutional changes through transfer of authority, reassignment of units, development of different relationships, devolution or concentration of authority that will have an important bearing on the achievement of goals. It may be necessary to ensure adequate attention to the matter by the creation of a lead agency with broad authority to coordinate – meaning control – the direction of activities in other agencies and to set priorities. Management of such a broad program might range from basic research (NSF) to applied research (Agriculture) to operational decisions for protecting the environment (EPA). Research on the appropriate organizational structure for dealing with climate change would be useful, both for the problem recognition phase and for later phases when the problem becomes one of prevention, cure or adaptation.

(5) *A Legislative or Judicial Strategy*. Climatic changes in their most drastic limits foreseen by scientists portend relatively extensive changes in the living conditions for entire regions of the country and for the world. In addition to whatever policies are adopted internationally, policies will be adopted domestically to ameliorate the impacts. Legislatures are frequently incapable or unwilling to legislate in sufficient detail to prescribe the behavior of bureaucrats who must execute their policies (Lowi, 1969). The result is bureaucratic discretion of sometimes staggering proportions.

The result of this legislative reluctance to specify both means and ends in detail is the intervention of the courts to protect interests that are unfairly dealt with by the bureaucracy. It is clear that environmentalists have made effective headway in challenging private and public actions detrimental to the environment through the courts, under the color of private injury or statutory law (Sax, 1971). Federal and state statutes have opened up broad avenues for intervention by private groups. The National Environmental Policy Act provided leverage through the requirements that developers – either public agencies or private groups operating under federal permits – demonstrate the impacts on the environment (Liroff, 1975). While largely procedural requirements, these impact statements have opened the process up and forced the public and private interests to face the damaging consequences of their actions. The courts have been sympathetic and have imposed burdens on the developers to fulfill their impact statement requirements in good faith.

To what extent is this approach valid in dealing with as complex and difficult-to-prove causative relationships between the actions of individuals, firms, and public agencies and the level of concentration of CO_2 in the atmosphere? (Stewart, 1977) Research on the appropriate roles of legislatures, bureaucrats and the courts in dealing with such complex problems would be appropriate, particularly spelling out the extent to which individuals and groups have access to the courts in pressing their claims for redress or ameliorative action. Such an investigation would have meaning and significance far broader than the CO_2 issue and might be designed to incorporate research on the possible roles of central governmental institutions in dealing with a variety of long-term, cumulative problems.

(6) *Long-range Planning to Deal with Climatic Impacts*. The nature of the climatic threat is long-term, gradual and cumulative. These characteristics of the impacts make planning both necessary and difficult. It is difficult because the nature and the timing of the impacts are uncertain and the resources available in the future are unclear. It is necessary because, as the scientists at the International Workshop on Climate Issues stated, it is necessary to "make use of and build in resilience to climate factors" (International Workshop on Climate Issues, 1978, p. 25). Such planning requires examination of alternative policies, evaluation of the ability of society to accommodate and respond to change and estimation of uncontrolled natural background.

The issue is not whether such planning should occur but the character and quality of such planning. The Workshop concluded that "prudent policy must not rigidly adhere to a particular line but must keep options open." This suggests a long-term flexible planning strategy, involving careful establishment of baseline conditions, both physically and socially, consistent monitoring, and periodic examination of the options at a given stage.

Planning in itself is not necessarily redistributive in character but it does hold out the possibility of broad controls over future economic development that may portend significant shifts in allocation of resources and therefore wealth accruing to given sectors of society. The implications of such planning might be readily foreseen if plans led to recommendations and then to concrete steps to *prevent* further CO_2 build-ups. Such steps would include the drastic reduction in the burning of fossil fuels on a world-wide basis (Kellogg, 1978). The disruption in energy production would inevitably spread from the energy sector itself to other sectors of society and would involve massive shifts in investments.

The term "planning" is virtually anathema in some circles within the United States although widely accepted here in practice and in both theory and practice elsewhere. The view of the sceptics are important reminders of the limitations of such efforts in achieving any level of predictive capacity, both in the United States and elsewhere (Wildavsky, 1973). Nevertheless, planning to deal with climatic change and other long-term threats to the environment would add strength to the movement toward centralized planning in this country, given the relatively broad range and depth of the social impacts that might be encountered. The quality of the planning effort required for undertaking actions to deal with climatic change would be a worthy subject for a research effort. Such research would be appropriate in every phase of problem consideration and with reference to whatever strategy was adopted.

10.2. Research on Adjustment Mechanisms for Those Injured by Climatic Changes

It is assumed – but the assumption may not be warranted – that the federal government will take responsibility for the consequences of climatic change. Both legally and politically it will be argued that increased CO_2 in the atmosphere is the result of deliberate government policy to encourage the burning of fossil fuels, especially if such policy is the consequence of slowing down the growth of nuclear power development and expansion of the nuclear industry. Having contributed to the accelerated growth of the problems associated with atmospheric change, the federal government will be seen as obliged to take measures to mitigate the injury to those whose lives and livelihoods are affected by the changes.

On the other hand, it is conceivable that (1) societal changes associated with climatic change will be slow and difficult to distinguish from natural cyclical climatic phenomena and (2) that the consequences are so widespread and severe that it will be difficult if not impossible for political institutions to assume the burdens associated with these changes. These consequences may be comparable to the problems of decay in the inner cities of major metropolitan areas of the United States. Some of the decay is the consequence of economic forces that have gradually shifted employment to the suburbs. But some of the decay can be attributed to policies to encourage suburban housing and interurban travel. It is not clear the extent to which public institutions can or should bear the burden of such changes. To a very considerable degree, the cost is borne by society in general in the form of lower standards of living.

The following subjects for research are clearly not the exclusive domain of the student of political institutions but they have important institutional components whose investigation would bring benefits in terms of a realistic understanding of the possibilities of achieving the goals of various policies. Implementation strategies must be integral parts of any adaptive approach or they are doomed to fall short of these goals.

(1) *Adjustment Mechanisms for Agriculture.* The interests most directly affected will be farmers and ranchers. A warming trend and reduced rainfall may severely injure farming and ranching that are dependent on rainfall. Some agriculture, notably that in the Western Great Plains and Rocky Mountains, is marginal at best and farming may be simply untenable in those regions. The spectre of drought, always on the horizon, now may become a permanent feature of these regions. Agriculture in the Southwest will also be affected as water supplies are reduced and less water is available for irrigation. To protect against erosion it may be necessary to impose more stringent controls over use of this marginal agricultural and grazing land. The government may wish to pay compensation for the loss of property values associated with such adjustments. Such efforts involve important questions of national economic efficiency and equity as well as regional development. Both the regulatory and compensatory mechanisms for providing such adjustments would require careful analysis to determine their costs and feasibility. Historically, such programs have been anemic and unfair because they have been underfunded and/or because payments were made to parties who were legally entitled to them but did not merit them by any standard of equity. If such mechanisms were to transcend national boundaries, the political problems would be seriously exacerbated.

(2) *Adjustment Mechanisms for Tourist and Recreation Industries.* A second major interest affected by climate change would be the tourist and recreation industries. Depending on some quality related to climate – snow, sunny, clear skies, wildlife habitats, forested land, water bodies – these industries may feel similarly aggrieved and deserving of public compensation and assistance in making necessary adjustments. Similar efforts will be required to sort out the efficiency, equity and administrative implications of such assistance.

(3) *Other Interests.* Additional interests that may be directly affected are industries

that depend on water supplies such as power plant and extraction industries. The inducements, subsidies, and relaxations of utility regulation structures may all be appropriate devices for assisting industries suffering such detrimental impacts.

(4) *Secondary Interests*. In addition to the interests affected directly by climate and rainfall changes, there are secondary interests that will be materially affected. These include virtually the entire range of private firms and public agencies that give support to primary industries: banks and other lending institutions, merchants, professionals of all kinds, local governments in all of their varieties, school systems, etc. Each will suffer significant losses as the primary industries are affected by climatic change. These losses will occur over long periods of time and it may well be that their investments have long since been amortized, thus reducing justification for public asistance. These secondary effects nevertheless suggest that entire communities and their political and economic infrastructures will be affected by climatic change and federal assistance to deal with these changes may have to be undertaken of a community-wide basis.

Given the impulse for energy development in the Western part of the United States, there is considerable experience with the fashioning of federal-state-local mechanisms for assisting in the transition from one kind of economy to another. Nevertheless, careful attention will have to be paid to the political and administrative requirements of what might be a massive assistance program dealing with entire regions of the country.

(5) *Research and Planning Water Supplies*. In many places throughout the nation there will be increased pressure on water supplies that already are inadequate at given places and times. Droughts may become more frequent and prolonged, necessitating additional storage and large-scale schemes for transferring water from one location to another. These will be increased pressure to develop water at places that have heretofore been protected as national parks and monuments. Alternatively, or in conjunction with water development, there will be increased demand for water conservation. It will be necessary for the government to create incentives or to impose regulations on water usage. Such regulations may have to do with limitations in water applications in agriculture to some strict notion of beneficial use, the creation of water markets in which water is applied to its highest economic use, requirements for dry cooling towers, reductions in instream or low-flow standards, etc.

Increasing pressures on water supplies inevitably bring with them increasing political controversy over water rights and water law. At least in the United States, states jealously guard their prerogatives to allocate water, define rights, and determine priorities. The intrusions by the federal government, which might be required under conditions of nation-wide climatic shifts, would require careful assessment of the necessary federal authoritiy while ensuring recognition of the legitimate interests of the states in this policy area.

(6) *Research on Public Works and/or Other Actions Required to Protect Coastal Cities*. Warming trends could lead to collapse of portions of the polar ice caps, thereby increasing sea levels throughout the world. If sea levels were to increase, many coastal cities would face gradual inundation (Schneider and Chen, 1980). It may be expected that the initial approach will be construction of barriers to the sea, construction of facilities along the

coast that are more resistant, and various other public works. Alternatively, it might be necessary to abandon certain facilities and activities that lie along the seacoast. Entire residential communities, port facilities, transportation systems, resorts and other commercial activities might be either protected, abandoned or a combination of both.

The staggering size and cost of such efforts would require the resources of all levels of government but it may be expected that the federal government would play a major role in such undertakings. It would be important to design a comprehensive strategy for dealing with this predictable inundation problem, encompassing not only the steps necessary along the coastline and within coastal cities and metropolitan areas but also a national strategy that might involve relocation of certain activities that are not dependent on the coast, to inland locations. Measures to deal with this problem could undoubtedly be various: financial inducements, regulatory measures, direct public works, exhortation, and voluntary efforts on the part of the private sector. Research on an appropriate mix of these measures including appropriate incentives and administrative mechanisms is necessary to anticipate the need and to meet the contingencies that would undoubtedly arise.

(7) *Research on Storage of Agricultural Products to Counter Perturbations in Rainfall Patterns*. This is the so-called "Genesis" strategy in which the United States and other nations with agricultural production in excess of domestic needs endeavor to protect their citizens and the people of other nations of the world against serious shortfalls of food supplies (Schneider, 1976). As the changes in rainfall and temperature occur as a result of increased CO_2 in the atmosphere, these nations would have to embark on programs of stockpiling food supplies. There has been considerable experience with stockpiling programs in the United States associated with price support and farmer subsidy programs. Not all of this experience has been positive in that there was enormous wastage, depressing effects on the market for farm products, and high administrative costs. Stoarage programs are therefore not popular and would be adopted over considerable opposition. Research would be useful on the character of such a program, its costs and potential benefits, the parameters of the program itself, methods of releasing food from storage and the conditions that would trigger releases, coordination with means of transportation and distribution, etc.

It has also been argued that the only sure way for adequate supplies of food to be available in times of shortages is to increase the production and storage of grain in the developing nations themselves (Schnittker Associates, 1978). Both U.S. and international assistance could be designed to accomplish this goal. Further investigation would be important to understand not only the physical boundaries of such policies but the political boundaries as well.

10.3. Regulatory Policy

The following policy issues are likely to be taken up in a regulatory framework, i.e., the sorting out of conflicting sectoral interests and the delineation and enforcement of rules that reflect the priorities and values of the national or regional communities. Because they are regulatory, these policies are likely to be more conflictive and more difficult to carry out. They involve not the mere expenditure of money but the careful sorting out of burdens and benefits and the direct imposition of authoratative control of human behavior.

(1) *Research on the Potential Need for Population Controls.* As part of a national strategy for dealing with shifts in agricultural production and coastal zone problems, it may be necessary to consider the desirability of national population policies. Such policies raise important questions of constitutional law as well as social equity. The policies of the United States government have always directly and indirectly affected population movements by public investments such as assistance to railroads in the West, mortgage guarantees for homeowners in the suburbs, reclamation of land, and location of defense and research installations. It may be that similar measures would meet the need to direct population away from those areas most severely impacted by climatic change. The measures would be controversial as in the past but they would not raise constitutional issues such as direct restrictions on mobility. Other nations have sometimes used more direct and forceful methods of population management and control owing to different constitutional traditions.

Research on the constitutional, legal and policy issues would be important in elucidating the assumptions, standards and consequences involved in such policy. To what extent would such national policies collide with various mechanisms of population management such as inducements to locate in given locations of the country or in urban or rural settings? To what degree and in what forms should the government assist individuals to make such location changes? To what extent would national population policy upset the balance of power between centralized government and local governments?

(2) *Research on Environmental Impacts of Climate Change.* Climatic change will have a direct impact on the environment in the form of altered patterns of rainfall, increased temperatures, changed growing seasons, changes in wildlife habitats, and innumerable complex relationships of an ecological kind. Decisions must be made regarding remedial measures that may be considered desirable or standards of human usage necessary to protect the environment. Such measures and standards will require careful scientific assessment of consequences, both in the short and long run. Because science seldom provides definitive answers, decisions will have to be made under conditions of uncertainty and therefore will be controversial.

Many of the issues that are likely to arise in this category are relatively familiar issues that may take on new meaning in the context of the CO_2 impacts problem. These include regulation of activities on the public domain: the balance among competing interests such as wildlife, cattle and sheep, recreation, and timber management; regulation of stream flows and water quality and the extent to which diminished (in some areas) and augmented (in other areas) streamflows may serve competing interests; regulation of strip mining, particularly with respect to the possibilities of reclamation of the mined land.

Here again there is an important political and administrative component that should be investigated. All too often, the best intentions founder on the failure to recognize the political and administrative measures required to carry out such protective programs. The resources provided may be inadequate. There may be a failure to achieve common agreement on standards, creating conflicting expectations regarding the scope and depth of the program. There may be too little recognition of competing values, particularly those that emanate from the localities where such measures are to be taken.

(3) *Research on Appropriate Policy for the United States with Nations Suffering Similar Consequences.* Climatic change will be global in character but the relationships among nations to deal with the phenomenon may be more regional in character. Unquestionably, the relationships with Canada and Mexico will be the crucial ones for the United States. It is predicted that the regions of the higher latitudes will feel the most dramatic impacts. A warming climate for Canada may make arable and susceptible of development land to the northward of presently cultivated regions. Canada's agricultural production, already substantial, may increase greatly. Its water supplies, currently in excess of its needs and therefore coveted by the United States interests, may be required to meet the needs of increased agricultural production. Such alterations in this one sector alone portend major policy issues with respect to world and North American food markets.

With Mexico, the situation may be even more complex. Increased desertification of the North American continent will have a serious impact on Mexico in its quest for national development. Reduced water supplies may imperil its developing agriculture in Northern Mexico and may endanger industrialization as well. Difficult relationships between the United States and Mexico with respect to the water supplies of the Colorado River and the Rio Grande River will be exacerbated. The immigration problem along the United States-Mexican border may become more acute (*Natural Resources Journal*, 1977).

The consequences of increasing CO_2 in the atmosphere will be worldwide, but perhaps more concentrated in the Northern Hemisphere. Thus, the fate of most of the industrialized nations will be affected by climate change. The effects may be more or less benign or damaging in Western Europe, the Soviet Union and China. Thus, while the effects of climate change may appear more apparent in the northern half of the Western Hemisphere, the relationships with none of the Northern Hemisphere nations can be safely ignored.

Given the past record, it is perhaps too much to expect a North American or a truly international approach to be taken but it may be hoped that the interrelationships among the three nations of North America as well as the other nations of the world and the various policy issues that are intertwined with climatic changes will be looked at in a more comprehensive way than has been the case in the past. Research on mechanisms for accomplishing this goal would be profitable.

(4) *Social-Order Consequences of Changed Precipitation and Temperatures.* Tobin has examined the political and social consequences of changing climate for people living in urban areas (Tobin, 1976). It can be argued that inclement weather might reduce political activity generally while making those who participate even more active because they have hurdled one significant barrier to participation. On the other hand, there appears to be a relationship between the incidence of riots and long, humid, hot and uncomfortable weather. One study concluded that there is a correlation between increased police activity and temperature increases and atmospheric pressure decreases.

Temperature and rainfall may also be correlated with mortality and morbidity rates. Increased heat may contribute to temperature inversions which, when combined with severe air pollution, may seriously damage the well-being of people living in the region, especially those with serious lung and respiratory problems.

Suffice it to say that such relationships are little understood and should be investigated more thoroughly. These kinds of effects have direct consequences on the burdens of government, especially local governments and private facilities concerned with health and welfare. The cost of providing police protection or health services may go up measurably as the climate changes. Police activity may be stepped up to control anti-social behavior, perhaps dividing a political community that is already riven with conflict owing to ethnic and economic inequities. The growth of these costs, unless ameliorated by some other political means, may have significant redistributive effects.

11. Alternatives to Government Intervention: The Private Economy

The foregoing approaches to public policy-making emphasize direct government intervention in the economy either through funding of projects and programs, regulation or imposition of taxes and distribution of benefits. One major alternative is the achievement of public policy goals through the private economy, largely by the imposition of penalties and the offering of incentives to those whose behavior may affect the climate or affect the behavior of those who are affected by the climate. Many critics of current air and water pollution control policy argue that such a system of effluent charges would provide a more effective, efficient and economical method of achieving pollution goals than the current regulatory approach (Anderson *et al.*, 1977).

Given the policy overload that arguably exists already in the federal bureaucracy it is essential to examine alternatives to policies that impose further burdens on it. Regulatory measures will undoubtedly be necessary but some combination of market incentives and regulatory requirements may be more effective in achieving the adjustment goals than any rigorous application of either. In any case, it is an important area of institutional design and the design effort would benefit from careful research and exploration of alternative mixtures.

It would be useful to examine the opportunities for utilizing this approach in achieving goals in response to potential or actual climatic change. As a part of a preventive strategy, it is clear that such a policy would work essentially like the policies recommended for dealing with air pollution. Firms would be charged fees in accordance with their production of effluent; the firms in turn would be in a position of selecting the precise means of achieving reductions of the costs associated with the charges. In some cases, it might be preferable to pay the charges; in most cases, firms would seek to reduce the charges by controlling output of the pollutant – in this case CO_2.

It is somewhat more speculative to assert the form that a market-oriented strategy might follow in the context of an adaptive strategy. One could envisage incentive payments to individuals to assist them in migrating from regions affected detrimentally by climatic change but these are hardly market-oriented incentives. On the other hand, it is conceivable that grazing fees, already a part of the structure of the government land-management agencies, might be designed less to obtain revenue or to compensate the treasury for public investment than a device to achieve certain environmental goals in the light of climatic change. Individual entrepreneurs could evaluate the charges in terms of their structure of costs and make the necessary adjustments to reduce the impact of the charges on their operations.

Still another approach that would emphasize the private sector would be the inclusion of the insurance industry as a major actor in dealing with climate change. Either with or without government involvement, perhaps through some reinsurance arrangement, the private insurance industry might participate by insuring against the potential effects of the consequences of climatic changes (Kunreuther, 1973). The private sector presently insures against hail and other weather events, and it has all-hazard insurance as well. Just as rates are structured to reflect actual event and claim experience, and the probabilities that claims will occur, the insurance industry could insure those who are endangered by fluctuations that will occur as climate changes. The lack of scientific certainty is a major hindrance to this approach, as insurers tend to avoid such situations or charge prohibitive rates. The accumulation of additional scientific evidence regarding the climatic trends as well as the actual evidence from experience will determine the rates farmers as well as others would have to pay. To the extent that public policy wishes to encourage this approach, the federal and state governments may provide financial support for such an approach.

12. Conclusions and Recommendations

The conclusions and recommendations below are basically general observations about the priorities and principles that should govern the formulation of a research agenda in political science. The study of political responses to climatic change or political strategies for dealing with the threat of climatic change raise new substantive problems for public institutions but they are clearly related to the traditional objects of research in the discipline of political science. Thus, the study of political responses or strategies could conceivably encompass virtually every sub-field within political science: governmental structure and processes, constitutional law, international relations, political philosophy and ideology, etc. It is inconceivable that sufficient resources would be available to undertake a comprehensive study of the political implications of this one phenomenon and thus it is necessary to relate the investigation of political implications to a broader investigative strategy. The following are an effort to accomplish that goal.

(1) *Research on the political implications of climatic change should be incorporated into larger research projects involving both the other social science disciplines and those of the natural sciences*. The studies in the fields of psychology and anthropology, for example, have their focus on individual and group perceptions of climatic phenomena and the behavioral responses to those perceptions. Such perceptions and behavior often and predictably would have political and institutional consequences that would warrant investigation. The careful design of comprehensive projects that begin with the expectation that the results of one project would dove-tail with others would maximize the benefits to be derived from the overall effort. Such an effort toward an integrated research design would have the advantage of allowing investigators to develop a more precise focus rather than to respond to the tendency to study the implications of climatic change from every aspect of a given discipline.

(2) *Given the fact that the predicted climatic change will be cumulative and gradual, it is important to study the preventative techniques first, leaving the adaptive techniques to a later time.* As the discussion of potential projects makes clear, there are a number of tasks of immediate importance to undertake if an understanding of the role of political institutions in dealing with climatic threats is to be well understood. These projects have to do with scientific consensus, planning and organization, and communication and learning with respect to climatic change. These investigations have to do with general governmental processes for dealing with complex and conflicting values and demands upon society. But the problem of climatic change and its effects is hardly unique in the sense that there are a large number of long-term cumulative problems that society must address. Thus, a study of governmental response to climatic challenge, accompanied by studies of governmental responses to similar challenges, would have cumulative value in their own right. Obviously, studies of techniques of adaptation could wait until funds were available, and perhaps more importantly, until the parameters of climatic impacts are more clearly discernible than they are at the present time.

(3) *As this research agenda has repeatedly pointed out, the investigation of domestic political institutions is crucial for success of any preventative or adaptive program, but such investigations must be made wherever the climatic challenge is encountered.* The appropriate framework for political research is comparative, i.e., the study of political institutions across cultural and national boundaries and their comparative capacity for dealing with similar social phenomena. Control of and adaptation to climatic impacts will inevitably require an international effort but such efforts will be implemented through domestic institutions within each country. International processes and institutions may be instrumental in arriving at agreements and making fundamental decisions about the duties, obligations and prerogatives of each nation but each nation will in turn transform those qualities into domestic law and program to accomplish the purposes of those agreements and decisions. Domestic institutions respond to a wide variety of forces that may or may not be present in the international arena. Any reasonable expectations of achieving the goals upon which international agreements are predicated rely upon assumptions about the responsiveness and responsibility of those domestic institutions. The comparative advantage of various alternative institutional arrangements at the domestic level should receive thorough evaluation.

(4) *The focus of political science research should be on decision-making arrangements and implementation strategies.* Decisions with respect to responses to climatic change impacts will undoubtedly be made through the traditional processes of decision-making and administration within each society. The actual actors and agencies that participate in the process will depend on the specific issue. Part of the strategy with respect to decision-making, however, depends on the formulation of the specific issue because that formulation may, in effect, determine who the decision-makers are. If made an issue of general redistribution of the benefits and costs of society, the issue may incite leading decision-makers such as presidents and premiers and leading members of legislative institutions. Treated as a technical issue of modest importance, it may excite the interest of scientists and lower level bureaucrats only. Similarly, will the issues related to climatic impacts be drawn narrowly by specific sectors or will they constitute a broad frontal

attack requiring extensive coordination? The formulation of the issues, then, may be a task requiring more than mere acceptance of fate.

In addition, those who are concerned with policy achievements must evaluate the alternative institutional arrangements through which decisions will be carried out. These issues concern such matters as assignments of principal tasks, coordination with agencies that have marginal but real stakes in the issue, ordering of priorities and achievement of agreement on them, procedural requirements to ensure conformity with laws and regulations of a more general sort, appropriate levels of funding, and other more technical matters. Realistic assessments should be made of such institutional arrangements in order to avoid the fate of much policy-making: symbolic statement of intention with little accomplishment. Part of this assessment clearly must involve the expectations of the behavior of both private and public actors and institutions. To what extent will private interests be encouraged to participate and to what extent will they resist efforts to obtain their cooperation?

(5) Finally, political science research should have a focus on improving the capacities of societies to learn, to modify behavior and to gain from cumulative experience. Public policy-making is not an exact science at best, is often experimental, and at worst is virtually a random series of efforts to deal with problems that are little understood and whose solutions can only be guessed at. This is true of present problems that are all too real, present and urgent. The climatic change problem is even more speculative in nature: its nature is little understood and controversial, and the effects that flow from it, both physical and social, can be the subject of intelligent estimation at best.

Institutional analysis should incorporate within it an awareness of the danger of early and false closure: the conclusion that the problem has been fully identified and the options narrowed before adequate scientific evidence is available. This may require a sensitivity to the need for building in redundancy, competition, as well as alternative approaches to research and policy. Such sensitivity is less concerned with neatness and order, with doctrine and ideology, than it is with gradual accomplishment of the nation's, and the world's, goals with respect to protection against the consequences of climatic change.

Dept. of Political Science,
University of California

Notes

* I wish to express my appreciation to Jack Corbett, J. Clarence Davies and Lester Milbrath for their searching and detailed critiques of a draft of this manuscript. While they bear no responsibility for the final version, their suggestions resulted in improvements throughout the manuscript.

[1] Edith Brown Weiss addresses the international, legal and institutional structure for dealing with the CO_2 in the following paper.

[2] See the Weiss paper for an extensive discussion of international institutions for negotiation of differences.

[3] I am grateful to Lester Milbrath for suggesting the appropriateness of this classification.

References

Anderson, Frederick R. *et al.*: 1977, *Environmental Improvement Through Economic Incentives*, Baltimore: The Johns Hopkins University Press.

Anderson, James E., Brady, David W., and Bullock, Charles III: 1978, *Public Policy and Politics in America*, North Scituate, Mass.; Duxbury Press.

Andrews, Richard N. L.: 1979, 'Environment and Energy: Implications of Overloaded Agendas', *Natural Resources Journal* **19**, 3, 487–503.

Apter, David: 1965, *The Politics of Modernization*, Chicago: The University of Chicago Press.

Bardach, Eugene: 1977, *The Implementation Game*, Cambridge, Mass: MIT Press.

Bertsch, Gary K., Clark, Robert P., and Wood, David M.: 1978, *Comparing Political Systems: Power and Policy in Three Worlds*, New York: John Wiley and Sons.

Burnham, Walter Dean: 1970, *Critical Elections and the Mainsprings of American Politics*, New York: W. W. Norton and Co.

Convery, Frank J., Royer, Jack P., and Stairs, Gerald R. (eds.): 1979, *Reorganization: Issues, Implications and Opportunities for U.S. Natural Resources Policy*, Durham, North Carolina: Center for Resource and Environmental Policy Research.

Cooley, Richard A. and Wandesforde-Smith, Geoffrey: 1970, *Congress and the Environment*, Seattle: University of Washington Press.

Corbett, Jack: 1979, 'Developing Response Strategies for Climate Change', paper prepared for AAAS-DOE Workshop on Environmental and Social Consequences of CO_2-Induced Climate Change, Annapolis, Maryland.

Dahlberg, Kenneth A.: 1979, *Beyond the Green Revolution: The Ecology and Politics of Global Agricultural Development*, New York: Plenum Press.

Daly, Herman E.: 1979, 'Entropy, Growth and the Political Economy of Scarcity', in Smith, V. Kerry (ed.), *Scarcity and Growth Reconsidered*, Baltimore: The Johns Hopkins Press.

Dolbeare, Kenneth M. and Dolbeare, Patricia: 1976, *American Ideologies: The Competing Political Beliefs of the 1970s*. Chicago: Rand McNally.

Domhoff, G. William: 1978, *The Powers That Be*, New York: Random House.

Downs, Anthony: 1972, 'Up and Down with Ecology – the "Issue-Attention Cycle"', *Public Interest* **28**, 38–50.

Dye, Thomas R.: 1975, *Understanding Public Policy*, 2nd edition, Englewood Cliffs, N.J.: Prentice-Hall.

Dye, Thomas R.: 1976, *Who's Running America: Institutional Leadership in the United States*, Englewood Cliffs, New Jersey: Prentice-Hall.

Edelman, Murry: 1964, *The Symbolic Uses of Politics*, Urbana: University of Illinois Press.

Falk, Richard A.: 1971, *This Endangered Planet: Prospects and Proposals for Human Survival*, New York: Random House.

Fiorina, Morris P.: 1977, *Congress: Keystone of the Washington Establishment*, New Haven, Conn.: Yale University Press.

Glantz, Michael H.: 1979, 'A Political View of CO_2', *Nature* **280**, 189–190.

Greenwald, Carol S.: 1977, *Group Power*, New York: Praeger Publishers.

Gribbin, John: 1978, *The Climate Threat*, Fontana: Collins.

Hanrieder, Wolfram: 1978, 'Dissolving International Politics: Reflections on the Nation-State', *American Political Science Review* **72**, 4, 1276–1287.

Hardin, Garrett: 1968, 'The Tragedy of the Commons', *Science* **162**, 1243–1248.

Ingram, Helen M. and Mann, Dean E.: 1978, 'Environmental Policy: From Innovation to Implementation', in *Public Policies in America*, Lowi, Theodore J., and Stone, Alan (eds.), Beverly Hills, California: Sage Publications.

Ingram, Helen M. and Mann, Dean E.: 1980, *Why Policies Succeed or Fail*, Beverly Hills, California: Sage Publications.

International Workshop on Climate Issues: 1978, *International Perspectives on the Study of Climate and Society*, Climate Research Board, National Academy of Sciences, Washington, D.C.

Jones, Charles O.: 1974, 'Speculative Augmentation in Federal Air Pollution Policy-Making', *Journal of Politics* **36**, 438–463.

Jones, Charles O.: 1975, Clean Air: *The Policies and Politics of Pollution Control*, Pittsburgh: Universiry of Pittsburgh Press.

Kelley, Donald R., Stunkel, Kenneth R., and Wescott, Richard R.: 1976, *The Economic Superpowers and the Environment: the United States, The Soviet Union, and Japan*, San Francisco: W. H. Freeman and Co.

Kellogg, W. W.: 1979, 'Potential Consequences of a Global Warming', in *Man's Impact on Climate*, Bach, Wilfrid, Pankrath, Jurgen and Kellogg, William (eds.), Developments in Atmospheric Science 10, Amsterdam: Elsevier Scientific Publishing Company.

Kunreuther, Howard: 1973, *Recovery from Natural Disaster: Insurance or Federal Aid?*, Washington, D.C.: American Enterprise Institute.

Landsberg, M. E.: 1975, 'Man-Made Climatic Changes', in *The Changing Global Environment*, Singer, S. Fred (ed.), Dordrecht-Holland: D. Reidel Publishing Company.

Liroff, Richard A.: 1976, *A National Policy for the Environment: NEPA and Its Aftermath*, Bloomington, Indiana: Indiana University Press.

Lowi, Theodore J.: 1972, 'Four Systems of Policy, Politics and Choice', *Public Administration Review* **XXXV**, 4, 288–300.

Lowi, Theodore J.: 1969, *The End of Liberalism*, New York: W. W. Norton and Co., Inc.

Mann, Dean E.: 1975, 'Political Incentives in U.S. Water Policy: Relationships Between Distributive and Regulatory Politics', in *What Government Does*, Holden, Matthew, Jr., and Dresang, Dennis L. (eds.), Beverly Hills, CA: Sage.

Mann, Dean E.: 1979, 'How We Might Respond to a Different Climatic Regime', paper at the DOE-AAAS Workshop on Environmental and Societal Consequences of CO_2-Induced Climate Change, Annapolis, Maryland.

Meyer-Abich, Klaus M.: 1980, 'Socio-economic Impacts of CO_2-Induced Climatic Changes and the Comparative Chances of Alternative Political Responses: Prevention, Compensation, Adaptation', *Climatic Change* **2**, 373–385.

Mitchell, Robert Cameron: 1978, 'The Public Speaks Again: A New Environmental Survey', 60 *Resources* **1**.

Mitchell, Robert Cameron: 1979, 'Silent Spring/Solid Majorities', 2 *Public Opinion* **16**.

Mitchell, Robert Cameron: 1980, 'How "Soft", "Deep", or "Left?" Present Constituencies in the Environmental Movement for Certain World Views', *Natural Resources Journal* **20**, 2, 345–358.

Morganthau, Hans J.: 1978, *Politics Among Nations: The Struggle for Power and Peace*, fifth edition, New York: Alfred A. Knopf.

Moss, Thomas H.: 1979, 'Can Our Political Institutions Deal With Climate Change'. (mimeo)

Niskanen, William A.: 1971, *Bureaucracy and Representative Government*, Chicago: Aldine, Atherton.

Ophuls, William: 1979, *Ecology and the Politics of Scarcity*, San Francisco: W. H. Freeman and Co.

Pressman, Jeffrey and Wildavsky, Aaron: 1973, *Implementation*, Berkeley: University of California Press.

Prewitt, Kenneth and Stone, Alan: 1973, *The Ruling Elites: Elite Theory, Power and American Democracy*, New York: Harper and Row.

Price, Don K.: 1965, *The Scientific Estate*, Cambridge, Mass.: The Belknap Press of Harvard University Press.

Sax, Joseph L.: 1971, *Defending the Environment*, New York: Alfred A. Knopf.

Schneider, Stephen H. with Mesirow, Lynne E.: 1976, *The Genesis Strategy*, New York: Plenum Press.

Schneider, Stephen H. and Chen, Robert S.: 1979, 'Carbon Dioxide Warming and Coastline Flooding: Physical Factors and Climate Impact', *Ann. Rev. Energy* **5**, 107–40.

Schnittker Associates: 1978, *Reducing the Climatic Vulnerability of Food Supplies in Developing Countries: Public and Private Alternatives*, Washington, D.C.

Schware, Robert: 1980, 'Toward a Political Analysis of the Consequences of a World Climate Change Produced by Increasing Atmospheric Carbon Dioxide', paper presented at the Southwestern Political Science Association Meeting, Houston, Texas.

Science: 1979, p. 883.

Science: 1980, **207**, 1462–3.

Segal, M.: 1972, *The Supersonic Transport: A Legislative History*, Library of Congress, Congressional Research Service.
Seidman, Harold: 1975, *Politics, Position and Power*, New York: Oxford University Press.
Sprout, Harold and Sprout, Margaret: 1971, *Toward a Politics of the Planet Earth*, New York: Van Nostrand Reinhold.
Stern, Arthur C.: 1977, *Air Pollution: Air Quality Management*, Vol. V, New York: Academic Press.
Stewart, Richard B.: 1977, 'Paradoxes of Liberty, Integrity and Fraternity: The Collective Nature of Environmental Quality and Judicial Review of Administrative Action', *Environmental Law* 7, 3.
Sundquist, James L;: 1973, *Dynamics of the Party System*, Washington, D.C.: The Brookings Institution.
'Symposium on U.S.-Mexican Transboundary Resources': 1977, *Natural Resources Journal* 17, 4.
Tobin, Richard J.: 1976, 'Climate Changes, Local Governments and Political Problems', in Ferrar, Terry A. (ed.), *Urban Costs of Climate Modification*, John Wiley and Sons, New York, pp. 239–258.
Tobin, Richard J.: 1979, *The Social Gamble*, Lexington, Mass.: Lexington Books.
Truman, David B.: 1971, *The Governmental Process: Political Interests and Public Opinion*, second edition, New York: Alfred A. Knopf.
Walker, Richard and Michael Storper: 1978, 'Erosion of the Clean Air Act of 1970: A Study of the Failure of Government Regulation and Planning', *Boston College Environmental Affairs Law Review* 7, 2, 189–257.
Wildavsky, Aaron: 1973, 'If Planning is Everything, Maybe It's Nothing', *Policy Sciences* 4, 127–153.
Wildavsky, Aaron: 1979, *Speaking Truth to Power: The Art and Craft of Policy Analysis*, Boston: Little Brown and Co.

INTRODUCTION TO: INTERNATIONAL LEGAL AND INSTITUTIONAL IMPLICATIONS OF AN INCREASE IN CARBON DIOXIDE: A PROPOSED RESEARCH STRAEGY

How can national preferences and choices translate into international agreement and action? Although this is certainly a key question for almost any international issue, it is especially crucial in relation to a long-term, global environmental problem such as the possibility of climate change. In particular, what institutions and procedures currently exist at the international level that can, firstly, help develop sufficient consensus for actions to be taken and, secondly, help orchestrate or even enforce such actions among enough nations for them to be effective?

Weiss takes the first necessary step of reviewing in detail key aspects of society's "experience-to-date" in attempting to deal with global resource and environmental problems. By drawing extensively on the Western legal literature, she reveals that both *principles* and *precedents* do indeed exist that are relevant to the possibility of climate change. The question then becomes one of assessing how such principles and precedents would likely be applied (or ignored) given either the prospect or the onset of climate changes.

To illuminate this question, Weiss proposes research on nine distinctive facets of the CO_2/climate problem that may strongly influence the "regime" or context in which the problem is viewed and treated. Each facet suggests some specific examples drawn from a variety of arenas that should inspire useful insights into how society might perceive and respond to CO_2-induced climate changes at the international level. Although the nine areas Weiss proposes may not be entirely comprehensive, they nevertheless provide a much-needed framework for organizing information about past societal experience and for developing more sharply focused *prescriptions* for international action.

R. S. CHEN

R. S. Chen, E. Boulding, and S. H. Schneider (eds.), Social Science Research and Climate Change: An Interdisciplinary Appraisal, 147.

EDITH BROWN WEISS

INTERNATIONAL LEGAL AND INSTITUTIONAL IMPLICATIONS OF AN INCREASE IN CARBON DIOXIDE: A PROPOSED RESEARCH STRATEGY

1. Summary

The problem of a build-up in carbon dioxide (CO_2) needs to be viewed as a problem in developing the appropriate transition strategies for moving from a fossil fuel to a non-fossil fuel economy in the next fifty to one hundred years. We need to develop processes for managing carbon dioxide emissions which will delay the warming of the earth's temperature from carbon dioxide sufficiently to give time to develop new technologies for storing and recycling the carbon dioxide and to adapt to any changes in climate. Central to this management strategy are scientific assessments as to how much carbon dioxide input will produce how much change in the global environment, when and where.

The carbon dioxide problem is foremost a problem in the management of fossil fuels. States have the sovereign right in international law to control the exploitation of their natural resources, and will resist measures which are seen as intruding upon this sovereignty. We need to explore the feasibility and desirability of establishing an international regime for setting a global ambient air guality standard for carbon dioxide, and for implementing it at the national/regional level through emission limitations designed to meet this standard. This system could involve measures for allocating emissions between competing producers, and perhaps provisions for the lease and sale of emission allocations.

It is essential to involve the private sector directly in the discussion of strategies for mitigating carbon dioxide emissions. We need to explore with industry appropriate incentives for developing controls and for developing new ways to store carbon dioxide.

Presently, the major contributors to carbon dioxide pollution are a handful of developed countries. The Organization for Economic Co-Operation and Development, the European Economic Community and the new Convention on Long-Range Transboundary Air Pollution offer appropriate forums for initiating discussions regarding strategies for monitoring, limiting, and allocating carbon dioxide emissions.

The increase in CO_2 is also a deforestation problem. The precise contribution of deforestation is scientifically unknown, but does not appear to be nearly as significant as from fossil fuels. Problems of deforestation involve a wide community of states, including many in the developing world, and a number of multi-national corporations. The problem for the international community is much broader than limiting the CO_2 emissions from deforestation: namely development and implementation of environmentally sound strategies for sustained yield management of the forests and the soils. Any progress in controlling CO_2 emissions is likely to be a by-product of progress in this area. CO_2 emission limitations are not a useful approach. Efforts to raise "international consciousness" about the implications of forest and soil management for carbon dioxide levels and global temperatures are a necessary and useful step.

R. S. Chen, E. Boulding, and S. H. Schneider (eds.), Social Science Research and Climate Change: An Interdisciplinary Appraisal, 148–175.

The atmosphere is a global resource, which states need to manage. International law indicates a growing obligation of states to develop processes for an equitable use of common natural resources and for preventing harm to other states from activities which arise within their own jurisdiction or under their control. It is essential that we begin the research necessary to explore the processes for minimizing the impact on the global temperature from the production and development of fossil fuels during this transition period. This is what the author refers to as the "CO_2 transition strategy" for managing the atmosphere.

2. Resource Management Approach to CO_2 During the Century of Transition

Partly as a result of the world's increasing use of fossil fuels, carbon dioxide is accumulating in the atmosphere at a rate estimated to double the present concentrations by about the year 2050.[1] Carbon dioxide in the atmosphere traps infra-red radiation coming from the earth's surface and prevents it from escaping into outer space, thereby raising the temperature of the earth's surface. This phenomenon is commonly known as the "greenhouse effect". If the concentration of carbon dioxide were to double over present levels, global climate models predict an average global surface warming between 1.5 °C and 3.5 °C, with greater temperature increases at higher latitudes.[2]

This increase in the global surface temperature would have major impacts on the world's climates, ocean currents, and growing seasons. It would significantly disrupt agricultural production and water supplies in some areas.[3] At some point, increased concentrations of carbon dioxide could cause the floating ice in the Arctic to disappear or trigger the disintegration of the West Antarctic Ice Sheet, which might ultimately raise sea level as much as fifteen to twenty-five feet.[4] The President's Council on Environmental Quality has concluded that this projected increase in carbon dioxide poses "one of the most important contemporary environmental problems" and "threatens the stability of climates worldwide and therefore the stability of all nations."[5]

The rapid increase in the atmospheric concentration of carbon dioxide to date is well-documented. Scientists do not agree, however, upon the rates of projected increase, the climatic effects of such projected increase, the sources of a carbon dioxide build-up, or the capacity of existing reservoirs in the global system to absorb future increases in carbon dioxide. Yet there is a clear warning from the scientific community that we need to begin now to manage the build-up of carbon dioxide in the atmosphere and to design and implement strategies for adapting to its impact. The problem becomes particularly important in light of United States policy favoring rapid development of our coal resources. In 1980 the President's Commission on Coal concluded that in order to reduce dependency on foreign oil, coal must replace oil and natural gas as the primary energy source.[6] Coal releases much more carbon dioxide than oil or natural gas.[7]

The build-up of carbon dioxide in the atmosphere challenges the international community to break new ground to handle the unique blend of political, economic, legal and scientific issues that it raises. How, in face of serious scientific uncertainties should/can States manage the emissions and release of carbon dioxide into the atmosphere, when many States contribute to the problem in widely varying degrees, when all States will be affected by the resulting climate changes but in different ways, when the activities contributing to a carbon dioxide build-up are central to energy and/or land-use practices of States, and when costly preventive strategies, to be effective, would have to be initiated at least a decade or more before the full effects of CO_2-induced climate change would be felt? What alternative strategies are available to States in adapting to CO_2-induced climate change? This paper outlines a research program designed to answer these questions.

2.1. Description of the CO_2 Problem

Conceptually, the global increase in carbon dioxide is a problem in the management of a scarce natural resource: the quality of the global atmosphere. Part of this is an international pollution problem in which states dump carbon dioxide into the atmosphere, which can lead to a decrease in the quality of the atmosphere. Since fossil fuel resources are limited and likely to be replaced by alternative energy supplies in the future, the steady build-up of carbon dioxide is a phenomenon which will probably take place only for the next fifty to one hundred years. The problem is thus one of managing the atmosphere during the transition from a fossil fuel economy to a non-fossil fuel economy.

Carbon dioxide is a gaseous by-product of the use of fossil fuels, of deforestation, or of other activities. It may thus be viewed as a pollutant.[8] In economic terms, countries that develop fossil fuel and emit carbon dioxide into the atmosphere are using the atmosphere as a "free good" in developing their own resources. They are not internalizing the cost of the diseconomies which they are inflicting upon the atmosphere by developing these resources. One can approach this problem either by regulatory mechanisms which seek to limit the amount of emissions, or by economic incentives, including taxes, which prompt contributors to take measures to limit the amount of emissions. Nationally, we have some experience in managing air pollution,[9] water pollution,[10] noise pollution,[11] and other forms of pollution.[12] Internationally, we have experience in dealing with pollutants which destroy the ozone layer,[13] with ocean pollution,[14] with international river pollution,[15] with outer space pollution,[16] and with air pollution.[17] Conceptually, the issues are not altogether new. What makes the carbon dioxide problem so difficult is that it is caused by many point sources of pollution and that the pollutants emerge as by-products of the use of critical natural resources – the consumption of fossil fuels and, to an extent yet unknown, the harvesting of forests and the misuse of soils. Moreover, the problem develops slowly with no immediate health or environmental effects, making it all the more difficult to convince decision-makers to take immediate action.[18] Approaching the CO_2 problem as a pollution problem reveals its basic nature – namely that it is a problem in energy management.

There appear to be two major sources for the build-up of carbon dioxide: (1) fossil fuels and (2) deforestation. The first is well-dcoumented, but there is considerable uncertainty about the net contribution of the latter.[19]

2.2. International Law and Carbon Dioxide Pollution

International law contains two principles which apply to the management of carbon dioxide accumulations: (1) a principle of "equitable use" which applies to countries using shared natural resources; and (2) a principle which makes states responsible for damage caused to the environment of other states in areas beyond their jurisdiction.

The Declaration on the Human Environment which was adopted by the United Nations Stockholm Conference on the Environment in 1972, provides that "States have, in accordance with the Charter of the United Nations and the principles of international law ... the responsibility to insure that activities within their jurisdiction or control do not cause damage to the environment of other states or areas beyond the limits of national jurisdiction."[22] A detailed analysis of the development of both legal principles can be found in Handl (equitable use)[23] and Weiss (State responsibility)[24] and is not repeated here. These principles have been adopted in some international agreements governing air and water pollution and the use of international rivers. They have been used to develop processes for information exchange, coordination, consultation and compensation for harm suffered.

Research is needed to attempt to formulate principles that could provide an acceptable basis and justification for international cooperation. This means examining the rather far-reaching implications of the principles of equitable use and state responsibility as applied to the carbon dioxide problem, and the likely problems that would be encountered in trying to implement them. We need to define specifically what equity means in the CO_2 context. Research should also identify and analyze alternative legal principles which may *de facto* come to govern the problem. It would be useful to study the development and application of international legal principles in two closely allied areas: management of international rivers and control of air pollution.

2.2.1. *International Rivers*. The underlying legal standard embodied – either explicitly or implicitly – in arrangements for managing international rivers and river basins is that of "equitable utilization". The principle is implemented primarily through water allocation and quality control mechanisms and compensatory schemes.

The Helsinki Rules on the Uses of the Waters of International Rivers adopted by the International Law Association in 1966 impose a duty on states to prevent future pollution of international drainage basins[25] and requires violators to cease the wrongful conduct and compensate the injured co-basin state.[26] The rules also suggest that states should take "all reasonable measures" to abate existing problems[27] and enter into negotiations to achieve this end.[28] The 1909 Treaty Between the United States and Great Britain Relating to Boundary Waters and Questions Arising Between the United States and Canada, while focusing on preventing boundary disputes and overseeing water diversion plans, prohibits transboundary water pollution.[29] The 1978 Agreement Between the United States and Canada on Great Lakes Water Quality picks up on this provision and requires the parties to use "best efforts" to ensure water quality standards are met.[30] The International Joint Commission, a fact-finding body established to implement the 1909 Treaty,[31] assists in implementing the 1978 Agreement.[32]

As between the United States and Mexico, the 1889 Convention[33] and the 1944 Treaty on the Utilization of Waters[34] govern water allocation and boundary line disputes. An International Boundary and Water Commission implements the agreements.[35] The United States and Mexico are expanding their efforts to include measures for water quality control. The Memorandum of Understanding between the Subsecretariat for Environment Improvement (SMA) of Mexico and the United States Environmental Protection Agency (EPA) for Cooperation of Environmental Programs and Transboundary Problems, signed on June 6, 1979, provides for consultation and an exchange of experts to resolve environmental problems.[36] It calls for periodic meetings and parallel efforts at research and monitoring of pollution. These two agencies are also to devise an early warning system for potential environmental problems.[37]

There are a number of river basin agreements designed to promote the harmonious development of the region, such as the Treaty on the Plate River Basin of 1969,[38] to which Argentina, Bolivia, Brazil, Paraguay and Uruguay are contracting parties. The agreements traditionally provide for a committee to coordinate use of water resources.[39] The problem is that these agreements usually exist only in form.

Several Western European countries – France, West Germany, Luxembourg, Switzerland and the Netherlands – have undertaken measures to prevent and abate pollution of the Rhine.[40] Under the Convention on the Protection of the Rhine Against Chemical Pollution, states agree to eliminate or reduce to agreed-upon emission standards the discharge of certain enumerated pollutants.[41] The International Commission for the Protection of the Rhine Against Pollution coordinates implementation.[42] The more recent agreement, the Convention on the Protection of the Rhine Against Pollution by Chlorides, has yet to go into effect.[43] This Convention calls for the reduction of the chloride content in the Rhine waters along the Netherlands/German border through the injection of chloride wastes into the sub-soil of Alsace.[44] Potassium mining in that region is the primary source of this pollutant.[45] The French safeguard is its unilateral right to cease injections when such injections appear to pose a "serious danger to the environment".[46] Nonetheless, this Convention has met with considerable opposition within France. Consequently, the French government has been unable to ratify the agreement in Parliament, much to the distress of the Dutch.[47]

It would be useful to survey briefly the international river and river basin management agreements – especially those focusing on water quality preservation – for the lessons that they provide for efforts to manage carbon dioxide accumulations. In general, these agreements have shown that upstream users put downstream users at their mercy. Nevertheless, past experience between the United States and Canada and between the United States and Mexico might yield insights useful in designing an appropriate framework to control CO_2 pollution. A review of experience with the Rhine Conventions would demonstrate the political obstacles that arise to implementation of multilateral efforts to control pollution.

2.2.2. *Air Pollution*. Existing conventions concerned with air pollution, like water pollution agreements, break down into two basic groups. The first imposes an obligation on the contracting party to inform and consult with another state when activities within the jurisdiction of that state may adversely affect the environment of the other state. It also contains arrangements for joint and/or coordinated research programs and data exchanges. Classic examples include the 1979 U.N. Economic Commission for Europe (ECE) Convention on Long-Range Transboundary Air Pollution[48] and the 1976 Agreement on Monitoring of the Stratosphere.[49] The second kind of agreement goes one step further to include provisions for dispute resolution, such as a right of access to domestic courts or administrative bodies or establishment of an impartial fact-finding commission. The Nordic Convention on the Protection of the Environment[50] and the United States – Canadian International Joint Commission's Michigan-Ontario Air Pollution Board[51] are good examples of the latter kind.

The most recent international agreement is the Convention on Long-Range Transboundary Air Pollution, which has been signed by thirty-five countries including the United States, Canada and most of Eastern and Western Europe.[52] The agreement imposes an obligation on contracting parties to consult upon request about activities which affect or pose a "significant risk" of long-range transboundary air pollution.[53] The legal duty to combat the discharge of air pollutants is admittedly weak. States are merely required to act "without undue delay."[54] To its credit it contains lengthy provisions on research, montoring and information exchange, which are aimed at mitigating SO_2 emissions.[55]

The 1976 Agreement on Monitoring of the Stratosphere – to which the United States, France and the United Kingdom are contracting parties – concentrates solely on increasing scientific understanding of the ozone layer.[56] It requires the parties to collect, exchange and analyze information on the stratosphere, and to fully integrate their activities with the existing international networks of the World Meteorological Organization and the UN Environment Programme.[57] One purpose of the agreement is to demonstrate the feasibility and utility of collaborative international action in this area.[58]

The Nordic Convention, in contrast to the above-mentioned agreement, provides a framework for abatement and compensatory relief for persons injured by transboundary air and water nuisances.[59] It requires the contracting parties – Denmark, Finland, Norway and Sweden – to accord to non-citizens equal access to administrative agencies and domestic courts and non-discriminatory treatment.[60] It accords the right to any person who is affected or may be affected "by environmentally harmful activities" to ask for measures to prevent damage.[61] In theory, this process might be applied to the carbon dioxide problem to enjoin activities until carbon dioxide standards are met. The Nordic Convention also allows compensatory relief[62] but such a mechanism is probably not practical for the carbon dioxide situation.

Coordinated monitoring and emission regulations, on the other hand, is not only viable but has been utilized in other transboundary air pollution situations. The International Joint Commission (IJC) between the United States and Canada undertook in the early 1970's to mitigate air pollution problems between Michigan and Ontario.[63] To assist its efforts, the IJC established, in 1976, the International Michigan-Ontario Air Pollution Board.[64] The function of this board is to coordinate the implementation of air pollution control programs including setting a minimum basis for emission standards.[65] As a result of this effort, the air quality in the area seems to have improved with the percentage of air quality readings failing to meet the IJC objectives declining throughout the region.[66]

International agreements concerned with managing air pollution offer at least some limited positive experiences to draw upon in developing a framework for managing carbon dioxide accumulation. It would be useful to analyze them further to define the extent to which they are in fact valuable models. They suggest that international scientific cooperation in the gathering of data and in monitoring CO_2 levels is a desirable and feasible step. We should explore the possibility of including the monitoring of CO_2 within existing networks for monitoring sulphur dioxide, chlorofluorocarbons, and other air pollutants.

In addition to the formal agreements concerned with air and water pollution, there are a number of arbitral decisions and negotiated settlements which incorporate the principles of equitable use and state responsibility for environmental harm. The Trail Smelter Arbitration is one of the most frequently cited cases. In this case an arbitral tribunal held Canada liable for damage in the State of Washington from the fumes emitting from a Canadian smelting company. The final decision of the tribunal in 1941 declared that "The tribunal, therefore, finds .... under the principles of international law, as well as of the law of the United States, no state has the right to use or permit use of its territory in such a manner as to cause injury by fumes in or to the territory of another or the properties of persons therein, when the case is of serious consequences or the injury is established by clear and convincing evidence."[67] Decisions of arbitral tribunals are important indicators of international law, but they are not binding and derive much of their force from the extent to which the principles enunciated therein are incorporated in future decisions or agreements.

A number of diplomatic settlements at least implicitly recognized the state responsibility for the consequences of pollution inflicted upon other countries. For example, Mexico complained that the water received from the Colorado River in the United States was too saline to be useful to Mexico and hence in violation of the 1944 treaty between the two countries. The Mexicans argued that the Treaty included a water quality standard, a point disputed by the United States. In settling this dispute with Mexico, the United States agreed to provide compensation in the form of assistance to rehabilitate the Mexicali Valley for the damages suffered from the saline pollution, although it did not actually acknowledge any obligation to do so in international law.[68]

The resolution of the United Nations General Assembly on Natural Resources Shared by Two or More States calls for an exchange of information and for prior consultation in using these resources.[69] States engaging in activities which potentially may cause pollution harmful to other states may have a duty to consult with those states or at least offer them an opportunity to engage in consultation. The value of consultation is that it offers a process by which to minimize harm and political conflict, and in the best of cases, to negotiate mutually acceptable arrangements between the concerned states. Research should explore the implications of these duties in the context of the CO_2 problem and attempt to define what they would require.

2.3. Preventive Management Strategies

The increasing concentration of carbon dioxide in the atmosphere may be viewed as a global pollution problem caused primarily by point sources of pollution. If it turns out that the main polluters (i.e. the main producers of CO_2) are also the countries that would be most hurt by a change in climatic conditions triggered by a CO_2 build-up, then it would be in their interest to band together to develop an allocation regime for CO_2 emissions in order to delay and/or avert such climatic change. If, on the other hand, there are important contributors to the CO_2 concentration in the atmosphere that stand to benefit from CO_2-induced climate changes, it would be very difficult to develop an effective allocation regime.

This is complicated by the fact that most of the scientific factors are still unclear: the relative importance of different sources for the higher CO_2 concentration in the atmosphere, the eventual fate of CO_2 in the global system, and the impact of a given level of CO_2 on the climate of specific regions and countries. What is more, the elucidation of these underlying phenomena will probably require at least a decade of scientific research.

For this reason, predictions of the political consequences of future climatic change are useful as illustrations of the kinds of political alignments that are likely to take place as the world comes to grips with the carbon dioxide problem. Fortunately, the impact of carbon dioxide accumulation is still distant enough so that we may be able to wait for better scientific information before taking significant political action.

Bearing in mind this caveat, preliminary data indicate that the United States and Europe are major contributors to the carbon dioxide increase, and that CO_2-induced climate change could impact them adversely.[70] In particular, U.S. agricultural production is predicted by some calculations to decrease as a result of warmer, drier weather in the grain belt,[71,72] while portions of the U.S. could be permanently submerged in the less likely event of the melting or collapse of the polar ice caps.[73] These predictions depend on uncertain scientific theories, and do not take into account the ability of the U.S. to respond by technological innovation, i.e. by developing new crop varieties or species adapted to the new ecological conditions. Still, the possibilities enumerated above do suggest that the U.S. has considerable incentive to reduce CO_2 accumulations if this is possible. From this projection, it appears to be in the interest of at least the United States and Europe to explore immediately strategies for managing carbon dixoide emissions during the coming century.

Under some calculations, countries located in monsoon regions are predicted to benefit as a warmer global temperature will trigger more favorable rains with accompanying improved crop production.[74] If these predictions are accepted, the possibility of such benefits can significantly affect the formation of an international consensus to control CO_2 emissions. Moves by the United States to limit CO_2 "pollution" would understandably be seen as another example of "environmental imperialism", especially to those beneficiaries who would not view the problem as "pollution" at all. Moreover, as noted earlier, Rotty has predicted that developing countries may be significant contributors of CO_2 emissions by the year 2025.[75] Unless the United States and other developed countries responsible for most of the CO_2 emissions have initiated measures to limit the CO_2 build-up in the atmosphere, maximum U.S. remedial initiatives later may have only minimal effect in abating the problem, if any.

One of the major contributors of CO_2 emissions today, the U.S.S.R, is a possible net gainer from the projected climate change. Warming global temperature is predicted to increase agricultural productivity by lengthening the growing season for some now unproductive lands.[76] China, which holds a major share of the earth's coal, would also benefit agriculturally. Rice yields are predicted to increase and multiple growing seasons are possible.[77] If these predictions are accepted, there may be few incentives for the U.S.S.R and China to engage in international cooperative action to limit CO_2 accumulations.

Predictions as to the likely effect of a given increase in carbon dioxide on specific regions are still far from certain and more research is needed. If we assume that the United States and other developed countries in the Northern Hemisphere either would benefit from the climatic change or could easily adapt to it and that areas in the developing world will become increasingly arid and agriculturally unproductive, then there may be significant incentives for international cooperation flowing from the demands of the developing world for a redistribution of the weath.[78]

An analysis of possible effects of an increase in carbon dioxide on key CO_2 contributors – those emitting now and in the future – is essential. But we should also have an "interest-analysis" for all countries or regions to assess the incentives of States to participate in international cooperative actions.

As a first step in devising appropriate strategies for managing carbon dioxide emissions, we need to obtain data which will give us a fuller understanding of who the important actors are and what proportion they are contributing to the pollution problem. This means that we need data which give a breakdown by states of major contributors to the increased carbon dioxide pollution, both now and projected into the future, and a breakdown which shows how much of the pollution is from fossil fuels, and how much from deforestation and how much from still other sources. Within the fossil fuel category, we need to know which fuels contribute most to the carbon dioxide build-up. Rotty (1979)[79] has published a table which gives global carbon dioxide production by world segments for the year 1974 (and projects it to the year 2025), while Steinberg, Albanese and Vi-Doung (1978)[80] have prepared a table showing carbon dioxide generation as a function of fuel source. Such tables are essential for later assessments of appropriate institutional arrangements. It will also be useful to have a breakdown in the use of fossil fuel as to residential heating, gasoline, petrochemicals, and other fossil fuel products, and what part of this use involves coal.

Carbon dioxide as a pollution problem is foremost an energy policy problem. This means that it is important to analyze the workings of the current oil economy, to investigate the private sector and the public sector outlook for the development and marketing of coal resources, and to consider and involve the private sector in the management of the carbon dioxide problem. The international oil companies have substantial investments in coal and oil shale, so they will be important actors in the future as well as today. More importantly, we need to explore with industries that use fossil fuels, the feasibility of technological innovations to reduce or recycle CO_2 emissions and the incentives needed to stimulate innovation. To date little attention has been given to how to involve private industry in a resolution of the carbon dioxide problem.[81]

Much of the literature tends to discuss the problem in policy terms that suggest an immediate switching to a non-fossil fuel economy or leaps to a discussion of managing the impacts of future climatic change.[82] Certainly these are policy elements that ought to be evaluated. The question that has been largely overlooked and that needs to be addressed is how we go about limiting carbon dioxide emissions from fossil fuels to a "manageable" amount that will delay the projected warming and/or allow time to develop new methods for preventing the release of CO_2 or recycling it, mitigating its climatic effects and adapting to the resulting climatic change. This means we need to know what kind of output of carbon dioxide would result in what kind of build-up in the system, and what rate of build-up of carbon dioxide can be absorbed over what periods of time. We can work from there to ask what measures may be necessary to ensure that states do not exceed these amounts in their production and use of fossil fuels. Given these calculations, it would then be possible to prepare for worst case scenarios, such as the utilities do in planning for fuel consumption during the winter season.

The management of carbon dioxide pollution requires processes for risk assessment and impact evaluation, and preventive strategies at the national and international level.

2.3.1. *National Preventive Strategies.* Carbon dioxide pollution raises the possibility of significant, adverse and irreversible climatic impacts, which could be catastrophic to certain parts of the economies of some countries. Yet, as discussed earlier, most of the important factors are still scientifically unknown and in dispute. This requires states to balance the economic risks of taking action at the present time in the face of scientifc uncertainty with the economic risks of not acting in the face of that scientific uncertainty. Since the former is usually easier to quantify, decisions are usually struck for the latter. It would be useful to examine analogous situations, such as Project Westford, which dispersed very fine copper needles into space, or the early NASA space experiments where officials had to decide what confidence levels they required in assessing potential risks from the experiments. We need to examine the sufficiency of the existing network for monitoring carbon dioxide, and to explore measures for disseminating the information in a meaningful way to concerned decision-makers.

The United States Federal Government can take steps to control carbon dioxide emissions, if needed.[83] It would be useful to have an identification of the measures potentially available, and an analysis of whether or not they are desirable or feasible and at what cost. Certainly costs would need to be carefully assessed before initiating any action. Substantive measures which might be taken include the following: (a) requiring any federal environmental impact statement concerned with a project that relies on fossil fuels to identify explicitly the impact of carbon dioxide emissions into the atmosphere;[84] (b) exploring the feasibility of the development and utilization of scrubbers to reduce carbon dioxide emissions or of methods to recycle carbon dioxide and analyzing the experience with the provisions of the U.S. Clean Air Act as they affect innovation in industry; (c) providing incentives to industry to engage in research and development which could lead to measures to reduce carbon dioxide emissions, and analyzing the export market potential for such measures; (d) imposing substantive limits on the production and consumption of coal (or other fossil fuels) and on the export of this resource to other countries, particularly Western Europe: (e) investigating other measures, such as provisions in the terms of lease for development of coal deposits and oil shale, which could provide for public intervention if needed, to manage the level of carbon dioxide emissions; and (f) rationing of fuels on the basis of carbon dioxide generating power, with exemptions for plants with scrubbers or other means of limiting CO_2 emission. Whether or not CO_2 emission control is a technically feasible goal is still unclear and controversial. Such controls will be adopted only if the costs to economic growth are acceptable, and these costs still need to be defined.

Any analysis of measures to prevent carbon dioxide emissions from exceeding "acceptable" limits must include an analysis of the relevant decision-makers within a country. In the United States these would include the various branches of the federal government (i.e., Department of Energy, Environmental Protection Agency, Council on Environmental Quality, National Oceanic and Atmospheric Administration, Department of Interior, National Aeronautics and Space Administration, Department of Defense), Congress, private industry, state governments, and consumers. An analysis of how environmental issues have developed in the context of energy resource development and management would be a useful input into any analysis of possible preventive strategies for carbon dioxide emissions.

2.3.2. *International Preventive Strategies.* Since the major contributors to the carbon dioxide build-up are presently only a handful of developed countries,[85] it should be possible initially to focus discussion among these primary contributors. This means the United States and Western Europe, and if possible, the Soviet Union and Eastern Europe. Initial consultations might begin with the United States and Western European countries under the umbrella of the European Economic Community or of the Organization for Economic Co-operation and Development, one component of which is the International Energy Agency. Again, it would be necessary to analyze the role that private industry could play. The oil market is an international one, and the coal market promises to be likewise. Discussions might focus on emission controls, such as they have focused on emission controls in controlling air pollution in the United States[86] and in Europe.[87]

The OECD's Environment Committee at its meeting in May 1978 recognized the potential concern of a carbon dioxide build-up from coal production.[88] The Council on Coal and the Environment specifically recommended that "member countries, in the light of appropriate research results, seek to define acceptable fuel qualities, emission levels or ambient media qualities, as appropriate for carbon dioxide."[89] This could encompass a scheme to control carbon dioxide emissions similar to that employed in the United States for air pollution: an international ambient standard for carbon dioxide, together with emission limitations designed to meet that ambient standard, which could be implemented at the national level.[90] Such a scheme could also be patterned after national programs to control water pollution, which focus mainly on effluent limitations and use of best available control technology.[91] Controls on emissions should be keyed to acceptable carbon dioxide limits which have been determined scientifically to be linked with given levels of temperature increases in climate. Admittedly this requires considerable advances in our scientific knowledge, but obtaining sufficent knowledge to establish such limits should be a primary objective. Allocation of emission limitations between countries would be a highly political decision, one which could prove to be intractable. Applying emission limitations within each country would entail difficult political decisions about who had the right to contribute what amount of carbon dioxide to the atmosphere.

Predictions by Rotty (1979) indicate that by the year 2025, 40% of the carbon dioxide build-up may be coming from the developing countries, who now contribute less than twenty percent to the problem.[93] Even if the projected contributions of the developing countries were considerably less, it would be essential to involve the participation of the rest of the world in any system of ambient standards and controls that is established. Initially this means, at a minimum, efforts to build international consciousness of the problem and to elicit participation of other countries at conferences which discuss the problem. Any system that is developed for limiting carbon dioxide concentrations in the atmosphere will eventually need to be carried forward with the participation of the developing countries. We will then be faced with issues similar to those that arose at the Stockholm Conference on the Environment in 1972, in which developing countries expressed legitimate concern over the potential conflict between development strategies appropriate for their countries and the concerns of the international community which would make those strategies either more expensive or would otherwise impede their development.[94]

Several international institutions are concerned with the carbon dioxide build-up, and an in-depth assessment of their potential role in the future should be done. These institutions include the World Meteorological Organization, the United Nations Environment Program (UNEP), the U.N. Educational, Scientific, and Cultural Organization (UNESCO), the International Council of Scientific Unions (ICSU), the U.N. Food and Agriculture Organization (FAO), the International Institute for Applied Systems Analysis (IIASA), the Organization for Economic Co-Operation and Development (OECD), the European Economic Community (EEC), and the NATO Committee on the Challenges of Modern Society. To date their role has been limited to identifying the problem, monitoring and evaluating scientific data, collecting and exchanging information regarding the build-up, and the facilitation and coordination of national and international programs related to it.[95] It would be useful to assess the extent to which these organizations, or an appropriate mix of them, might be employed in analyzing the impact of various policy choices regarding the control of carbon dioxide pollution.

2.3.3. *Preventive Strategies for Deforestation.* In order to develop appropriate management strategies for a carbon dioxide build-up from deforestation, it is necessary to identify those forests in the world which are likely to be deforested, and to assess the amount to which deforestation does, in fact, contribute to the carbon dioxide build-up. The problem is to determine the direction and magnitude of the biospheric signal. Are the forests of the world being cut down at a rate that exceeds reforestation and that exceeds the progressively faster growth rates of forests that may result from increasing levels of carbon dioxide? The dispute within the scientific community regarding the contribution of deforestation makes it especially difficult to develop strategies to manage it. After we have identified the geographical areas of concern, we need to identify and assess alternative ways to slow down the rate of deforestation, or to counteract its effects. Some argue that it is not deforestation *per se* that produces CO_2, but the destruction of the organic matter of the soil that follows misuse of cleared land. If true, this would be the issue for international attention.

The problem for international attention is the development and implementation of environmentally sound strategies for sustained yield management of the forests and the soils. Any success in controlling CO_2 emissions is likely to come as a by-product of progress on this general problem. A regime of separate emission limitations for CO_2 from the deforestation of tropical forests is not likely to be a useful or viable approach.

Any effort to control deforestation frontally assaults cherished notions of national sovereignty over the exploitation of a state's natural resources. For this reason, states can be expected to resist such measures. A first and necessary step will be to encourage efforts to raise international consciousness about the implications of forest and soil management strategies for levels of carbon dioxide in the atmosphere, the impact of these levels on global temperatures, and the long-run implications of a rise in temperature for the economy and well-being of individual countries.

The community of actors in deforestation differs from that of fossil fuels. Most of the forests are in the developing world. Multi-national companies and governments of at least some of the developed countries are frequently responsible for the clearing of forests. Thus, the appropriate forum for discussing this problem would be a broader based community then either the EFC or OECD, perhaps UNEP.

Some international attention has already been focused on the problem of deforestation. In 1979, the Governing Council of the United Nations Environment Program adopted a resolution, introduced by the United States, which called for a meeting of experts to develop proposals for an integrated international program on the conservation and wise utilization of tropical forests and to report to the UNEP governing council in April 1980.[96] UNEP sponsored a conference on tropical forests in February 1980 in Kenya.[97]

In the United States, former President Carter's Environmental Message to Congress on August 2, 1979, addressed two global environmental problems – acid rains and deforestation of tropical forests.[98] The President referred to estimates that the world forest could decline by 20% by the year 2000 and noted, "Forest loss may adversely alter the global climate through the production of CO_2. These changes and their effects are not well understood and are being studied by scientists but the possibilities are disturbing and warrant caution."[99] Subsequently, the United States introduced a resolution to the UNEP Governing Council calling for UNEP to discuss the problem.[100] In a memo to the Secretary of Agriculture, the President directed that "high priority be given to: (1) improved monitoring of world forests; (2) research on preservation of natural forest ecosystems; (3) research on tropical forest multiple use management; (4) studies on increasing yields of tropical agriculture; (5) demonstration of integrated projects of reforestation, efficient fuel wood use and alternative energy sources; and (6) examination of how U.S. citizens and corporations can be encouraged to follow sound forest management practices."[101] An interagency task force was established in November 1979 on deforestation to issue a report in 1980.[102] It allegedly recommended that deforestation assessment be a component of environmental impact statements and that any overseas U.S. projects by federal agencies and private institutions contain assessments of the effects of the planned activities on tropical forests.[103]

Efforts can also be made at the regional level to address the problem of deforestation. In South America, the new Amazon Pact[104] offers a possible umbrella for such discussions and for an exchange of data. It is important to assess the potential value of existing institutional arrangements for managing the problems associated with carbon dioxide and to analyze in detail the applications of international legal principles to the issues.

2.3.4. *Preventive Strategies for Chlorofluorocarbons.* The experience of the U.S. in attempting to limit emissions of chlorofluorocarbons, which may deplete the ozone layer, offers useful insights into the problems associated with limiting carbon dioxide emissions. In March 1978, the Federal Drug Administration and Environmental Protection Agency (EPA) issued regulations banning the manufacturing and shipping of fluorocarbons for non-essential aerosol uses.[105] The ban is expected to reduce total chlorofluorocarbon emissions by considerably less than 25% globally.[106] The Clean Air Act requires EPA *inter alia*, to conduct studies of the ozone problem in an effort to determine what further regulation of clorofluorocarbons is necessary.[107] In October 1980 EPA proposed a set of rules which would impose mandatory emission controls for users of chlorofluorocarbons and would limit chlorofluorocarbon production through distribution of marketable production permits.[108] In response to comments to the proposed rulemaking, EPA has backed off from these approaches and is now trying to develop with NASA an early warning system to detect ozone depletion. The Clean Air Act also requires the National Oceanic and Atmospheric Administration to establish a contrinuing program of research and monitoring of changes in the stratosphere and of climate effects resulting from such change.[109] The results to date indicate that chlorofluorocarbon concentration is increasing at five to ten percent per year, but do not show positive evidence of chlorofluorocarbon depletion of the ozone layer.[110] The impact of the aerosol ban may not be visible for a number of years.

Action has also been taken in the European community to limit the use of chlorofluorocarbons in aerosol sprays. On December 17, 1979, the Environment Ministers of the EEC adopted a proposal which would reduce chlorofluorocarbons in aerosol sprays by 30% of 1976 levels over the next two years (December 1979 to December 31, 1981), and which would limit production of chlorofluorocarbons, F-11 and F-12, to present levels.[111] Several delegations pushed for a stricter standard, since the standard is less strict than in some existing national laws. Support for a stricter standard also came from two other directions: (1) the European Parliament's Environment, Public Health and Consumer Protection Committee, which adopted a resolution at a meeting in Brussels, Noevember 1979, calling for a 50% reduction in the use of chlorofluorocarbons by 1981, and a total ban by 1983,[112] and (2) the conclusion of UNEP's International Scientific Community at its third session in Paris, November 1979, that aerosol propellants, such as hydrofluorocarbons, could destroy 15% of the ozone layer within one hundred years if present levels are maintained.[113] The EEC called for a re-examination of the problem "in light of available scientific and economic data" in order to adopt necessary and new measures by June 30, 1981. By contrast, the OECD has limited its involvement to studies of the economic impact of chlorofluorocarbon regulation. UNEP has assumed the role of co-ordinating scientific research on the subject for the international community.

Some countries have acted unilaterally. Sweden banned aerosol manufacture and importation in December 1977.[114] West Germany, through cooperative government-industry consultation, has been able to get German industries to agree to reduce aerosol chlorofluorocarbon use by 25% by 1979 and 50% by 1981.[115] Canada initiated "voluntary" reductions of aerosol use of chlorofluorocarbons.[116]

Developments in the regulation of chlorofluorocarbons offer some insight into the issues that will be raised in regulating carbon dioxide emissions. On the one hand, it should be possible to develop further and maintain the scientific network necessary for proper monitoring of carbon dioxide. On the other hand, it will be very tough to get the international agreement necessary to manage a global pollution problem. The countries that have been most willing to regulate chlorofluorocarbons, except the United States, have been those which contribute least to the problem. Canada is responsible for only 2% of the world's chlorofluorocarbons and Sweden does not manufacture any.[117]

The central problem in the transnational regulation of chlorofluorocarbons has been the inability of countries to agree upon a strict standard limiting the use of certain aerosol sprays containing these chemical compounds. In many ways the problems are the same. The number of countries which are major contributors are small – the developed countries. The effect of polluting the ozone layer with chlorofluorocarbons is the depletion of a global resource. The effect of an increase in carbon dioxide is a global rise in temperature. Yet the modest success both in the United States and most recently in the EEC in achieving at least some form of regulation on the emission of chlorofluorocarbons offers some hope for building such measures to control the emission of carbon dioxide. Carbon dioxide poses a much tougher problem. Unlike chlorofluorocarbons, its source is at the heart of a country's economy – the production and use of fossil fuels for energy, or the removal of its forests.

2.3.5. *Adaptive Strategies for Carbon Dioxide Pollution.* If climate change occurs as a result of temperature increases from carbon dioxide, many of the adaptive strategies are likely to focus on preventing the problem from getting worse. Our experience with responding to problems of air and water pollution suggests that our adaptive response to the impacts of pollution is to try to diminish the amount of pollution. Thus, one component of the adaptive strategy is likely to be an intensified search for technical fixes, such as ways to expand the capacity of oceans to absorb carbon dioxide. But we will also need to anticipate the effects of possible climate change, particularly upon water supplies and migration patterns, and upon the general dislocation of a country's economy.

How states adapt to climatic change will depend on whether or not the changes are beneficial or adverse to their economic and strategic interests. We need to know who is likely to be affected how and when by what degree of climatic change. In the absence of hard data, probably the best contribution is to explore the consequences of various credible climate scenarios. It will be in the interest of those countries that are hurt to join together in measures to alleviate the stress caused by climate changes. The international and legal institutional literature concerned with management of scarce natural resources and with responses to natural hazards should offer useful insights into alternative strategies and into the shortcomings of such strategies. We need to address the implications of possible CO_2-induced climate change for our international economic and monetary system, and for international political stability. This means research into strategies for adapting to CO_2-induced climate change, such as policies to increase more efficient use of water resources, compensatory mechanisms to help economies vulnerable to climate change and programs to manage the disruption of large-scale population migrations. It also suggests research into measures to encourage a stable international political and legal order that can respond to long-term CO_2-induced climatic change.

3. Proposed Research Strategy

The objective of the research outlined in this strategy is to promote an understanding of the principles of law and of institutional design which must underlie any international regime to deal with climate effects of carbon dioxide emissions. This problem displays a unique set of characteristics and hence poses a novel problem to the international system. On the other hand, there are instructive parallels in other problems in international law and politics. A study of the unique aspects of the CO_2 problem and of the instructive elements from other problems comprise a small, but important, research program.

From the point of view of the international lawyer, a regime to handle the problem of CO_2-induced climate change must consider the following (1) CO_2 as a pollutant, the emission of which by one country may, on balance, harm another; (2) CO_2 as a non-point source pollutant produced by a multitude of relatively small emitters; (3) "clean air" as a global atmospheric resource to be managed internationally; (4) CO_2 emission control as a neglected technology for which innovation should be encouraged by public policy at the national and international levels; (5) climatology as a relatively undeveloped science of general interest to the world community at national and international levels; (6) climatology as a rapidly developing science, the results of which should be fed into relevant policy-making bodies at national and international levels; (7) climate warming as a long-term development affecting precipitation patterns and agricultural productivity; (8) climate warming as a long-term development affecting migration patterns and economic comparative advantage; and (9) climate warming as a long-term development affecting balance of political power between countries.

(1) *CO_2 as a pollutant, the emission of which by one country may harm another.*

The situation is familiar in transboundary pollution cases, such as the Trail Smelter Arbitration between the United States and Canada, and in current national and regional air and water pollution problems. In this context we should examine U.S. experience with an ambient air quality standard and pollutant specific emission standards to control air pollution and European efforts to control water pollution on the Rhine. The acid rain problem in Europe and between Canada and the United States raises similar issues.

(2) *CO_2 as a non-point source pollutant produced by a multitude of small emitters.*

This is familiar in a national context – e.g. ground water pollution, storm run-off and smog control. It is familiar in the international context through the threatened depletion of the ozone layer.

(3) *"Clean Air" as a global atmospheric resource to be managed internationally.*

One of the best examples of international regulation of a common resource is the World Administrative Radio Conference (WARC) for frequency allocation and the regime for orbital space. We also have considerable experience with managing scarce living natural resources, such as fisheries, polar bears, fur seals, whales and endangered species. For specific problems of data collection, see reports of the U.N. Food and Agricultural Organization and the new Agreement on Monitoring the Stratosphere.

(4) *CO_2 emission control as a neglected technology for which innovation should be encouraged by public policy at the national and international levels.*

Innovation at the national level is frequently encouraged by direct incentives (e.g. the French nuclear or computer programs), or by regulation (e.g. U.S. approach to automobile mileage and emissions). Examples of international collaboration for technological innovation are found in The Consultative Group on International Agricultural

Research (CGIAR) and in various OECD programs (such as that for barnacle control). International administrative regimes can also be a major influence on the direction of innovation, as exemplified by the WARC.

(5) *Climatology as a relatively undeveloped science of general interest to the world community at national and international levels.*

Examples of large-scale international cooperation on mission-oriented basic research include the Global Atmospheric Research Program (WMO-ICSU), the training and research program on tropical diseases (World Health Organization) and international cooperation on fusion research.

(6) *Climatology as a rapidly developing science, the results of which should be fed into relevant policy-making bodies at national and international levels.*

Examples are the impact of advances in stratospheric chemistry on the use of aerosols and the ongoing controversy over the need for guidelines on recombinant DNA research.

(7) *Climate warming as a long-term development affecting precipitation patterns and agricultural productivity.*

The World Food Conference and the institutions created there are an example of global reaction to climate-created crisis in the agricultural sector. By contrast, the World Water Conference had little impact on national or international practice, despite the need to treat water resources as a scarce resource. A few states, such as Israel, have been developing methods for expanding water supplies by using water more efficiently and turning to less water-intensive technology. In most countries drastic changes in the administrative framework or pricing system and in some cases major investments in physical superstructure would be needed to implement such strategies.

(8) *Climate warming as a long-term development affecting economic comparative advantage and population migration patterns.*

Here the politics of the U.N. Industrial Development Organization (UNIDO), the U.N. Conference on Trade and Development (UNCTAD), and the General Agreement on Tarrifs and Trade (GATT) show the world's hitherto unsuccessful efforts to deal with the North-South implications of shifts in global comparative advantage. There are many elaborate codes on trade practices, which have been set up to cope with consequences of shift in comparative advantage. There are also domestic arrangements – e.g. dislocation allowances – to cope with human consequences of obsolescence. States have considerable experience in dealing with interstate population migrations occasioned by wars, natural disasters (including droughts), and the search for economic opportunity.

(9) *Climate warming as a long-term development affecting balance of political power between countries.*

To study the consequences of shifts in relative power among countries is to study the whole history of geopolitics. One example of substantial shifts in political power following economic change is found in the greater influence of OPEC countries after 1973, as reflected not only in voting power in a few international institutions, but in the new status of Saudi Arabia and Iran as global powers. There may be examples of countries seeking to forestall long range technological or other changes through international diplomacy, because of the fear that it will diminish their political powers.

This research would provide systematic and comprehensive analysis of our experience with a variety of legal, institutional and political measures in partially parallel situations. This rich background would be intended to offer insights useful in the analysis of the issues as they arise specifically in the CO_2 context. Given that the CO_2 problem is *sui generis*, each of the parallels should not be explained exhaustively, but only in the detail necessary for this purpose. For some it will be enough to note obvious points; a few areas will deserve detailed investigation, perhaps with case studies.

The proposed research should then proceed to in-depth analysis of the unique blend of issues raised by the build-up of carbon dioxide in the atmosphere. This means direct reasoned consideration of the following issues.[119] What are the incentives and disincentives for international cooperation in the specific context of the CO_2 problem, and what means are available to induce "abstainers" to cooperate? Should we attempt to apply generally accepted principles of international law to this problem and if so, what should the content be? What is the desirability and feasibility of alternative preventive and adjustment strategies under varying scenarios of national interest? What institutional framework would be most appropriate for the various arrangements that may be necessary? What is the process for developing the necessary institutional framework? To what extent can we build upon existing national and international institutions? How do we prepare ourselves to make reasoned policy decisions in the face of scientific uncertainty, when such decisions to be effective must be initiated far in advance of when the impact of climate change will be felt? Given the divergence between national and global interests, how might a global consensus realistically be obtained? If we view CO_2 in a broader international policy framework, what are the important trade-offs with energy policy, land use practices, and still other policies which will affect the choice of policy instruments for the carbon dioxide problem?

If the research proposed can begin to help us understand the answers to these questions, we will be better prepared to determine what we should and can do to respond to the rising levels of carbon dioxide in the atmosphere.

For this research program, personnel should be experienced in interdisciplinary work, so that they can easily pose questions of relevance to the scientific community and to the other related disciplines. Facilities should be such that there is ready access to the most current national initiatives in controlling air and water pollution and to the initiatives and thinking of the international community, including the EEC. Research personnel should also have sensitivity to the policy process, both national and international, in order that the assessment of alternative strategies be maximally useful to those concerned with decision making during the next century.

It is essential that a comprehensive research strategy for the CO_2 problem involve social scientists from the beginning. The goal should be not only to increase our knowledge about the carbon cycle and the effects of carbon dioxide on climate, but also to understand the human problems associated with rising levels of CO_2 and to try to do something about them. This requires a multi-disciplinary research effort. The research proposed is one component.

Georgetown University Law Center

Notes

* The author wishes to thank Ms. Michele Giusiana for her research assistance in the preparation of these materials.

[1] Climate Research Board, National Research Council, U.S. National Academy of Sciences, *Carbon Dioxide and Climate: A Scientific Assessment* (1979), [hereinafter cited as National Research Council Report]; G. Woodwell, G. MacDonald, R. Revelle and C. Keeling, *The Carbon Dioxide Problem: Implications for Policy in the Management of Energy and Other Resources* (1979), [hereinafter cited as CEQ Report], *reprinted in* Senate Comm. on Governmental Affairs, 96th Cong., 1st Sess., Symposium on Carbon Dioxide Accumulation in the Atmosphere, Synthetic Fuels and Energy Policy 44 (1979) [hereinafter cited as CO_2 Symposium].

[2] National Research Council Report, *supra* note 1, at 1. For a general description of the "greenhouse effect", *see* CEQ Report, *supra* note 1, *reprinted in* CO_2 Symposium, *supra* note 1, at 45–47.

[3] S. Schneider, *So What If the Climate Changes? reprinted in* CO_2 Symposium, *supra* note 1, at 78. *See also* S. Schneider *The Genesis Strategy* (1976).

[4] S. Schneider and R. Chen, *Carbon Dioxide Warming and Coastline Flooding: Physical Factors and Climatic Impact Ann. Rev. Energy* 5: 107–40 (1980).

[5] CO_2 Symposium, *supra* note 1, at iii.

[6] President's Commission on Coal, Recommendations and Summary Findings (1980).

[7] MacDonald has concluded that coal releases 2.5 (x 10^{15} g.) of carbon dioxide per 100 quads of energy, while oil releases 2.0 and gas 1.45. CEQ Report, *supra* note 1, *reprinted in* CO_2 Symposium, *supra* note 1, at 50.

[8] While there is evidence that increased CO_2 concentrations in the atmosphere will have a positive impact on agriculture in some parts of the world, the author believes the predicted adverse consequences of the "greenhouse effect" justifies its classification as a pollutant. *See also* text accompanying notes 4–5, *supra*, and notes 70–80 *infra* on incentives for cooperation.

[9] Clean Air Act, 45 U.S.C. § 7401 (1980).

[10] Clean Water Act, 33 U.S.C. § 1251 (1980), Safe Drinking Water Act, 45 U.S.C. § 3000(f) – (j – 10) 1976 and Supp. 1 (1980).

[11] Noise Control Act, 45 U.S.C. Sec. 4901 (1976).

[12] E.g., Toxic Substances Control Act 15 U.S.C. Sec. 2601 (1976); Resource, Conservation and Recovery Act, 42 U.S.C. Sec. 6901.

[13] *See, e.g.*, Agreement on Monitoring of Stratosphere, May 5, 1976, United States-France-United Kingdom, 27 U.S.T. 1437, T.I.A.S. No. 8255, *reprinted in* [1978] 1 [Reference File] Int'l. Envir. Rep. (BNA) § 21: 2501. For further discussion, see notes 56–68 *infra* and accompanying text.

[14] *See, e.g.*, International Convention for the Prevention of Pollution of the Sea by Oil, *open for signature*, May 12, 1954, 12 U.S.T. 2989, T.I.A.S. No. 4900, 327 U.N.T.S. 3, *as amended* Apr. 11, 1962, 17 U.S.T. 1523 T.I.A.S. No. 6109, 600 U.N.T.S. 322, *as amended* Oct. 21, 1969, 28 U.S.T. 1205, T.I.A.S. No. 8505, *reprinted in* [1978] (1 Reference File) Nt'l. Envir. Rep. (BNA) 21: 0301, Convention on the Prevention of Marine Pollution by Dumping of Wastes and Other Matter, *done* Dec. 29, 1972, U.S.T. 2403; T.I.A.S. No. 8165, *reprinted in* [1978] (1 Reference File) – Nt'l. Envir. Rep. (BNA) 21: 1901.

[15] *See, e.g.*, Helsinki Rules on the Uses of the Waters of International Rivers, U.N. Doc. A/CN. 4/274. Vol. 11, at 29 (1966), *reprinted in* The Report of the Fifty-Second Conference of the International Law Association 484, August 20, 1966 [hereinafter cited as Helsinki Rules]; Treaty Relating to Boundary Waters Between the United States and Canada, January 11, 1909, United States-United Kingdom, 36 Stat. 2448 T.S. No. 548 [hereinafter cited as 1909 U.S. – Canada Treaty]; Convention on the Protection of the Rhine Against Chemical Pollution, *done* December 3, 1976, *reprinted in* 16 Int'l Legal Mat. 242 (1977) [hereinafter cited as Rhine Chemical Convention]. For further discussion, see notes 25–47 *infra* and accompanying text.

[16] Treaty on Principles Governing the Activities of States in the Exploration and Use of Outer Space, Including the Moon and Other Celestial Bodies, *done* January 27, 1967, art. IX, 18 U.S.T. 2410, T.I.A.S. No. 6347, 610 U.N.T.S. 205.

[17] Convention on Long-Range Transboundary Air Pollution, *done* November 13, 1979, 7. U.N. *ECE*, Annex 1, U.N. Doc. E/ECE/HLM–1/2 (1979), *reprinted in* 18 Int'l Legal Mats. 1442 (1979)

[hereinafter cited as TBAP Convention]; Convention on the Protection of the Environment, *done* February 19, 1974, *reprinted in* 13 Int'l Legal Mats. 591 (1974) [hereinafter cited as Nordic Convention]. For further discussion, see notes 48–66 *infra* and accompanying text.

[18] In contrast, the fact that scientific studies linked depletion of the ozone layer to increased incidence of skin cancer undoubtedly contributed to prompt adoption of legislation banning the use of chlorofluorocarbons in aerosol spray cans. *See* text Section 2.3.4. *infra*.

[19] *See generally* Adams, Mantovani & Lundell, *Wood Versus Fossil Fuel as a Source of Excess Carbon Dioxide in the Atmosphere: A Preliminary Report*, 196 *Science* 54–56 (1977); Bolin, *Changes of Land Biota and Their Importance for the Carbon Cycle, id* at 253–58 (1978); Woodwell, Whittaker, Reiners, Likens, Delwiche, and Botkin, *The Biota and the World Carbon Budget, id*. at 141–46.

[20] CO_2 Symposium, *supra* note 1, at 150. Rotty projects that by the year 2025, the primary contributors of carbon dioxide accumulation will shift from developed to developing countries. He estimates that, in 50 years, the LDC's will comprise 40% of the global emissions of carbon dioxide while the emissions from the U.S., U.S.S.R., and European nations will total 28%. *Id*. For detailed analysis of projected rates of energy use by region, *see* R. Rotty & G. Marland, Constraints on Carbon Dioxide Production from Fossil Fuel Use (May 1980) (Institute for Energy Analysis, Oak Ridge).

[21] The total picture narrows when the carbon wealth of nations is viewed exclusively in terms of coal. In such an analysis, the U.S., U.S.S.R., and China are the main contributors.

[22] Stockholm Declaration of the United Nations Conference on the Human Environment, Principle 21, *adopted* June 16, 1972, U.N. Document A/Conf.48/14 (1972), *reprinted* in 11 Int'l Legal Mats. 1416 (1972) [hereinafter cited as Stockholm Declaration]. The final report of the Stockholm conference is conveniently reprinted in a Senate Foreign Relations Committee publication: Senate Comm. on Foreign Relations, 92b Cong. 2d Sess., United Nations Conference on the Human Environment – Report to the Senate by Senator Clairborne Pell and Senator Clifford Case 12–90 (Comm. Print 1972). A House Public Works Committee print includes a summary of the Conference's recommendation and outlines the position taken by the United States on the matters discussed at the Conference. Staff of House Comm. on Public Works, 2d Sess. Report on the United Nations Conference on the Human Environment (Comm. Print 1972).

[23] *See* Handl, *The Principle of "Equitable Use" as Applied to Internationally Shared Resources: Its Role in Resolving Potential International Disputes over Transfrontier Pollution*, Revue Belge de Droit International (1978–79). *See also* Agreements for the Full Utilization of the Nile Waters, November 8, 1959. U.A.R. Sudan, 453 U.N.T.S. 51; The Indus Water Treaty, September 19, 1969, India–Pakistan–I.B.R.D., 419 U.N.T.S. 125; Treaty Relating to the Uses of Waters of the Niagara River, February 27, 1950, United States–Canada, 1 U.S.T. 694, T.I.A.S. No. 2130, 132 U.N.T.S. 223; Treaty Relating to the Utilization of the Waters of the Colorado and Tijuana Rivers, and of the Rio Grande, February 3, 1944, United States–Mexico, 59 Stat. 1219, T.S. 994, 3 U.N.T.S. 313.

[24] Weiss, *International Liability for Weather Modification*, 1 Climatic Change, 267 (1978).

[25] Helsinki Rules, *supra* note 15, art. X(1) (a).

[26] *Id*. art. XI(1).

[27] *Id*. art. X(1) (b).

[28] *Id*. art. XI(2).

[29] 1909 U.S.–Canada Treaty, *supra* note 15, art. IV.

[30] Agreement on Great Lakes Water Quality, November 22, 1978, United States–Canada, art. II, T.I.A.S. No. 9257 [hereinafter cited as 1978 Great Lakes Agreement].

[31] 1909 U.S.–Canada Treaty, *supra* note 15, art. VII.

[32] 1978 Great Lakes Agreement, *supra* note 30, art. VII.

[33] Convention to Facilitate the Carrying Out of the Principles Contained in the Treaty of November 12, 1884. March 1, 1889, United States–Mexico, 26 Stat. 1512, T.S. No. 232 [hereinafter cited as 1889 U.S.–Mexico Convention].

[34] Treaty Relating to the Utilization of Waters, February 3, 1944, United States–Mexico, art. 2, 3. 59 Stat. 1219, T.S. No. 944 [hereinafter cited as 1944 U.S. Mexico Treaty].

[35] 1889 U.S.–Mexico Convention, *supra* note 33, Art. II; 1944 U.S.–Mexico Treaty, *supra* note 34 Art. 1.

[36] Memorandum of Understanding for Cooperation on Environmental Programs and Transboundary Problems, June 19, 1978, United States–Mexico, T.I.A.S. 9265.

[37] *Id.* ¶ 8.
[38] Treaty on the Plate River Basin, *signed at* Brasilia, April 23, 1969, art. 1, *reprinted in* 8 Int'l Legal Mats. 905 (1960).
[39] The Plate River Basin Treaty establishes the Inter-Government Committee to perform this function. *Id.* art. III.
[40] Rhine Chemical Convention, note 15 *supra*. Convention on the Protection of the Rhine Against Pollution by Chlorides, *done at* Bonn, December 3, 1976, *reprinted in* 16 Int'l Legal Mats. 265 (1977) [hereinafter cited as Rhine Chloride Convention].
[41] Rhine Chemical Convention, *supra* note 25 arts. 1–4.
[42] *Id.* art. 2.
[43] [1979] 2 Int'l. Envir. Rep. (BNA) 975.
[44] Rhine Chloride Convention, *supra* note 40, art. 2.
[45] [1979] 2 Int'l. Envir. Rep. (BNA) 975.
[46] Rhine Chloride Convention, *supra* note 40, art. 2.
[47] The Dutch, the downstream recipients of pollution in the Rhine, recalled their ambassador from Paris after the French government withdrew its bill for ratifying the Convention from Parliament. Moreover, the Dutch requested the EEC to pressure the French to ratify the agreement. [1979] 2 Int'l. Envir. Rep. (BNA) 975.
[48] TBAP Convention, note 17 *supra.*
[49] Agreement on Monitoring of the Stratosphere, note 13 *supra.*
[50] Nordic Convention, note 17 *supra*.
[51] 1974 Michigan/Ontario Memorandum of Understanding on Transboundary Air Pollution Control (June 26, 1975) (available from the International Joint Commission); Reference from the Governments of the U.S. and Canada (July 8, 1975) (available from the International Joint Commission).
[52] [1979] 2 Int'l. Envir. Rep. (BNA) 976–77.
[53] TBAP Convention, *supra* note 17, art. 5.
[54] *Id.* art. 3.
[55] *Id.* art. 6–9.
[56] Agreement on Monitoring of the Stratosphere, *supra* note 13 arts. Ii–Iv.
[57] *Id.* art. VI.
[58] *Id.* art. I.
[59] Nordic Convention, note 17 *supra. See also* OECD Council Recommendation on the Equal Rights of Access to Information, Participation in Hearings and Procedures by Persons Affected by Transfrontier Pollution, *adopted* May 11, 1976, *reprinted in* in Int'l Legal Mat. 1218 (1976).
[60] *Id.* art. 3.
[61] *Id.*
[62] *Id.*
[63] See note 51 *supra.*
[64] International Joint Commission, Second Annual Report on Michigan/Ontario Air Pollution, 2 (1977) (available from the IJC).
[65] *Id.* at 1–2.
[66] *Id.* at 2–4.
[67] Trail Smelter Case (United States v. Canada), 3 R Int'l. Arb. Awards 1905 (1941).
[68] Agreement Confirming Minute 242 of the International Boundary and Water Commission, August 30, 1973, United States–Mexico, 24 U.S.T. 1968, T.I.A.S. No. 7708.
[69] G. A. Res. 3129, 28 U.N. Gaor. Supp. (No. 30) 49 U.N. Doc. A/9030 (1973), *reprinted in* 13 Int'l. Legal Mat. 232 (1974).
[70] R. Rotty, Growth in Global Energy Demand and Contribution of Alternative Supply Systems (Institute for Energy Analysis, Oak Ridge Associated Universities, 1979); R. Rotty, Past and Future Emission of Carbon Dioxide (Institute for Energy Analysis, Oak Ridge Associated Universities, 1980).
[71] W. Bach, *Impact of World Fossil Fuel Use on Global Climate: Policy Implications and Recommendations, reprinted in* CO_2 Sumposium, *supra* note 1, at 121.
[72] Copper, *What Might Man-Induced Climate Change Mean?* 56 Foreign Aff. 500, 513 (1978).
[73] Schneider, note 3 *supra*.

[74] Cooper, *supra* note 72, at 507. For a complete analysis, *see* Bryson, *A Perspective on Climate Change*, 184 Science 753 (1974).
[75] See note 20 *supra* and accompanying text.
[76] Cooper, *supra* note 72, at 505. While Canada will experience a similar lengthening of its growing season, this change will be less beneficial since much of the land affected would be located on the Laurentian Shield. *Id.* at 513. *See* Rotty, note 71 *supra.*
[77] Bach, *reprinted in* CO_2 Symposium, *supra* note 1, at 138; Cooper, *supra* note 72, at 514–15.
[78] *See* Report of Thomas C. Schelling *et al.*, Ad Hoc Study Panel on Economic and Social Aspects of Carbon Dioxide Increase, to Philip Handler, National Academy of Sciences (Climate Board, April 18, 1980).
[79] CO_2 Symposium, *supra* note 1 at 158.
[80] *Id.* at 159.
[81] No representatives from private industry were invited to the Senate Governmental Affairs Committee's carbon dioxide symposium. The West German government, however, is pursuing intensive research in conjunction with the German coal industry to develop a technical fix, scrubbers, for the carbon dioxide problem. CO_2 Symposium, *supra* note 1, at 28. In a study for the U.S. Department of Energy, Albanese and Steinberg of Brookhaven National Laboratory investigated the practicability of alternative CO_2 control systems. They concluded that most potential CO_2 control systems reduce power generation efficiency and require significant energy input. Of the several options studied, the most promising is to store CO_2 in the deep ocean. Albanese and Steinberg, Environmental Control Technology for Atmospheric Carbon Dioxide (May 1980) (Brookhaven National Laboratory).
[82] *See, e.g.*, Cooper, note 72 *supra.*
[83] Some action has already been taken, for example, the July 30, 1979, Senate Committee on Governmental Affairs hearings exploring the relationship between carbon dioxide accumulation and synthetic fuels production. CO_2 Symposium, *supra* note 1. As a result of these hearings, Senator Ribicoff introduced an amendment authorizing the National Academy of Sciences to undertake a two-year study of the carbon dioxide problem. That amendment, with the bill, is now part of the Synthetic Fuels Act.
[84] On the requirement of federal environmental impact statements, see National Environmental Policy Act of 1969, 42 U.S.C. § 4332(2) (c) (1976).
[85] The Soviets have moved to convene a ministerial meeting of the United Nations Economic Commission for Europe to discuss the world energy situation. A Soviet initiative last November resulted in the Convention on Long Range Transboundary Air Pollution. While such a meeting would be an appropriate forum for discussing the carbon dioxide problem (the U.S., Canada, and East and West Europe are all members of ECE), the chances for such a meeting are probably slim as long as Afghanistan and other situations remain unresolved. [1980] 339 Envt'l Users Rep. (BNA) 15. *See* note 20 and accompanying text. *See also* note 21 *supra.*
[86] Clean Air Act, 42 U.S.C. Sec. 7401 (1980).
[87] For a survey on national European Air laws, *see* appendix to this article.
[88] OECD Environment Comm. Press Release (Paris, May 8, 1979).
[89] Recommendation of the Council on Coal and the Environment, *id.* at 12.
[90] The U.S. Clean Air Act embodies a two-step approach to maintaining and enhancing air quality. First, the federal government establishes a primary (for the protection of public health) and a secondary (for the protection of public welfare) national ambient air quality standard for each pollutant listed in the Act. 42 U.S.C. § 7401(b) (1980). Second, the states, through state implementation plans, formulate emission standards defining specific quantitative limits on the amount of the pollutants individual sources may release. 42 U.S.C. § 7410 (1980). For a discussion, see Rodgers, Environmental Law 254–59 (1977).
[91] *E.G.* Clean Water Act, 33 U.S.C. § 1251 (Supp. 1. 1976).
[93] *See* note 71 *supra* and accompanying text.
[94] *See* Stockholm Declaration, note 22 *supra.*
[95] Cain, *CO_2 and the Climate: Monitoring and a Search for Understanding*, in *Environmental Protection Activities of International Organizations*, (Kay and Jacobson, ed.), in press. The article provides a useful review of the involvement of international organizations to date with the carbon dioxide problem.

[96] [1979] 2 Int'l. Envir. Rep. (BNA) 841.
[97] *Id*. The UNEP meeting in Kenya advocated that the global community should give "urgent consideration" to the building up of alternative energy supplies such as solar power and windpower, which should take over from the carbon dioxide producing fuels. [1980] 3 Int'l. Envir. Rep. (BNA) 241.
[98] 15 Weekly Comp. or Pres. Doc. 1353, 1371–73 (August 2, 1979).
[99] *Id*. at 1371.
[100] [1979] 2 Int'l. Envir. Rep. (BNA) 841.
[101] *Id*.
[102] *Id*.
[103] [Current] Int'l. Envir. Rep. (BNA) 49.
[104] Treaty for Amazonian Cooperation, *done at* Brasilia July 3, 1978, *reprinted in* 17 Int'l Legal Mat. 1045 (1978).
[105] 43 Fed. Reg. 11.301–26 (1978). Two states – Oregon and Michigan – have passed legislation banning non-essential aerosol use of chlorofluorocarbon compounds in an effort to protect the ozone layer. Or. Rev. Stat. § § 468.605, 468.996 (6); Mich Comp. Laws. Ann. § § 336.103, 335.107 (effective Mar. 31, 1977). Maine and Minnesota have also enacted legislation prohibiting aerosol sprays containing chlorofluorocarbons for certain uses. [1977] 8 Envir. Rep (BNA) 179, 666. The New York Department of Environmental Conservation adopted regulations requiring warming labels on aerosols with chlorofluorocarbons. *Id*. at 135. At one time, 20 states were considering some type of legislation. *Id*. at 1693–94.
[106] Stoel, Compton & Gibbons, *International Regulation of Chlhorofluoromethanes*, 3 Env. Pol'y and Law 130 (1977).
[107] 42 U.S.C. § 7453 (1980).
[108] Advance Notice of Proposed Rulemaking, 45 Fed. Reg. 66726 (1980).
[109] 42 U.S.C. § 7454 (1980).
[110] Conversation with Dr. Lester Machta, Director of Air Laboratories, NOAA (Jan. 13, 1980).
[111] [1980] 3 Int'l. Envir. Rep. (BNA) 3.
[112] *Id*.
[113] *Id*.
[114] Amendment of Ordinance 1973.334 on Products Hazardous to Health and the Environment adding Section 486. [1977] Swedish Fed. Stat. 1095, *codified in* [1973] Swedish Code of Statutes 329.
[115] T. Stoel, International Regulation of Fluorocarbons V (1–2) (Apr. 17, 1979) (draft paper available from the Natural Resources Defense Council).
[116] 11 Can. Stat. Chapter 72 (1975).
[117] Stoel, *supra* note 115, at VII (6–7).
[118] The author thanks Professor Oscar Schachter of Columbia University Law School for his helpful comments on this section.

Appendix A: Selected International Agreements: An Outline

(I) Air Agreements.
  (A) International.
  (B) National.
(II) International Rivers and River Basin Treaties.
(III) Living and Scarce Resources Treaties.
(IV) General International Legal Principles Concerning the Environment.
(V) Miscellaneous Treaties.

(I) Air Agreements
(A) International

(1) Long-Range Transboundary Air Pollution Convention (done November 13, 1979); signatories: thirty-five countries including the U.S., U.S.S.R., and most Eastern and Western European nations;

purpose: to arrange for concerned research and monitoring of transboundary air pollution, in particular sulphur dioxide. O.J. No. C 281, 10, 11, 1979.

(2) Agreement on Monitoring of the Stratosphere (signed 3 May 1976); signatories: U.S., U.K., and France; *purpose*: to pursue research on the stratosphere, specifically the ozone layer, and coordnate national efforts with the activities of international organizations working in the area. (Reference File) *Int'l Envir. Rep.* (BNA) 21: 2501.

(3) Convention on Protection of the Environment (done February 19, 1974, entered into force October 5, 1976); signatories: Denmark, Finland, Norway and Sweden; *purpose*: to guarantee equal right of access to domestic courts for citizens of foreign states injured by transboundary air and water pollution. 13 *Int'l Legal Materials* 591.

(4) Canada – U.S. Agreement in exchange of information on weather modification activities (done March 26, 1975, entered into force March 26, 1975); *purpose*: to exchange information on weather modification and to consult each other as to the impact of such activities. 14 *Int'l Legal Materials* 589.

(B) National

(1) Belgium Law of 28 December 1964 with Regard to the Control of Atmospheric Pollution (Moniteur Belge, 14 January, 1965, No. 9, pp. 345–347) (Reference File) *Int'l Envir. Rep.* (BNA) 211: 1001.

(2) Canada: Clean Air Act, Statutes of Canada 1970–1971, Chapter 47 (effective November 1, 1971) (Reference File) *Int'l Envir. Rep.* (BNA) 51: 1901.

(3) Denmark: Environmental Protection Act (Act No. 372) assented to 13 June, 1973 (Reference File) *Int'l Envir. Rep.* (BNA) 221: 0201.

(4) France: Decree Concerning the Control of Polluting Emissions into the Atmosphere and Certain Uses of Thermal Energy Decree No. 74–415 of 13 May 1974 (Journal official de la Republique Francais 15 May 1974, No. 115, pp. 5178–5179) (Reference File) *Int'l Envir. Rep.* (BNA) 231: 3201.

(5) Ireland: Control of Atmospheric Pollution Regulations 1970 S.I. No. 156 of 1970 (Reference File) *Int'l Envir. Rep.* (BNA) 251: 0501.

(6) Italy: Law Embodying Provisions as to the Control of Atmospheric Pollution (The Anti-Smog Law) Law No. 615 of 13 July 1966 (Gazzetta Ufficialle della Repubblica Italiana, Part I, 13 August 1966, No. 201, pp. 4091–4096) (Reference File) *Int'l Envir. Rep.* (BNA) 261: 0501.

(7) Japan: Air Pollution Control Law as amended (Reference File) *Int'l Envir. Rep.* (BNA) 91: 0901.

(8) Netherlands: Act of 26 November 1970 Containing Regulations for the Prevention and Limitation of Air Pollution (Air Pollution Act) (Reference File) *Int'l Envir. Rep.* (BNA) 281: 1001.

(9) United Kingdom: Clean Air Act 1968 (1968 c. 62) (Reference File) *Int'l Envir. Rep.* (BNA) 291: 1301.

(10) United States: Clean Air Act, as amended, 42 U.S.C. Sec. 7401 *et. seq.*

(11) West Germany: Federal Emission Control Law 15 March 1974 (BGBI I, pp. 721, 1193 as amended by the Law Amending the Laws to Introduce the Penal Code, 15 August, 1974 BGBI I, 9. 1942) (Reference File) *Int'l Envir. Rep.* (BNA) 241: 1001.

(II) International Rivers and River Basin Treaties

(1) Helsinki Rules on the Uses of the Waters of International Rivers (adopted by the International Law Association 1966) U.N. Doc. A/CN. 4/274 (Vol. 11) at 290, *purpose*: to identify state obligations and duties for the equitable use of international water.

(2) U.S. – Canada

(a) 1909 Treaty Between the U.S. and Great Britain Relating to Boundary Waters, and Questions Arising Between the U.S. and Canada, *purpose*: to establish a mechanism for resolving boundary, water allocation and water pollution disputes through the International Joint Commission (Reference File) *Int'l Envir. Rep.* (BNA) 31: 0401.

(b) 1978 Agreement Between the U.S. and Canada on the Great Lakes Water Quality, *purpose*: to coordinate parallel water quality programs with minimum emission standards (Reference Files) *Int'l Envir. Rep.* (BNA) 31: 0601.

(3) U.S. – Mexico

(a) 1899 Convention Between the U.S. and Mexico, *purpose*: to set a framework for resolving boundary disputes caused by changes in the river beds of the Colorado and Rio Grande (Rio Bravo) (Reference File) *Int'l Envir. Rep.* (BNA) 31: 0201.

(b) 1944 Treaty Between the U.S. and Mexico on the Utilization of Waters, *purpose*: to allocate water resources of shared rivers (Reference File) *Int'l Envir. Rep.* (BNA) 31: 0251.

(c) Memorandum of Understanding Between the Subsecretariat for Environmental Improvement of Mexico and the Environmental Protection Agency of the U.S. for Cooperation on Environmental Programs and Transboundary Problems (signed June 6, 1978 at Mexico City); *purpose*: to arrange for the exchange of experts and coordinate parallel programs on environmental problems. 17 *Int'l Legal Materials* 1056.

(d) Mexico – U.S. Agreement on the Permanent and Definitive Solution to the International Problem of the Salinity of the Colorado River (done at Mexico City and entered into force August 30, 1973). 12 *Int'l Legal Materials* 1105.

(4) Treaty on the River Plate Basin (done April 23, 1969 at Brasilia); signatories: Argentina, Bolivia, Brazil, Paraguay and Uruguay; *purpose*: to combine efforts to promote harmonious development of the basin through exchange of information and consultation coordinated by the Inter-Governmental Coordinating Committee. 8 *Int'l Legal Materials* 905.

(5) Convention on the Protection of the Rhine Against Chemical Pollution (done at Bonn, 3 December 1976); signatories: European Economic Community, France, West Germany, Luxembourg, the Netherlands and Switzerland; *purpose*: to take measures to protect the Rhine including the establishment of emission standards. 16 *Int'l Legal Materials* 242.

(6) Convention on the Protection of the Rhine Against Pollution by Chlorides (done at Bonn, 3 December 1976); signatories: France, West Germany, Luxembourg, Netherlands and Switzerland; *purpose*: to reduce the chloride content of the Rhine by injecting waste in the Alsatian sub-soil if environmentally safe. 16 *Int'l Legal Materials* 265.

(7) Agreement on the Implementation of an European Project on Pollution (Sewage Sludge) (done at Brussels 23 May, 1971, entered into force 1 August 1972); signatories: Denmark, France, West Germany, Italy, Netherlands, Norway, Sweden, Switzerland and United Kingdom; *purpose*: to coordinate research and development operations on methods for processing and disposing of sewage sludge. 12 *Int'l Legal Materials* 9.

(8) Treaty for Amazonian Cooperation (done at Brasilia, July 13, 1978); signatories: Bolivia, Brazil, Columbia, Ecuador, Guyana, Peru, Surinam and Venezuela; *purpose*: to undertake concerned actions to promote harmonious development of the Amazon through information exchanges and consultation. 17 *Int'l Legal Materials* 1045.

(III) Living and Scarce Resources Treaties

(1) Convention on Fishing and Conservation of the Living Resources of the High Seas (done at Geneva on 29 April 1958, entered into force 20 March, 1966); signatories: 22 countries have ratified. 559 *U.N.T.S.* 286.

(2) Convention on Fishing and Conservation of the Living Resources in the Baltic Sea and the Belts (done at Gdansk, September 13, 1973); signatories: Denmark, Finland, East and West Germany, Poland, Sweden, and U.S.S.R.; *purpose*: to expand and coordinate studies toward preserving and increasing the living resources of the Baltic Sea and the Belts and obtaining the optimum yield. 12 *Int'l Legal Materials* 1291.

(3) International Convention for the Regulation of Whaling (done at Washington, December 2, 1946, entered into force November 10, 1948); signatories: U.S., United Kingdom, Iceland, Norway, Australia, South Africa, U.S.S.R., the Netherlands, Panama, France, Sweden, Japan, Brazil, Denmark, New Zealand, Canada and Mexico: *purpose*: to develop effective conservation and development of whale stocks through regulation by the International Whaling Commission. *U.N.T.S.*, Vol. 161 at 74.

(4) Agreement Concerning an International Observer Scheme for Factory Ships Engaged in Pelagic Whaling in the Antarctic (signed at London, October 28, 1963); signatories: Japan, Netherlands, Norway, U.S.S.R., and U.K; *purpose*: to conserve whale stocks through the use of observers on board factory ships engaged in pelagic whaling to the implementation of the agreement. 3 *Int'l Legal Materials* 107.

(5) Convention for the Conservation of Antarctic Seals (done in London, February 3–11, 1972); signatories: same as Antarctic Treaty of 1959; *purpose*: to protect seals in water and on sea ice which the Antarctic Treaty of 1959 omitted. 11 *Int'l Legal Materials* 251.

(6) Convention on International Trade in Endangered Species of Wild Fauna and Flora (signed at Washington, March 3, 1973); eighty countries participated in the International Plenipotentiary Conference to Conclude an International Convention on Trade in Certain Species of Wildlife; *purpose*: to establish a system by which states may strictly control international trade of species in danger of becoming extinct and may monitor trade in species with the potential of becoming endangered. 12 *Int'l Legal Materials* 1085.

(7) Agreement on Conservation of Polar Bears (done November 15, 1973); signatories: Canada, Denmark, Norway, U.S.S.R., and U.S. 13 *Int'l Legal Materials* 13.

(8) Convention on Wetlands of International Importance Especially as Waterfowl Habitat (done at Ramsar, February 3, 1971, entered into force December 21, 1975); signatories: Australia, Bulgaria, Finland, Greece, Iran, Norway, South Africa, Sweden, Switzerland and United Kingdom. 11 *Int'l Legal Materials* 979.

(IV) General International Legal Principles Concerning the Environment

(1) U.N. Conference on the Human Environment (adopted in Stockholm, June 16, 1972); *purpose*: to enunciate common principles for the preservation and enhancement of the human environment, the most notable ones being Principle 21 (Recommending States act responsibly to ensure activities within their jurisdiction do not cause damage to the environment of other states) and Principle 22 (Recommending States to develop further the international law regarding liability and compensation for victims of transboundary pollution). 11 *Int'l Legal Materials* 1416.

(2) U.N. General Assembly Resolution on Permanent Sovereignty over Natural Resources (adopted December 18, 1972), (U.N.G.A. Doc. A/Res/3016 (XXVII) of January 15, 1973); *purpose*: to declare the right of states to permanent sovereignty over all their natural resources on land and in the seabed and sub-soil within their jurisdiction. 12 *Int'l Legal Materials* 226.

(3) U.N. General Assembly Resolution on Environmental Cooperation Concerning Natural Resources Shared by States (adopted December 13, 1973); *purpose*: to recommend that states sharing natural resources inform and consult with one another in joint and harmonious utilization of the resource. 13 *Int'l Legal Materials* 232.

(4) OECD Council Recommendation on the Equal Right of Access to Information, Participation in Hearings and Administrative and Judicial Procedures by Persons Affected by Transfrontier Pollution (adopted May 11, 1976); *purpose*: to recommend each state accord non-citizens injured by transboundary pollution equal access to and non-discriminatory treatment in domestic courts. 15 *Int'l Legal Materials* 1218.

(V) Miscellaneous Treaties

(1) Antarctic Treaty (signed at Washington, Dec. 1, 1959, entered into force June 23, 1961); signatories: Argentina, Australia, Belgium, Chile, France, Japan, New Zealand, Norway, South Africa, U.S.S.R., United Kingdom and U.S.; *purpose*: to establish that the Antarctic shall be used for peaceful purposes only and to promote international cooperation for scientific investigation of the region. 402 *U.N.T.S.* 71.

(2) Treaty on Principles Governing the Activities of States in the Exploration and Use of Outer Space, Including the Moon and Celestial Bodies (done January 27, 1967, entered into force October 10, 1967); signatories: twenty countries including the U.S. and U.S.S.R.; *purpose*: to provide for the non-restrictive and peaceful exploration of outer space. 610 *U.N.T.S.* 206.

Appendix B: Selected Bibliography

Books and Pamphlets

Brown, Cornell, Fabian, and Weiss: 1977, *Regimes for the Ocean, Outer Space and Weather.*

Carbon Dioxide Effects Research and Assessment Program: 1979, *Workshop on the Global Effects of Carbon Dioxide from Fossil Fuels*. Edited by Elliott and Machta.

Developments in Atmospheric Science 10; 1979, Bach, Pankrath, and Kellog (eds.).
European Environmental Law: 1977.
Kiss: 1976, *Survey of Current Developments in International Environmental Law*.
Law, Institutions and the Global Environment: 1972, L. Hargrove (ed.).
Nordhaus: 1979, *Efficient Use of Energy Resources*.
OECD: 1977, *Legal Aspects of Transfrontier Pollution*.
Rodgers: 1978, *Environmental Law*.
Schachter: 1977, *Sharing the World's Resources*.
Schneider, J.: 1979, *World Public Order of the Environment*.
Schneider, S. H. and Mesirov, L. E.: 1976, *The Genesis Strategy*.
Stoel, T., Miller, A., and Milroy B.: 1980, *Fluorocarbon Regulation*.
Thalmann: 1971, *Grundprincipien des Modernen Zwischenstaatlichen Nachbarrechts*.

Articles and Reports

Adams, J. A. S., Mantovani, M. S. M., and Lundell, L.: 'Wood Versus Fossil Fuel As a Source of Excess Carbon Dioxide in the Atmosphere: A Preliminary Report', *Science* **196**, 54–6.
Bolin, B.: 'Changes of Land Biota and Their Importance for the Carbon Cycle', *Science* **196**, 613–15.
Bramson: 'State Responsibility and International Environmental Law', in R. Stein, (ed.), *International Responsibility for Environmental Injury*. The American Society of International Law, (forthcoming).
Brownlie: 1973, 'A Survey of International Customary Rules of Environmental Protection'. 13 *Natural Resources Journal* **179**.
Cain. 'CO_2 and the Climate: Monitoring and a Search for Understanding', in Kay and Jacobson (eds.), *Environmental Protection Activities of International Organizations*. The American Society of International Law (in press).
Cooper: 1978, 'What Might Man-Induced Climate Change Mean?', *Foreign Affairs* **56**, 500.
Comm. on Governmental Affairs, 98th Cong., 1st Sess., Symposium on Carbon Dioxide Accumulation in the Atmosphere, Synthetic Fuels and Energy Policy, June 30, 1979.
Council on Environmental Quality: 1981, *Global Energy Futures and the Carbon Dioxide Problem*, 92pp.
Dyson and Marland: 1979, 'Technical Fixes for the Climatic Effects of CO_2', Elliott and Machta (eds.), *Workshop on the Global Effects of Carbon Dioxide from Fossil Fuels*. Carbon Dioxide Effects Research and Assessment Program.
Faturos: 'Basis of Liability and Standing to Complain for Transnational Environmental Injury', in R. Stein (ed.), *International Responsibility for Environmental Injury*. American Society of International Law, (forthcoming).
Goldie: 1965, 'Liability for Damange and the Progressive Development of International Law', 14 *International and Comparative Law Quarterly* **14**, 1189.
Gorbiel: 1978, 'Legal Status of Geostationary Orbit: Some Remarks'. *Journal of Space Law* **6**, 171–77.
Grave: 1979, 'Geostationary Orbit Issues of Law and Policy'. *American Journal of International Law* **73**, 444–61.
Handl: 1978–79. 'The Principle of 'Equitable Use' As Applied to Internationally Shared Natural Resources: Its Role in Resolving Potentially International Disputes Over Transfrontier Pollution'. *Revue Belge de Droit International*.
Handl: 1975, 'Territorial Sovereignty and the Problem of Transnational Pollution'. *American Journal of International Law* **69**, 50.
Hoffman: 1976, 'State Responsibility in International Law and Transboundary Pollution Injuries'. *International and Comparative Law Quarterly* **25**.
Iluyomade: 1975, 'Scope and Content of a Complaint of Abuse of Right in International Law'. *Harvard International Journal* **16**, 47.
Nanda: 1975, 'Establishment of International Standards for Transnational Environmental Injury'. *Iowa Law Review* **60**, 1039.
National Research Council, National Academy of Sciences: 1979, *Carbon Dioxide and Climate: A Scientific Assessment*. Ad Hoc Study Group on Carbon Dioxide and Climate.

Nordhaus: 1976, "Economic Growth and Climate: The Carbon Dioxide Problem". *American Economic Review* **66**, 341.
'Orbital Saturation: The Necessity for International Regulation of Geosychronous Orbits'. 1979, *Clifornia Western International Law Journal* 9: 139–56.
Recommendations and Summary Findings: 1980, President's Commission on Coal.
Report to the Senate by Senator Claiborne Pell and Senator Clifford Case, U.N. Conference on the Human Environment, 1972, 92nd Cong., 2nd Sess.
Rotty: 1979, 'Energy Demand and Global Climate Change'. *Man's Impact on Climate*, 269.
Rutkowski: 1979, 'World Administrative Radio Conference: The International Telecommunications Union in a Changing World'. *International Lawyer* **13**, 289–327.
Sands: 'The Role of Domestic Procedures in Transnational Environmental Disputes', in R. Stein and A. Chayes (eds.), *International Environmental Disputes*. American Society of International Law, (forthcoming).
Stoel, Compton and Gibbons: 1977, 'International Regulation of Chlorofluoromethanes'. *Environmental Policy and Law* **3**, 129.
Stoel: 1979, 'International Regulation of Fluorocarbons'. Draft prepared for the American Society of International Law.
Stuiver: 1978, 'Atmospheric Carbon Dioxide and Carbon Reservoir Changes." *Science* **199**, 253–8.
Symposium on U.S. – Mexico Transboundary Resources: 1977–78, *Natural Resources Journal* **17**, 543 and **18**, 1.
Weiss: 1978, 'International Liability for Weather Modification'. *Climatic Change* **1**, 267.
Wilson: 1973, 'Creating Mechanisms for International Environmental Action: Requirements and Response'. *Standford Journal of International Studies* **8**, 113.
Woodwell *et al.*: 1978, 'The Biota and the World Carbon Budget'. *Science* **199**, 141–46.

INTRODUCTION TO: PSYCHOLOGICAL DIMENSIONS OF CLIMATIC CHANGE*

Entering the psychological domain of perception of and response to a possible CO_2-induced global climate warming, Fischhoff and Furby boldly address those most difficult-of-all questions, ones that underlie this entire book: What does science tell us about the chances that climate warming will happen, how does the planner decide whether to do anything about it, and how will the layperson respond? The scholar confronts both riches and poverty in trying to apply knowledge from the science of psychology to these questions. On the side of lay perceptions alone, there is a vast social psychological literature on attitude change, fear communication, mass media, two-step communication flows, and the role of direct experience in perception; there is much survey research data on the ebb and flow of political alienation and the associated distrust of official statements about such things as oil shortages, nuclear power safety, cigarettes, cancer and fluoridation.

On the other hand, as Ola Svenson points out, applications of decision theory and the psychology of choice and decision are an elephant on a saucer of basic research facts. What Fischhoff and Furby are proposing to do is to enlarge the saucer, to discover more about how people use information to make decisions. Essentially their proposals are based on the following organization of the decision problem sequence:

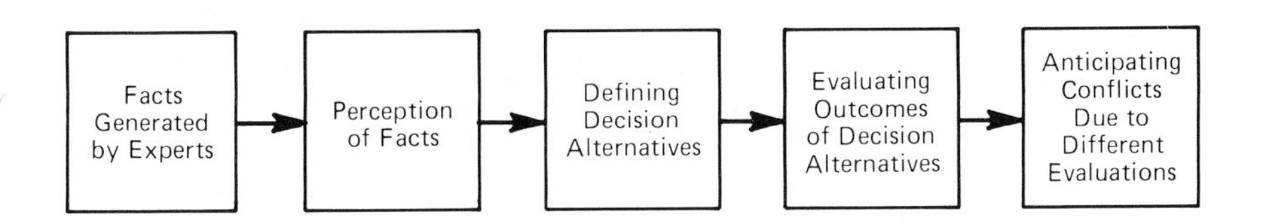

Nowhere is the essentially social nature of the climate change problem clearer than in this chapter. Ordinarily discussions of CO_2-induced climate change are hedged round by the constraints of a technological frame of reference. The solutions sought for possible crisis outcomes are technological ones. Fischhoff and Furby clarify the social nature of the process of arriving at scientific judgments, as well as the lay response. This removes the aura of infallibility from any solutions physical science alone might offer. The fact that scientists have been demonstrated to place more confidence in their own data than tests of scientific reliability allow, and that scientist and lay person alike appear to be subject to the Polyanna effect, making people more optimistic than they are justified to be suggests that the process of data evaluation needs a lot more attention.

Establishing the social character of the scientific perceptions of the climate change problem does not alter the physical realities of climate phenomena, or diminish the role of the physical sciences and technology in dealing with them. Fischhoff and Furby propose to work at the interface between technology and the human being, studying

R. S. Chen, E. Boulding, and S. H. Schneider (eds.), Social Science Research and Climate Change: An Interdisciplinary Appraisal, 177–179.

technology in human hands from a psychological perspective. The significance of the proposed research goes well beyond the climate change problem. It also is relevant to perceptions and decision making with regard to war, famine and energy crises. The hold of modern society on reality is far more tenuous than we thought. We don't *see* well, and we don't make good assessments of what we see. The feedback from the research proposed here could make us see better.

This chapter, like all the chapters in this book, is essentially about the future and how to deal with uncertainty, how to prepare for change. One aspect of that future which is relevant to the issue of response to the threat of climate change and which is not covered in this or any other chapter is the fact that every society, and the individuals in it, hold images of the future in their heads. They visualize, consciously or unconsciously, how things will be. Those visualizations, or images, act as magnets on people's behavior in the present, and draw them towards their envisioned (hoped-for or feared) future (Polak, 1973). At a very rudimentary level, this is the phenomenon of self-fulfilling expectations. This is a macrohistorical process, and the entire history of civilizations can be written in terms of their generative images operating through time. The action options Fischhoff and Furby focus on in their third and fourth proposals are in fact perceived in terms of those pre-existing images. If those images of the future are not analyzed and understood, conclusions about how people process information and choose options will be erroneous. Images of the future are not *values*, although values help shape them; they are coherent expectations about how life may be lived. If these images are bleak and negative, additional information about climate change may lead to hopelessness. If they are dynamic and positive, additional information about climate change will be taken as a challenge which can be met.

Because different societies have different images of the future, based on different values, and differ in their overall optimism or pessimism as a society, no study of the psychological dimensions of the response to climate change can be based in one society alone. The authors are well aware of that, and emphasize the importance of an international team of researchers working together. However, we are probably all in danger of underestimating the effort involved in moving from the research statements made in these chapters, largely out of the United States research experience, with some Canadian and European infusion, to a set of research concepts that will make sense in each of the major civilizational traditions that must be taken account of in global problem solving. While the problem of the commons will be widely understood, and the concept of justice appears to be universal, fairness is a peculiarly Anglo-Saxon construct, and democracy has as many meanings as there are languages containing the word. While the research proposed here comes out of the United States civic culture and the United States research culture, it is universalistic in its import and in the long run will help provide the basis for dealing with social perceptions of global problems in the world community.

E. BOULDING

Note

* Comments from the following readers of this chapter in manuscript form have been drawn on for this introduction: Harry C. Triandis, University of Illinois; Philip Brickman, University of Michigan; and Ola Svenson, University of Stockholm.

Reference

Polak, Fred: 1973, *The Image of the Future*, trans. and abridged., E. Boulding, Elsevier.

BARUCH FISCHHOFF AND LITA FURBY

PSYCHOLOGICAL DIMENSIONS OF CLIMATIC CHANGE

1. Coping with Climatic Change: A Decision Problem

The prospect of CO_2-induced climate change poses a series of interlocking decisions to be made by individuals and groups, national and international bodies. At each level, people must decide whether the problem is worth attending to and if so, should efforts be made to prevent the build-up from happening (e.g., by drastically restricting the consumption of fossil fuels), to implement curative schemes (e.g., massive reforestation programs), to adapt to the new world we are creating (e.g., by developing new crops or moving large populations) or to promote the build-up (for those who hope to benefit from the change). Each decision requires an assessment of what is happening, what the possible effects are and how well one likes them. The quality of these assessments at one level constrains the wisdom of the decisions made at others. Failure of the U.S. to adopt a coherent policy is likely to thwart any international effort. Absence of international cooperation may lead U.S. consumers to ask "why should we drive less when the Brazilians provide tax incentives for logging out the Amazon?" We are all in trouble if the climatologists seriously understate or overstate how much they know. How such assessments are made, by consumers, legislators, diplomats or scientists, would seem to be eminently psychological questions.

Like other decision problems, CO_2-related questions require a choice between alternatives. What distinguishes them is the magnitude of the stakes involved and the very difficult choices posed by the various decision options.

Whether done formally or informally, examination of the alternatives in a decision problem involves the following five interdependent steps:

(1) Specifying the objectives (i.e., what one wants),

(2) Defining the possible alternatives (including "do nothing"),

(3) Identifying the possible consequences of each alternative (including, but not restricted to, risks),

(4) Specifying the desirability of the various consequences and the likelihood of their being achieved, and

(5) Comparing the alternatives and selecting the best one.

The initial steps describe the problem as it is perceived by the decision maker, whereas the final step is prescriptive, in the sense of prescribing the option that should be selected (given the logic of the analysis).

Cost-benefit analysts, decision analysts, operations researchers, management scientists and others have devoted their careers to implementing this simple scheme in complex situations. Although based on an appealing premise and supported by a sophisticated methodology, these procedures have a number of characteristic limits on their usefulness as management tools. These limits arise when the mathematical formalisms confront the

R. S. Chen, E. Boulding, and S. H. Schneider (eds.), Social Science Research and Climate Change: An Interdisciplinary Appraisal, 180–203.

fallible individuals who must conduct, accept, or implement them. Be they technical experts, lay interveners or government regulators, these individuals all have, to some extent, limited capacity to process technical information, restricted resources to devote to the project at hand, irrational apprehensions about its consequences, intransigent prejudices about the facts of the matter, ulterior motives, and incoherent and unstable values on critical issues. Deliberately or inadvertently, these human properties tend to foil the best-laid plans of the purveyors of decision-making schemes (Fischhoff, 1979).

If such schemes are hard to implement when one tries to do so, it seems unlikely that they would be a very good description of unaided decision making (Fischhoff *et al.*, 1982). Predicting the decisions of others is nonetheless a crucial aspect of coping with the possibility of climatic change. The policy maker must anticipate the answers to questions like: will the public perceive this change as a possibility? Will they choose to adapt a conservation ethic? What economic incentives are available or efficacious? Will people respond to the international and intranational conflicts resulting from climate change cooperatively or belligerently? What will various groups perceive as their own best interests? Are people willing to make sacrifices for the common weal?

The following pages describe five major research projects designed to improve our ability to cope with possible CO_2-induced climatic change. Each promises to improve our ability to make deliberative decisions and to understand intuitive decisions. Understanding how various kinds of people make decisions naturally is a precondition for providing them with the information and aids needed to make better decisions. Although a decision-making framework is used as an expository device, these projects draw on the talents of individuals from a variety of areas in psychology (in addition to decision making). Moreover, each contains a description of the other disciplines that could most usefully be involved. A concluding section describes the interdisciplinary and international perspective we believe to be necessary to the success of the entire CO_2 research enterprise. The exposition draws examples from other contemporary problems, both because there is no corpus of research on these topics directed explicity at CO_2 issues and because the results of the proposed CO_2 research would have relevance for coping with those other problems.

The five projects are:

(1) Identifying and characterizing subjective aspects of the "facts" of CO_2-induced climatic change. Where does judgment enter into the work of experts and their communication of that work? Where do questions of fact and of value intermix? How good is expert judgment? How well are experts able to assess the definitiveness of their own work?

(2) Understanding and improving lay decision makers' understanding of the facts of CO_2-induced climatic change. How do they interpret (often conflicting) expert testimony? Is such testimony about climate consistent with their direct sensory experience with weather; if not, how are conflicts resolved? What kinds of information pose particular conceptual problems (e.g., very low probabilities, interaction of long-range cycles with varying periodicity)? How can such problems be remedied, so that best use is made of available scientific evidence?

(3) Clarifying and enriching the space of possible action options. What options naturally occur to (different groups of) people? How is feasibility judged? How can the set

be enlarged? What consequences (or side effects) tend to be overlooked? In what ways are decision makers prisoners of their own experience?

(4) Understanding how the alternative responses to climatic change are evaluated. How do people combine multiple and conflicting costs and benefits, on different dimensions with varying degrees of risk, and arrive at a single decision? How can people's values be elicited, so as to inform government officials (e.g., are traditional surveys adequate)? How can elicitation methods (e.g., survey techniques) distort the values expressed through them?

(5) Anticipating conflicts and developing means for their resolution. How will climate change pit nation against nation, group against group? What commons dilemmas exist today and will be created in the future? Can frameworks be developed to aid their resolution? What sorts of mistrust and misunderstanding might emerge and might be avoided?

2. Project 1. Identifying and Characterizing Subjective Aspects of the "Facts" of CO_2-Induced Climatic Change

2.1. Primary Research Questions

(1) Where do subjective judgments enter into scientific analyses?
(2) How valid are those judgments?
(3) How well are experts able to identify and assess such judgments?
(4) How can we make better use of our experts by having a better appreciation of the limits of their abilities?

2.2. Background

For the threat of climatic change to assume a respected place among the constellation of problems about which people are concerned, they must be convinced that it is a reasonably likely occurence. Unfortunately, assessing the probability of such extreme events can be a very difficult business. At times, it is possible to identify a population of events from which a sample may be drawn as a step toward assessing the probability of the event in question. The copious records of ice pack movements maintained in Iceland over the last millenium provide a clue to the probability of an extremely cold year in given future periods. The apparent absence of a full-scale meltdown in the 1000 or so years of nuclear reactor operation may allow setting some bounds on the probability of future meltdowns. Of course, extrapolation from any of these historical records is a matter of judgment. Changes in design, public scrutiny and federal regulation may render the next 1000 reactor years appreciably different from their predecessors. The new conditions created by increased CO_2 concentrations may change climatic variability in a way that amplifies or dampens yearly or daily fluctuations.

Even if experts were to agree on the relevance of these records, a sample of one thousand reactor- or calendar-years may be insufficient. Given the magnitude of possible consequences, a 0.0001 chance of a meltdown might be deemed unconscionable, but we will be well into the next century and irrevocably committed to nuclear power and its consequences before we will have enough hands-on experience to assess the probability of a meltdown to the desired accuracy. We know that meltdowns are unlikely (in the present sense), but whether they are unlikely enough may not be known until it is too late or may not be known at all.

When no historical record is available upon which to base conjectures, one is left with conjecture alone. In the scientific community, the more sophisticated conjectures are based upon models. General circulation models (GCM's) represent one such genre; the fault trees analyses of a loss-of-coolant accident upon which the "Rasmussen" Reactor Safety Study was based (NRC, 1975) represent another. Both focus on component processes and the interactions between them, instead of the problem in its full complexity.

The fault tree involves a logical structuring of what would have to happen for a core to melt down. If sufficiently detailed, it will reach a level of specificity for which we have relevant experience (e.g., the operation of individual valves). An overall probability of failure for the system is determined by combining the needed component failures. Unfortunately, some components are entirely novel or have never been used in these particular conditions; their performance parameters must be guessed. Furthermore, the logical structure and completeness of the tree are more or less matter of opinion.

GCM's share the same strengths and weaknesses as fault trees. They attempt to predict the unknown world of heightened CO_2 concentrations on the basis of related observables and their hypothesized interconnections. These are, respectively, recorded atmospheric and oceanographic conditions and generally accepted theories of their dynamic interaction. As with fault trees, some of the data are uncertain and some of the logic is disputable.

Thus, facts about climate are often revealed through the filter of formal analyses rather than through direct experience. One's faith in the results so revealed depends on the success of the analysts in identifying all relevant components, assessing their values, and understanding their interrelations. Recent psychological research suggests some likely bounds on their success and our faith. People apparently have limited ability to recognize the assumptions upon which their judgments are based, appraise the completeness of problem representations, or assess the limits of their own knowledge. Typically, their inability encourages overconfidence (Fischhoff *et al.*, 1977, 1978).

One might hope that the results of previous research conducted on lay people could not be generalized to technical experts, that somehow the latter's substantive knowledge and training would lead to improved judgment when forced to go beyond the available data. Unfortunately, a modicum of systematic data and many anecdotal reports suggest that this is not the case. As a case in point, a high level peer review found that the Recactor Safety Study had greatly overstated the precision of its conclusions (NRC, 1978). The unpleasant surprise at Three Mile Island demonstrated that it had not included all pathways to disaster nor even explicitly raised a number of critical and erroneous assumptions (e.g., that trained personnel would always be available). For their part, GCM's necessarily omit some aspects of the environment believed to be relatively unimportant (for the sake of manageability) and incorporate untested assumptions provided by other disciplines (e.g., that the rate of increase in CO_2 production of the last 20 years will continue unabated in the future, in a world that may have more or less nuclear power, war, recession, and environmental awareness than its predecessor). They seem to be poorly suited for even providing guesses at their accuracy.

If one reads such analyses and the rare subsequent evaluations with an eye to the psychology of the analyst, there seem to be generic sources of error and omission. These include (a) failure to consider the imaginative ways in which human error can mess up a system (e.g., the Browns Ferry fire in which the world's largest nuclear power plant almost melted down due to a technician checking for an air leak with a candle in direct violation of standard operating procedures); (b) insensitivity to the assumptions an analysis makes about constancies in the world in which the system is embedded (e.g., no major changes in government regulatory policy); (c) overconfidence in current scientific and technological knowledge (e.g., assuming that there are no new chemical, physical, biological, or psychological effects to be discovered); (d) failure to see how the system functions as a whole (e.g., a system may fail because a backup component has been removed for routine maintenance).

2.3. *Research Plan*

For judgments to be evaluated, they must first be identified. Since individuals often have very little insight into the workings of their own intellectual processes, it is not enough to ask someone, "How did you arrive at that answer?" or "What unstated assumptions guided you?" The first step in this project would be a joint effort by substantive experts and judgment experts to answer those questions with respect to scientific analyses of CO_2-induced climate change. The second step would be an analysis of the resultant answers according to the principles of cognitive psychology, and a review of extant literature to see if there is a basis for trusting or doubting such judgments. A moderate amount of additional empirical work will undoubtedly be necessary. The third would be to devise ways to enhance the performance of technical experts who are involved in assessing the CO_2 climate-change prospects.

At the moment, the intellectual processes of the highly trained are rarely studied. There are, however, research methodologies that could readily be extended to this problem. Some critical questions this research should address are: Are experts any different from lay people in their basic cognitive functioning (i.e., can one generalize to experts from research conducted with lay people)? Does professional training encourage or discourage particular misperceptions? Do technical specialists tend to isolate aspects of the CO_2 phenomenon and its impact rather than integrate their results to a system-wide analysis that includes possible compensating, exacerbating or masking effects? How independent can the opinions of two experts be when they have gone through similar training and specialization? How well do experts understand the limits of their own knowledge? Further research questions arise if one considers experts not as dispassionate interpreters of results, but as individuals strongly motivated to confirm pet theories or satisfy employers.

The knowledge possessed by experts may not always be organized in their minds in the form desired by decision makers or the risk analysts paid to help them. For example, an experienced mechanic who sees problems as they come in to the garage may be ill-equipped to estimate break-down rates or the likelihood of malfunctions co-occurring. Theoretically appealing summary measures of complex situations are of little use if no one can produce them, or if none of the relevant decision makers can understand them. Development of ways to elicit from experts what they know about climate or society will have to be a joint effort of substantive experts and experts in information processing.

A variety of behavioral assumptions underlies many climatological, economic, agronomic and other theories. Examples might be: the public can be viewed as passive impactees, doing little to shape their own world; energy consumption will continue to grow at historical rates; people will not respond to altruistic appeals. Once spotted, such assumptions are subject to empirical tests. Given the large cumulative impact of small changes in, say, energy consumption rates on the conclusions of analyses, such tests and corrections can markedly change the climate picture.

Experts bring with them to any problem an image of its dimensions. This problem definition mixes issues of fact and value. By ignoring some topics and giving little weight to others, it can largely determine the subsequent decision. One would like to know what consequences and strategies they consider (or reject)? Where do they turn for advice on feasibility? What control strategies are they likely to neglect? In what ways are they captive of untested theories or the failure of basic researchers to study potentially useful topics? Studying any of these topics presumes that outsiders can somehow contribute to the wisdom of the recognized experts in a field. The basis for that presumption is the possibility that although experts have a near-monopoly on the best facts, they may not see problems in the full richness that could be obtained by considering the perspectives of a diverse group of others.

2.4. Research Outcomes

The following products can be visualized: (1) Technical papers reporting the results of research on the nature and quality of expert judgment in assessing the facts about CO_2-related issues. (2) Guides translating the conclusions of these technical reports into a form useful for decision makers outside the expert community (i.e., government, the lay public, intervenors, social critics). (3) Practical procedures for better exploiting the educated intuitions of experts. (4) Regular presentations at meetings to disseminate the most useful results to both experts and lay decision makers (and to assess their perceived needs). The research itself should foster better communication between the experts and the public they serve. The participating discipline should include psychology, statistics, cognitive science and relevant substantive professions.

3. Project 2. Understanding and Improving Lay Decision Makers' Perceptions of the Facts of CO_2-Induced Climatic Change

3.1. Primary Research Questions

(1) How do lay decision makers interpret the fact presented to them by experts?
(2) Is this testimony about climate consistent with their direct sensory experience with weather; if not, how are conflicts resolved?
(3) What kinds of information pose particular conceptual problems?
(4) How can such problems be remedied, so that decision makers can make the best use of available scientific knowledge and the wisdom of their own experience?

3.2. Background

The facts of climate change reveal themselves to experts through the filter of various research methods, formal models, and professional prejudices, each with their strengths and weaknesses. They reveal themselves to non-experts through unsystematic experience and reports from the front by experts, seers, and the news media that traffic in such reports.

To make use of what the experts report, one must understand both the substance of their message and the qualifications that (should) accompany it. An obvious limit on our ability to understand substance is having the report couched in unfamiliar technical terms. These can mislead (say, when technical terms have common language counterparts with different meanings), confuse (perhaps leading us to think that we understand when we really do not) and dissuade us from even attempting to understand.

Obviously, most scientific problems afford opportunities for asserting some sort of elite control. However, even well-meaning attempts to inform the public may go astray. CO_2 issues make a terrific chalk talk, but their impact may be lost if care is not taken to draw causal links between its parts (Tversky and Kahneman, 1980), particularly those links connecting human behavior and climatological consequences. Without such explicit ties, a CO_2 crisis may appear implausible as well as improbable. On some level, it may be hard to believe that global cataclysm might be the result of such innocuous and sensible acts as lighting home fires and burning leaves. The CO_2 problem represents a global commons dilemma in which seemingly inconsequential individual decisions combine to produce universally adverse consequences in the long run. Although moralizing might lead to more prosocial behavior (Dawes, 1980), it is likely to have little effect until recipients are convinced that a dilemma exists.

Even if people are willing to listen, it may be difficult to present low probabilities to them comprehensibly. Is, for example, the difference between 0.001 and 0.0001, so stated, meaningful to people? Scattered evidence suggests that people may ignore or exaggerate probabilities in that range (Slovic *et al.*, 1977; Lichtenstein *et al.*, 1978). One alternative is to provide a concrete referent in the form of a familiar event with an accurately judged probability of similar magnitude. The efficacy of this (or any other) procedure for communicating low probabilities has yet to be demonstrated.

As a guide to action, the uncertainty surrounding the experts' best guess may be as important as the substance of the guess. One wants to know "just how high could it be?" and "do these experts know enough for me to take their best guess seriously?" A good deal of evidence (e.g., Gettys *et al.*, 1973; Kahneman and Tversky, 1973) suggests that were such qualifications provided, they would not be used properly. In particular, people seem to be as confident making inferences from highly unreliable data as from reliable data, rather than less confident, as statistical theory dictates. If, as suggested above, there is also a propensity for experts to exaggerate how much they know, one should expect a gap between the credibility afforded to scientific analyses and that which they merit.

Another form of credibility problem arises when the integrity of the source is threatened. Most people probably have learned to discount what they see on TV because of its tendency to sensationalize. Whether they are aware of the subtle biases that can enter into scientific analyses may be another question. For example, the very raising of CO_2 questions rather than those surrounding other hazards of potentially greater magnitude may reflect a desire to make life easier for one domestic energy industry (nuclear); not raising them may reflect a desire to obscure international energy issues (the fact that the industrialized countries are enjoying most of the benefits of creating the CO_2 imbalance whose costs will be borne by everyone). As a counterpoint, one might note that despite the enormous destructive potential of earthquakes in the U.S. and the fairly high likelihood of their occurrence, almost no research is going into improving human response. Seismological research designed to develop the capacity for earthquake prediction is, however, well-funded despite some serious suggestion (National Academy of Sciences, 1978) that the expected value of forecasts is negative, once one considers social reactions to them.

Unlike some environmental "problems", climate is directly experienced. That experience may set us wondering about the likelihood of major climatic changes (say, as did the recent West Coast drought and severe Northeastern winters). Once we are interested, that experience may support or contradict what the experts tell us with regard to CO_2 protections. In other cases, personal experience may be all we have to go on.

How good are we at assessing the likelihood of natural events? Lichtenstein *et al.* (1978) asked people to judge the likelihood of a randomly selected individual dying from a variety of recognizable, but not necessarily common, causes (e.g., botulism, tornadoes, cancer). They found that people (a) had a pretty good idea of the relative frequency of most causes of death, (b) substantially underestimated the differences in the likelihoods of the most and least frequent and (c) persistently misjudged the relative likelihood of those causes of death that are unusually visible (e.g., tornadoes) or invisible (e.g., asthma). Slovic, Fischhoff and Lichtenstein (1979) found a similar pattern of results in estimates of the fatalities from various technological hazards, although this work has yet to be extended to judgments of climate change.

Assessing people's knowledge about risks may be far from easy. A recent study asking people about the lethality of some causes of death (i.e., the probability of dying given that one was afflicted) found that formally irrelevant changes in response mode produced appreciable differences in assessed probability (Fischhoff *et al.*, in press). For example, death rates derived from responses to the question "For every 100 000 afflicted, how many die?" were roughly two orders of magnitude greater than those in response to "For every individual who dies, how many are afflicted but survive?" These differences seem due in part to the effect question format has one how people access their knowledge, and in part to variants of the well-known effects that the design of magnitude estimation experiments has on the results of those experiments (Poulton, 1968). Furthermore, even in situations where people have fairly accurate assessments of observable phenomena, their notion of underlying mechanisms may be quite in error. For example, an atypical period of rainy weather following the first agricultural settlements in the High Plains of the U.S. led to the belief (endorsed by the AAAS) that "rain follows the plow". The more normal drought years following the breaking of the sod resulted in tragic disruption of lives and loss of topsoil (Burton *et al.*, 1978; Opie, 1979).

Although many of the climatic fluctuations and meteorological events that may be affected by possible CO_2 changes have some natural, semi-observable frequency, the event itself does not. In fact, one directly sees little or nothing to indicate that some global dislocation may be on the way as a result of commonplace actions taken by all the earth's denizens. Those who have not heard the cry of alarmed climatologists (e.g., Bryson, 1974; Schneider and Mesirow, 1976) are doubtless worrying about other things. While everybody is doing something about the weather, no one is talking about it. Those who have heard the cry may "overinterpret" short-term climate fluctuations as evidence of long-term climate change.

3.3. Research Plan

To assess people's knowledge of climate change, one must first establish what it is that they need to know, and then characterize that message with regard to the kinds of information it embodies. For example, understanding climate requires a grasp of information that is surrounded by uncertainty, reflects complex interactions between different variables, can be overwhelming in its volume, often deals with time spans much longer than one's lifetime, expresses very low probabilities, and so on. For any kind of information, one should ask a series of questions: (a) What are its formal properties? (b) What are its observable signs? (c) How are those signs revealed to the individual? (d) Are they contradicted, supported or hidden by immediate experience? (e) Do people have an intuitive grasp of such information? (f) To the extent that they do not have such a grasp, what is the nature of their misunderstandings? (g) How great are such misunderstandings and how severe are their consequences? (h) Does natural experience provide feedback highlighting misunderstandings and inducing improvement?

If we hope to improve as well as predict performance, we must also ask: (i) Can understanding be enhanced, for example, by generating better evidence, developing superior presentations or altering basic approaches to knowledge?

These questions ask, in essence, how adequate people's cognitive skills are for coping with the information they receive. As the previous section indicated, there is an extensive body of psychological methods and knowledge about many of these questions. The application of that body to the climate arena must be systematically tested. In addition, we will need to develop more sophisticated techniques to establish what people know and how they think about climate risk. These techniques will reveal not just a snapshot of summary statistical knowledge, but an understanding of people's thought processes and potential for understanding properly presented risk information. Different procedures will be needed for populations differing in verbal and technical literacy (Whyte, 1977).

Once developed, these tools should be applied both to groups representative of the general population and to special-interest groups, each serving a different purpose. Surveys of the general public would show the potential for concern and misinformation; studies of interest groups would show how that concern is realized among people who have thought more about the issues. Each should, in turn, stimulate further elaboration of research instruments designed to find out: What do people know? What information do they want? What sources do they trust? What does climate mean to them?

The nature of the survey will depend in part upon why it is being conducted. At one extreme, it may be designed to establish how much the public already knows, as a guide to determining how far it should be allowed to make decisions in its own behalf. At the other extreme, perceptions would be studied as part of a concerted effort to enhance the public's decision-making ability. In that case, it might be embedded in an attempt to provide meaningful public participation in climate decisions, identifying areas of weakness with an eye to helping the respondents acquire competence, seeking a defensible basis for differences between lay and expert perceptions (Fischhoff *et al.*, in press, b).

This research on lay perceptions of the facts of CO_2-induced climate change must be international in scope if it is to be of maximum utility. People from different cultures often have very different ways of perceiving and understanding a given phenomenon, and a cross-cultural approach to this research will significantly enhance the eventual coordination of responses to CO_2 buildup on a global scale.

3.4. Research Outcome

The following products can be visualized: (1) Scientific papers extending existing judgment work to perception of climate change and opening new research areas. (2) Surveys of public knowledge and opinion on CO_2-induced climate change and its potential impact; results would guide both decision makers and communication specialists. (3) Guides for experts on presenting climate information, and bulletins to experts on what the public wants to know. The participating disciplines should include psychology, sociology, anthropology, plus some technical consulting from climate and survey researchers.

4. Project 3. Clarifying and Enriching the Space of Possible Action Options

4.1. Primary Research Questions

(1) What options naturally occur to people?
(2) How is feasibility judged?
(3) What consequences (or side effects) tend to be overlooked?
(4) In what ways are decision makers prisoners of their own experience?

4.2. Background

As every politician knows, controlling the agenda in a policy debate is part of a winning strategy. The agenda of a formal analysis, like that of any other decision making process, is embodied in its problem statement. Its terms formally foreclose some decision options by not raising them as possibilities. Other options are effectively eliminated by giving little or no weight to the consequences that they best serve. Experienced participants in technology sieges know the power of definitions. They fight hard to have their concerns reflected in the analytical mandate; failing that, they may fight dirty to impeach the resultant analysis. Comprehensiveness is the key not only to political acceptability, but also to conceptual soundness. Many analysts consider only one option (build the plant) or variants on one option (build it here or there or there), or only alternate forms of the same kind of solution (e.g., pesticide X or pesticide Y). Some neglect even the option of foregoing the project (and the risk that that entails). Ignored consequences do not go away; overlooked options may dominate considered ones. For initial analyses designed to enhance our intuitions by framing the overall decision problem, breadth is more important than depth. Guaranteeing minimal representation to all topics should precede elaborating any one topic with costly numerical or modeling exercises.

There is very little previous research on how individuals or groups formulate alternative action plans when faced with a problem to solve. What we do know from previous work is that the most successful and creative problem solvers are those who are not burdened by unnecessary assumptions, i.e., those who are able to break out of habitual patterns of thinking and see things from unusual perspectives. What we do *not* know is exactly what facilitates and what impedes creative option generation. Any attempt to predict, prevent, or mitigate the CO_2 effect must necessarily be based on some assumptions regarding human behavior, about what individuals will do (e.g., continue to consume energy at present rates), about what people will value (e.g., efficiency and cost-effectiveness over a clean environment), and about what other regions or nations will do (e.g., switch from fossil fuels to nuclear energy). Unrecognized assumptions are as much a handicap to lay decision makers as to experts, constraining the set of options generated without one's being aware of and able to evaluate that constraint. For example, generation of alternative responses to the predicted CO_2 increase requires some sort of causal interpretation of the phenomenon. If the burning of fossil fuels is considered to be the cause of climate change, then obviously one of the most effective options is to halt the use of fossil fuels. However, research suggests that once one sufficient explanation has been offered for an event, other possible causes are immediately and undeservedly seen to be less likely to have been involved (Shaklee and Fischhoff, 1979). Thus, if both fossil fuel burning and deforestation can cause an increase in CO_2, people may tend to focus on one cause and its management to the exclusion of the other. Witness the greater concentration on modifying the rate of burning fossil fuels than on modifying the rate of deforestation.

Another potential limiting factor on option generation is vested interests. The "logic" of one's own position may make it extremely difficult for Brazilians to conceive of, let alone advocate, halting deforestation; or for anti-nuclear environmental groups to suggest increased reliance on nuclear power; or for Americans to suggest giving up the automobile, as possible courses of action. We need research on what factors exacerbate and minimize the blinders of self-interest. Are people only able to generate action options that have obvious personal value, or are there at least some conditions under which they are likely to think of less self-serving solutions?

An important determinant of generating action options is how one views a change from the status quo. When is a "crisis" perceived as an opportunity or challenge, and when is it viewed as a hardship or disaster? Some agriculturalists see increased CO_2 not as a problem, but as an opportunity, since more CO_2 in the atmosphere increases photosynthesis as well as water use efficiency, and since adaptation to new climatic conditions is considered a stimulus to technological development. Many American settlers who moved west into unknown climate conditions also apparently viewed adaptation to the unknown as a challenge. Yet we know very little about (a) why some individuals and some cultures view change from the status quo as undesirable, while others view it as desirable, (b) why a given individual views certain "crises" negatively, but others positively, and (c) how these differing perceptions affect the way people generate action options.

4.3. Research Plans

Since the set of options is constrained by reality and imagination, a combination of projects is needed. One is a study of how individuals generate a set of possible and reasonable options. Particular attention should be given to cultural differences, showing how the conceptual space of different groups is limited. The eventual international cooperative effort in dealing with the CO_2 issue will require an understanding of the mental world within which others live. The second project is to exploit this understanding to produce the broadest range of possible responses to the CO_2 phenomenon. It should involve a vareity of disciplines and non-academics, poets, workers, clergy and so on, in hopes that their life experiences will reveal hitherto unconsidered possibilities.

4.4. Research Outcomes

The following products can be visualized: (1) Technical papers on the psychological and social processes governing the generation and evaluation of alternative solutions in problem situations in general and climate change problems in particular. (2) A broad set of possible responses to the CO_2 phenomenon for the consideration of policy and lay decision makers, along with an analysis of their feasibility and value assumption (i.e., the world outlooks they represent, the interests they favor). (3) Active participation of research in scientific and policy-making forums devoted to climate change. Such participation will facilitate accommodating research and scientific and political realities as well as changing those realities by expanding and clarifying the range of possible options. Research on options will be multidisciplinary. Psychologists will study individual and group processes in option generation; anthropologists and historians will examine cultural influences and historical examples of how people have viewed pending changes from the status quo, and how they generate action options under those conditions. Philosophers can help offer perspectives on cultural and historical assumptions influencing the generation of alternative responses, one of the most important assumptions being how we view our relationship to nature and the environment.

5. Project 4. Understanding How Alternative Responses to Climatic Change Are Evaluated

5.1. Primary Research Question

(1) How do people combine multiple and conflicting risks and benefits of various options into a single decision?
(2) How can people's opinions on these issues be accurately elicited so as to inform government officials?
(3) How can faulty elicitation methods distort the values expressed through them?

5.2. Background

The generation of creative responses to the CO_2 phenomenon does not assure their implementation. The strategy that is adopted depends upon how the various options are evaluated. This process is sometimes implicit, occurring right at the time of option generation (e.g., the possibility of moving the soil of Iowa to Minnesota may be discarded as unfeasible as soon as it is formulated), and sometimes it is explicit (involving the assignment of probabilities and values to various alternative outcomes). Our knowledge of the evaluation process is only rudimentary. We know little about how people combine multiple and conflicting costs and benefits, on different dimensions with varying degree of risk, and arrive at a single decision. For example, replacing fossil fuels with nuclear power increases the risk of radioactive contamination, but lowers the risk of climate change. Contamination is a low probability, catastrophic possibility. Climate change is a higher probability, less calamitous eventuality. How do people put these kinds of information together and weight the various options? Are these systematic biases in the evaluative strategies that people use (such as overestimating the likelihood of the most "available" scenarios), leading them to select alternatives that they don't "really" prefer? If so, are there ways of eliminating, or at least minimizing, these biases?

Since climate is part of our lives, we should, it would seem, have no trouble comperhending what the outcomes of CO_2-induced changes are and how much we would like them (e.g., what it means to have an average increase of 2 °C). There are, however, a number of reasons to doubt this presumption, all of which have analogs in the reasons for doubting the assumption that because we all live in society, we would be able to understand the meaning of a projected shift in one of its parameters (e.g., an increase in the median age or percentage of handicapped or price of fuel). One is that we do not experience our environment directly; rather, we have about us a series of defenses that regulate contacts so as to make them more pleasant and less demanding. Air conditioning and social norms are two obvious examples. We may have little idea of what life would be like if the conditions to which that veneer of civilization were adapted were changed.

A related reason for doubt is that we experience weather not climate, people not society. As a result, we seldom have to confront the complexity of the natural and social ecologies within which we live. We may not realize that an older world threatens the bankruptcy of the social security system or that a warmer world will eliminate the hard freezes that keep pests from destroying susceptible crops in some regions. Although the connections are straightforward and comprehensible when drawn, one should not expect either experts or lay people to recognize spontaneously the secondary or tertiary effects of projected changes.

Finally, no one knows how well people are able to imagine dramatic changes or, conversely, to what extent they are prisoners of their own experience. Do any of us who have not suffered that unmaskable pain of cancer know what it means? (If we did, would any of us be smoking?) What presumptions about unalterable aspects of human nature constrain our imaginations regarding, say, what awaits us in foreign countries or prison? Can we flesh out projections of climatic conditions outside of our species' experience? Can we really know what it will be like to live in the greenhouse? Without that experiential understanding, can we act appropriately to the possiblity? A related argument is used by some foes of nuclear power, who say that since we can't grasp the time span during which some radioactive wastes must be stored, we should avoid the whole business; without basic comprehension, wise decision making is infeasible.

Understanding effects requires not only factual knowledge, but also an evaluative assessment. Do we want this to happen? How badly? Such questions would seem to be the last redoubt of unaided intuition. Who knows better than an individual what he or she prefers? When one is considering simple, familiar events with which people have hands-on experience, it may be reasonable to assume that they have well-articulated opinions. Regarding the novel, global consequences potentially associated with CO_2-induced climatic change or nuclear meltdowns, that may not be the case. Our values may be incoherent, not thought through. In thinking about what are acceptable levels of risk, for example, we may be unfamiliar with the terms in which issues are formulated (e.g., social discount rates, miniscule probabilities, or megadeaths). We may have contradictory values (e.g., a strong aversion to catastrophic losses of life and a realization that we're not more moved by a plane crash with 500 fatalities than one with 300). We may occupy different roles in life (parents, workers, children) which produce clear-cut but inconsistent values. We may vacillate between incompatible, but strongly held, positions (e.g., freedom of speech is inviolate, but should be denied to authoritarian movements). We may not even know how to begin thinking about some issues (e.g., the appropriate tradeoff between the opportunity to dye one's hair and a vague, minute increase in the probability of cancer 20 years from now). Our views may undergo changes over time (say, as we near the hour of decision or the consequence itself) and we may not know which view should form the basis of our decision (Fischhoff *et al.*, in press, a).

The low rates of "no opinion" responses encountered by surveys addressing diverse and obscure topics suggest that most people are capable of providing some answer to whatever question is put to them. Where values are labile or absent, however, these responses may reflect a desire to be counted, rather than deeply-held opinions. The recently-commissioned National Academy of Sciences panel on "Survey Measurement of Subjective Phenomena" is one sign of the growing realization that existing procedures are not up to the tasks put to them. Just as decision makers confronted with climate-related problems cannot assume that an acceptable decision-making tool is available for the asking, they cannot assume that someone is able to find out what the public thinks about any and every question that comes to mind.

5.3. Research Plan

Two types of research projects will be needed. One will study the ways people make complex evaluative judgments, and how they decide to act or to continue to wait in situations of uncertainty. There are undoubtedly cultural differences in option evaluation and in the conditions under which preventive or corrective action will be undertaken. One of the most important determinants of cultural differences in evaluating the pros and cons of taking action is likely to be the degree of control over nature people perceive as possible and/or appropriate. Perceived control is known to be a major determinant of individual differences in many aspects of behavior within American culture. We need to know more about cultural differences on this and related dimensions if we expect to understand how various regions and countries around the globe will respond to information about the possibility of CO_2-induced climate change. In addition, we must examine whether people can be taught or influenced to evaluate response strategies in different (and more adaptive) ways. If so, what are the political and ethical implications of altering the evaluation process? For example, simply discussing low probability events (such as a nuclear plant meltdown) may increase their perceived probability and thereby affect the evaluation of response options involving those events. Research must assess such effects in the evaluation process and their policy implications. It must also address what parameters policy makers use in evaluating a set of alternative options. Are they likely to overemphasize certain aspects of a given course of action (e.g., the technical feasibility) and to neglect others (e.g., fears, attitudes, rivalries, etc., that may render a technically sound plan impossible to implement)?

A second program of research is needed to develop improved methods for surveying attitudes toward the issues raised by climatic change. Unlike the traditional survey with its philosophy of having impassive interviewers bounce stimulus questions off objectified respondents, these new methods may include structured interactions, designed to illuminate issues by presenting alternative perspectives for the respondents' consideration; they may use a variety of convergent methods; they may involve iterative procedures, in which the respondent goes through the issues several times until a feeling of closure is reached (or rejected, because no resolution seems possible). Formulating items would require the services not only of communications specialists, expert in expressing clearly the question that interests the sponsor of the study, but also substantive experts (e.g., philosophers, climatologists) able to tell whether the question itself was well conceived.

Many of the disparaging remarks one hears about the irrationality shown by "the public" in its responses to attitude surveys may reflect the inadequacy of survey design for the reasons just discussed. "Garbage in-garbage out" holds when addressing people as well as computers. Once developed, these newer, more sophisticated survey methods should be applied to find out what people want from their leaders in response to climate change and how they themselves intend to deal with the issues under their own control. "The people" is usually defined as those individuals represented by a probability sample of adults who can be found and will respond. For some novel issues, even the most sensitive interactive interview may not be able to generate enough understanding to make the results useful. In such cases, the public weal may be better served by questioning intact groups with some interest in the topic, or paying a representative group of citizens to follow the issues over a period of time, developing expertise.

When people do not have articulated opinions on specific risk issues, it may be the job of the responsible interviewer to help them develop positions consistent with their underlying values. One aspect of this aid is offering ways to think about a problem; a second aspect is working out together the implications of various policies that people might consider advocating.

Such analyses are not pulled from one's sleeve. A team of philosophers, economists, psychologists, sociologists, and others is needed to (a) articulate or speculate about the concerns motivating people's attitudes; (b) examine the implications of these positions; (c) offer alternative perspectives, e.g., how *might* people think about intergenerational equity issues or relations between people and other species? An unexploited source of potential insight would be working out the implications of various philosophies of life. Although only a minority of society might subscribe to these philosophies, all might learn something from exploring what a coherent libertarian, Marxist, Hindu, Christian, or Dadaist approach to nature and its challenges would be.

5.4. Research Outcomes

The following products can be visualized: (1) Basic methods for survey research into attitudes regarding the consequences of climate change and options for dealing with them. (2) Reports of empirical studies into how various population groups evaluate the research options generated in project 3. (3) Analyses interpreting what the public wants, what it might want if "better informed" and what would be the consequences of adopting policies consistent with those desires. The participating disciplines should include psychology, sociology, philosophy, anthropology, and economics.

6. Project 5. Anticipating and Clarifying Conflicts Created by the Inequitable Effects of CO_2-Induced Climate Change; Offering Paths of Resolution

6.1. Primary Research Questions

(1) How will climate change pit nation against nation, group against group?
(2) What commons dilemmas will be created (or exist already)?
(3) What sorts of mistrust and misunderstanding will emerge and can be avoided?
(4) Can frameworks or options be devised for conflict resolution?

6.2. Background

One of the major consequences of a CO_2-induced climate change is likely to be a significant change in distribution of resources. Intraregional, intranational, and international redistributions are likely to occur. However, nobody knows in advance exactly what those redistributions will look like. Local, national, and international communities will be deciding what to do, if anything, about possible redistributions. To anticipate and inform their decisions, we need to know how people make judgments about resource distribution, and what conditions might foster the most harmonious outcomes. How do people solve distribution problems, and how satisfied are they with their solutions? Which procedures and outcomes are considered fair or just; which promote cooperation and goodwill rather than conflict and resentment?

Perception of distributive justice is currently an active area of study in psychology. Extending this work to the topics raised by climatic change will require asking the following questions:

(a) Are distributional evaluations situation-specific, or do people apply general principles? For example, do people conceive of balancing inequities in different situations? If agriculture in certain regions or countries is hurt by the CO_2-induced climate changes, should they be given advantages in other areas (e.g., fewer trading restrictions)? Will they demand such advantages?

(b) What are the consequences of creating distributions that are judged as unjust? How do people react cognitively, emotionally, and behaviorally? Are over-reward and under-reward reacted to differently? Are unjust distributions less distressing when they are expected? How will regions and countries act if they feel the CO_2 issue is not dealt with equitably by local, national, and world decision-making bodies?

(c) How is anti-social behavior leading to inequities perceived and handled? If a country continues to burn a large amount of fossil fuel and that is perceived as contributing to the CO_2 problem, how does that influence the way other countries are willing to share resources with that country? Will CO_2 issues be seen in isolation or lost in the broader context of relations?

(d) What characteristics of the interaction between parties affect distributional behavior; e.g., do the parties presently in control of distributing certain resources do so differently depending on whether or not they expect other parties to be distributing them in the future? Would they, if long-term dependency were clarified?

(e) How do people judge the relative importance of equality of *opportunity* and equality of *outcome*? Is it enough to give different regions an equal "opportunity" to develop energy sources other than fossil fuels, or should they be assured equal amounts of future energy? Brickman (1977) argues that people will accept inequality in opportunity in order to achieve equality of outcome and that this preference for equality of outcome is greater when people are in Rawls' (1971) "original position" and don't know whether they will be advantaged or disadvantaged by the inequality in opportunity (see below).

(f) Is there a different in *satisfaction* versus *fairness* judgments of different-shaped distributions? Brickman (1975) has argued that positively skewed distributions are preferred to equal or negatively skewed ones, but that equal distributions are judged fairest. If satisfaction and fairness judgments are somewhat independent, the presumption that they are not needs to be challenged in the interests of generating more accepted solutions.

(g) Do *procedures* for distributing resources affect justice evaluations independent of the *outcomes* themselves? In solving this global problem, how important is widespread regional and international participation in the research and the decision process? Folger (1977) argues that procedures and outcomes interact in determining justice evaluations. We need to know more about how they interact.

(h) Do *public* judgments of fairness differ from *private* ones? Rivera and Tedeschi (1976) argue that people express much more satisfaction with being over-rewarded when their opinions are expressed privately than when they are in an experimental situation. Is this true for nations as well? If so, what effect does it have on worldwide cooperation and the structuring of negotiations?

To date, these kinds of issues have almost invariably been studied in a laboratory setting with tasks, rewards, and situational context determined by the investigator. Another major shortcoming of past research is that it has essentially imposed a simple-minded formulation of equity, according to which outcomes should be proportional to inputs. As a result, the typical study provides relative input information and asks for judgments of what are just distributions of outcomes, clearly implying that respondents should base their judgments on the relative input information. We need research examining the relevance of equity theory to CO_2-induced redistributions of resources. What are the "inputs" in this situation (e.g., proportional contribution to the CO_2 buildup over the past 100 years or over the next 10 years)? Are relative inputs of different regions and nations considered relevant to a global strategy to reduce suffering from redistributions? To what extent do the attitudes in existing studies simply reflect the particular historical, cultural, and economic conditions of Americans who participate in psychological experiments, rather than some fundamental characteristic of human nature? Sampson (1975) has argued that the desire for "equity" reflects the emphasis on agency and competition which presently dominates Western civilization, whereas "equality" reflects an emphasis on communion and cooperation, which is characteristic of other cultural and economic systems, both present and past.

Although the study of distributional justice is clearly relevant to understanding how people can and will handle the CO_2 issue, a number of innovations and modifications to current research paradigms will be necessary. An important lacuna in existing research is the study of judgments about changes in existing resource levels. Climate changes will necessarily involve both goods and bads, i.e., benefits for some, but costs for others. Previous research has generally been limited to the study of positive goods, or "rewards". One exception is Brickman and Bryan's (1975, 1976) studies of *transfers* of goods between two parties. Their approach could profitably be applied to the study of regional and national changes in resource distribution. Attention will also have to be given to who is the initiating agency of such transfers. Some people might believe that equity, like effective public participation in decision making, cannot be given, but must be taken. A very special kind of agency is nature. When are inequities viewed as naturally caused? What redress is asked for such inequities?

6.3. Research Plan

In planning relevant research on resource distribution, it must be recognized that the CO_2 phenomenon presents an unusually complex distribution problem requiring new methods and conceptualizations. First, there is not simply a single resource to be allocated among everyone, but rather different resources go to different people: for some people it is energy (by burning fossil fuels), for others it is fish supply (possibly affected by changes in ocean temperature and currents), and for still others it is crops (affected by changes in precipitation). In addition, each resource may have different values to different people. Previous research has dealt exclusively with the same resource being distributed to all.

Second, the potential climate changes are global in nature. People of all nationalities and cultures could be affected. Previous studies have focused almost exclusively on Americans' judgments of distribution fairness.

Third, given the uncertainties of the CO_2-induced climate changes (if they happen at all, what form will they take in any given locality, and how will they sum up across any given nation?), a collectivity of individuals, regions, and/or nations, must decide on a subsequent distribution of goods *without* knowing where any given individual (region or nation) will fall on that distribution. Previous research has generally been limited to resource allocation where each individual knows where he or she will stand in the various outcome distributions under consideration.

In this respect, making decisions about the most just way to handle CO_2-induced redistribution of resources presents us with a real-world analogue of John Rawls' "original position". One aspect of the present project would be a programmatic effort examining the empirical validity of Rawls' theory, expanding on Brickman's (1977) demonstration of the applicability of psychological research methods. His theory suggests that a consideration of the CO_2 buildup by people who find themselves under a "veil of ignorance" as to their future situation can result in just decisions that are recognized as such by all concerned. Knowledge of the empirical validity of Rawls' formulation, and the degree to which it applies to international cooperation on the CO_2 issue, would obviously be invaluable. One further question would be the effect of variations in degrees of risk and of uncertainty; although no one knows exactly how they will be affected by possible climate changes, some people will face much more uncertainty than others, and even among those confronted with the same level of uncertainty, some stand a chance to gain or lose much more than others.

A second subproject would involve applying Rawls' theory to groups (or their representatives) negotiating from an original position. To date, research has been limited to *individual* negotiators representing only themselves. Yet international cooperative decisions on how to handle the CO_2 issue will undoubtedly be made by a handful of people representing regions or entire nations. We know little about how such representatives make distributional decisions and resolve conflicts between their own interests and those of their group.

Research on *judgments* about the most just way to deal with CO_2-induced climate changes must be complemented by research on *behavior* of people facing various degrees of risk and uncertainty, and voicing conflicting claims. In many cases, the CO_2 situation qualifies as a "commons dilemma" or "social trap" (Dawes, 1980). In these situations, a group of individuals, each acting in a way apparently best personally, produce an effect that is bad for all of them. Since the effect of each individual's action is relatively small, it cannot be seen as either causing or potentially alleviating the problem. Rather, it is the sum of individuals' actions that creates the problem, and only their collective action can alleviate it. Referring to the CO_2 context, the energy and/or forestry policy of any given region or country may not discernibly affect the CO_2 levels, and thus it may be difficult for individual regions or countries to decide to curtail fossil fuel consumption or deforestation when their own impact on the problem seems negligible.

Both theoretical and empirical work on behavior in commons dilemma situations is in its infancy. We know very little about circumstances under which cooperative solutions are fostered. Early work has focused on such variables as group size, degree of discussion and communication among members, relative amounts to be gained and lost, etc. It needs to be expanded, with a greater emphasis on group behavior at the level of regions and nations, and address cognitive issues more directly, i.e., what determines when a situation is interpreted as a commons dilemma?

6.4. Research Outcomes

The following products can be visualized: (1) Innovative methodologies for studying resource distribution decisions; (2) Empirical studies of distributional judgments and behavior; (3) A comprehensive delineation of the alternative ways these issues are addressed in different cultures and different historical periods. The participating disciplines should include psychology, political science, sociology, economics, philosophy and history.

7. Guiding Principles

Several principles are fundamental to the success of the above-proposed research projects and to their utilization in dealing with possible CO_2-induced climate changes.

7.1. Interdisciplinary Focus

Psychological issues cannot be studied in a vacuum. When we ask how people perceive the world, how they make judgments and decisions, and how they behave as members of groups and/or nations, we are asking questions which can only be answered meaningfully by a multidisciplinary team including sociologists, anthropologists, historians, political scientists, and philosophers, with backup from climatologists and other technical specialists. In this perspective, we are echoing the sentiments of the World Climate Conference: "Efforts should be made to ensure that the environment in the institutions in which the projects will be carried out is favorable to interdisciplinary research which is a necessary condition for progress in such a complex field of investigation" (1979).

If interdisciplinary research is so good, why is there so little of it?

One reason is that no one is trained to do it. Rather, the interested parties are trained in their respective professions and are drawn to interaction via involvement in some substantive problem. The nascent fields that result tend to be strong on commitment and on the sort of fresh ideas produced by rubbing strange disciplines together. Weaknesses lie in decreasing quality control and conceptual clarity as one leaves traditional fields, with their strongly developed sense of "what good is" in the way of research. Thus, although the potential payoffs are large in interdisciplinary research, so are the problems and pitfalls.

A second reason why so few people take the interdisciplinary plunge is that there are often rather meager rewards for doing so. University departments like people who can teach the traditional courses and be evaluated by the usual criteria. Real-life problems calling for many perspectives are often in the lock of one discipline (i.e., economics, engineering), which is unwilling to give more than lip service to sharing attention or resources.

A final problem is the lack of persuasive models for how interdisciplinary research might be conducted.

The simplest mode of interaction would be to compare terminology to reveal the hidden assumptions in our frames of reference. If we do not clarify such assumptions, we risk ethnocentric misconceptualizations and the attendant dangers of (a) not realizing that the terms we used have different interpretations in the populations we are studying (and with whom we must communicate), (b) deluding ourselves into thinking that the focus of our research life is also the focus of our respondents' lives, (c) misinterpreting our subjects' lives by failing to see their internal logic. Clarifying the assumptions our psychological work makes about the world in which behavior is embedded is a first step toward establishing the generalizability of our results and developing a theory of context to complement our more evolved theories of the individual.

A higher level of contact can be seen in sorties across disciplinary boundaries, returning with bounty in the form of stolen methodologies. Many major advances have been the result of such appropriation. Kates (1962) and others changed geography by introducing attitude measurement, thereby freeing the field from reliance on purely physical measures.

A dangerous limitation to such contact is borrowing tools from another domain without the full appreciation of their limitations that comes from extended professional socialization in that domain. Hexter (1971) characterized historians borrowing notions from the analytic philosophy of science, just as philosophers were becoming disillusioned with analytic methods, as "rats jumping aboard intellectually sinking ships." Similar criticisms might be leveled against psychohistorians embracing psychoanalysis as an analytic tool just as psychologists were giving up on it as a research methodology, or cliometricians applying economic analysis to historical settings just as economists are questioning the validity of their measures.

The highest form of interdisciplinary work is actually working together with people from other disciplines. Although full collaboration is rare, its salutary effects are widely enough acknowledged for working together to be regarded as virtuous. Only by extended interaction can we learn to incorporate other disciplinary perspectives in our own work. Since most collaborative works are unique products of the interactions between the perspectives and personalities brought to bear on a particular problem, there are no firm standards or systematic means to ensure quality control. Disciplines progress by trial and error. Active collaborations attempt to create new, integrative disciplines in whole cloth at first crack. They can't always do it, and may not always be able to assess the validity of their attempt.

An alternative goal for collaboration is not to create a new discipline, with the capacity for getting the right answers to a newly, but narrowly, defined set of questions. Rather, one can acknowledge that there are no "right" answers (or at least no way to be certain that we have come across them) to questions rich enough to draw talents from a variety of fields. What one can hope for is to avoid getting the wrong answers, with each discipline helping to avoid particular kinds of errors.

7.2. *Cross-Cultural Emphasis in the Context of an International Research Effort*

Just as research on the psychological dimensions of possible CO_2-induced climate change must be cross-disciplinary, it must also be cross-cultural. The problems of ethnocentric research programs are nowhere more evident than in the current crisis in social psychology. As Triandis (1975) has argued, the study of human behavior in a single culture, by researchers from a single culture, results in "knowledge" with very limited replicability and generality. The resulting theories do not account for the complex interactions between person and context. Instead, we need cross-cultural studies in the tradition of Whiting (1964), or along the lines of the more recent "ecological functionalism" approach (Berry and Dasen, 1974). These cross-cultural research programs attempt to tie the nature of the physical environment to the nature of the social environment and to particular psychological phenomena. They often focus on new higher-order variables that account for much of the variance in social behavior, and that can be systematically related to culture-specific behavior (Whiting, 1968).

Since CO_2-induced climate changes will be worldwide, it is clear that the psychological studies proposed above must be carried out in the full range of impacted cultures if we are to understand and improve global response to this issue. In order to implement such a research program, an international effort is needed. Effective cross-cultural work cannot be achieved by United States researchers alone. Research design, data collection, analysis, and interpretation will all suffer if we do not achieve international collaboration from the very beginning. As a starter, the present proposal has been critiqued by well-known psychologists with different cultural perspectives. Later, selection of investigators and research centers to conduct the proposed projects should ensure that a variety of nations and cultures will be involved in planning and conducting the basic research.

It should be emphasized that the participation of the international research community is important not only for the quality of the studies themselves, but also for their acceptance and utilization by the international community of policy makers. If, for example, the Third World has not participated in a cooperative effort to research the societal impacts of possible climate change, the likelihood of their participation in any eventual cooperative response is greatly diminished. As Rep. George Brown has argued,

> It is a matter of efficiency in the sense of using the vast observation methods and data resources around the world. It's also a matter of political education in the sense that whatever joint world efforts might be required in climate – such as related to CO_2 – will inevitably only be viable if the world has jointly obtained and studied the data. . . . Lip service to this concept and recitation of previously formulated joint scientific efforts is not sufficient to fill the mandate intended here (1979, p. 3).

The National Climate Program Act has already recognized the importance of a coordinated international effort by stipulating that "measures for increasing international cooperation in climate research, monitoring, analysis, and data dissemination must be included as a basic element of the National Program" (National Climate Program, Preliminary 5-Year Plan, July, 1979, p. 68).

7.3. Combination of Basic and Applied Research

A third guiding principle is that one needs a mix of basic and applied research. Experience has shown that leaping into highly specific problems without a theoretical framework or carefully developed methodology tends to be unproductive. "As soon as you break a practical problem into its more basic elements you are faced with numerous fundamental questions, requiring basic research. You can *not* make progress in the solution of the practical problem unless you solve the *basic* problems" (Triandis, 1978, p. 385). For some topics, the basic research background already exists; for others, it will have to be developed. On the other hand, without a constant reminder of the applied focus, academics do tend to pursue their own agendas. With attention to this problem, we feel that on many topics, the path from basic research to application may be fairly short. Although it may be difficult for a mission-oriented agency to envision itself conducting basic research in the social sciences, a similar attitude by all agencies would mean that the basic infrastructure for solving applied problems would never be built.

Decision Research
A Branch of Perceptronics
Eugene, Oregon

The Wright Institute
Eugene, Oregon

References

Berry, J. W. and Dasen, P.: 1974, *Culture and Cognition: Readings in Cross-Cultural Psychology*. London: Methuen.

Brickman, P.: 1975, 'Adaptation Level Determinants of Satisfaction with Equal and Unequal Outcome Distributions in Skill and Chance Situations', *Journal of Personality and Social Psychology* **32**, 191–198.

Brickman, P.: 1977, 'Preference for Inequality', *Sociometry* **40**, 303–310.

Brickman, P. and Bryan, J. H.: 1975, 'Moral Judgment of Theft, Charity, and Third-Party Transfers that Increase or Decrease Equality', *Journal of Personality and Social Psychology* **32**, 156–161.

Brickman, P. and Bryan, J. H.: 1976, 'Equity Versus Equality as Factors in Children's Moral Judgments of Thefts, Charity, and Third-Party Transfers', *Journal of Personality and Social Psychology* **34**, 757–761.

Brown, G.: 1979, 'Implementation of the Climate Act', hearing before the Subcommittee of Natural Resources and the Environment.

Bryson, R. A.: 1974, 'A Perspective on Climatic Change', *Science* **184**, 753–759.

Burton, I., Kates, R., and White, G. F.: 1978, *The Environment as Hazard*, New York: Oxford University Press.

Dawes, R. M.: 1980, 'Social Dilemmas', *Annual Review of Psychology* **31**, 169–194.

Fischhoff, B.: 1979, 'Behavioral Aspects of Cost-Benefit Analysis', in G. Goodman and W. Rowe (eds.), *Energy Risk Management*, London: Academic Press.

Fischhoff, B., Goitein, B., and Shapira, Z.: 'The Experienced Utility of Expected Utility Approaches', in N. Feather (ed.), 1982, *Expectations and Actions: Expectancy-Value Models in Psychology*. Hillsdale, N.J.: Erlbaum.

Fischhoff, B., Slovic, P., and Lichtenstein, S.: 1977, 'Knowing with Certainty: The Appropriateness of Extreme Confidence', *Journal of Experimental Psychology: Human Perception and Performance* **3**, 552–564.

Fischhoff, B., Slovic, P., and Lichtenstein, S.: 1978, 'Fault Trees: Sensitivity of Estimated Failure Probabilities to Problem Representation', *Journal of Experimental Psychology: Human Perception and Performance* **4**, 330–344.

Fischhoff, B., Slovic, P., and Lichtenstein, S.: 1980, 'Knowing What You Want: Measuring Labile Values', in T. Wallstein (ed.), *Cognitive Processes in Choice and Decision Behavior*, Hillsdale, N.J.: Erlbaum.

Fischhoff, B., Slovic, P., and Lichtenstein, S.: 1981, 'Lay Foibles and Expert Fables in Judgments about Risk', in T. O'Riordan and R. K. Turner (eds.), *Progress in Resource Management and Environmental Planning, Vol. 3*. Chichester: Wiley.

Folger, R.: 1977, 'Distributive and Procedural Justice: Combined Impact of "Voice" and Improvement on Experienced Inequity', *Journal of Personality and Social Psychology* **35**, 108–119.

Gettys, C. F., Kelley, C. W., and Peterson, C. R.: 1973, 'Best Guess Hypothesis in Multi-Stage Inference', *Organizational Behavior and Human Performance* **10**, 364–373.

Hexter, J. H.: 1971, *The History Primer*, New York: Basic Books.

Kahneman, D. and Tversky, A.: 1973, 'On the Psychology of Prediction', *Psychological Review* **80**, 237–251.

Kates, R. W.: 1962, *Hazard and Choice Perception in Flood Plain Management*, Chicago: University of Chicago, Department of Geography, Research Paper No. 78.

Lichtenstein, S., Slovic, P., Fischhoff, B., Layman, M., and Combs, B.: 1978, 'Judged Frequency of Lethal Events', *Journal of Experimental Psychology: Human Learning and Memory* **4**, 551–578.

National Academy of Sciences: 1978, *Committee on Socio-Economic Effects of Earthquake Prediction*. Washington, D.C.: The National Academy of Sciences.

Nuclear Regulatory Commission: 1975, *Reactor Safety Study: An Assessment of Accident Risks in U.S. Commercial Nuclear Power Plants*. WASH 1400 (NUREG–75/014). Washington, D.C.: The Commission.

Nuclear Regulatory Commission: 1978, *Risk Assessment Review Group Report to the U.S. Nuclear Regulatory Commission*, NUREG/CR–0400, Washington, D.C.: The Commission.

Opie, J.: 1979, *America's Seventy-Year Mistake: Settlement and Farming on the Arid Great Plains, 1870–1940*, Paper presented for Panel on Social and Institutional Responses, AAAS-DOE Workshop on Environmental and Societal Consequences of a Possible CO_2-Induced Climate Change, Annapolis, Maryland.

Poulton, E. C.: 1968, 'The New Psychophysics: Six Models for Magnitude Estimation', *Psychological Bulletin* **69**, 1–19.

Rawls, J.: 1971, *A Theory of Justice*, Cambridge, Ma.: Harvard Univ. Press.

Rivera, A. N. and Tedeschi, J. T.: 1976, 'Public Versus Private Reactions to Positive Inequity', *Journal of Personality and Social Psychology* **34**, 895–900.

Sampson, E. E.: 1975, 'On Justice as Equality', *Journal of Social Issues* **31**, 45–64.

Schneider, S. H. and Mesirow, L. E.: 1976, *The Genesis Strategy*, New York: Plenum.

Shaklee, H. and Fischhoff, B.: 1979, *Strategies of Information Search in Causal Analysis*, Decision Research Report 79–1.

Slovic, P., Fischhoff, B., and Lichtenstein, S.: 1979, 'Perceived Risk', *Environment* **21**, 14–20, 36–39.

Slovic, P., Fischhoff, B., Lichtenstein, S., Corrigan, B., and Combs, B.: 'Preferences for Insuring Against Probable Small Losses: Implications for the Theory and Practice of Insurance', *Journal of Risk and Insurance* **44**, 237–258.

Triandis, H. C.: 1975, 'Social Psychology and Cultural Analysis', *Journal for the Theory of Social Behaviour* **5**, 81–106.

Triandis, H. C.: 1978, 'Basic Research in the Context of Applied Research in Personality and Social Psychology', *Personality and Social Psychology Bulletin* **4**, 383–387.

Tversky, A. and Kahneman, D.: 1980, 'Causal Schemata in Judgment under Uncertainty', in M. Fishbein (ed.), *Progress in Social Psychology*, Hillsdale, N.J.: Erlbaum.

Whiting, J. W. M.: 1964, 'Effects of Climate on Certain Cultural Practices', in W. H. Goodenough (ed.), *Explorations in Cultural Anthropology*, New York: McGraw-Hill, pp. 496–544.

Whiting, J. W. M.: 1968, 'Methods and Problems in Cross-Cultural Research', in G. Lindzey and E. Aronson (eds.), *Handbook of Social Psychology*, Vol. 2. Reading, Mass.: Addison-Wesley, pp. 693–728.

World Climate Conference, Geneva: World Meteorological Organization, 1978.

Whyte, A. V. T.: 1977, *Guidelines for Field Studies in Environmental Perception*, Paris: UNESCO.

INTRODUCTION TO: ANTHROPOLOGICAL PERSPECTIVES ON CLIMATE CHANGE

In moving from the contributions of psychology to climate change to those of anthropology, we are moving from Fischhoff and Furby's focus on perception and the more cognitive aspects of decision-making to Torry's focus on behavioral responses to actual climate change. The Fischhoff and Furby paper was prospective in its time orientation. Climate change might or might not come. Torry, as an anthropologist, knows that climate-change-surrogates such as soil depletion, drought, flood and frost are part of the folk experience in all parts of the world, and considers how to examine the actual adaptive behavior that takes place under environmental stress.

Torry chooses an interesting approach. Rather than giving a general overview of anthropological contributions to the comparative study of human response to environmental stress, he has chosen to construct a method of systems analysis. Using the household as a core concept, he shows how the tools of anthropology can be used by policy makers in facilitating adaptation of societies to climate change. Readers looking for a literature review of relevant anthropological studies can turn to Torry's excellent review articles, "Anthropological Studies in Hazardous Environments: Past Trends and New Horizons" (Torry, 1979a) and "Anthropology and Disaster Research" (Torry, 1979b).

The value of the present chapter lies in its presentation of a systems analytic framework for analyzing past adaptive behavior and for forecasting future adaptive behaviors. The approach offered in the "household organizational environment system" (HOES) is essentially that of human ecology. Sociologists will recognize its kinship to Otis Dudley Duncan's POET (people, organization, environment, technology). The uniqueness of HOES as an analytic construct lies in its focus on aggregates of households, rather than on "people" or "community". The household is indeed, as Torry says, society's most elemental unit of collective resource management. An aggregate of households enmeshed in an organizational environment is a more dynamic concept than community, accessing a richer, more complex set of interactions and institutional patternings. Furthermore, the organizational environment construct can account for as many levels of the larger social environment as the analyst wants it to. HOES is a social construct. Unlike Dudley's POET, it does not include the physical environment, only what human beings have created to deal with that environment. It can therefore be studied in every type of climatic setting.

Because Torry is describing a systems-analytic tool, the illustrative concreteness of specific cases of HOES within particular cultural settings and under particular climatically induced stresses is lacking. There is no assumption that there is one grand system for dealing with climate stress, only suggestions for identifying the system any given society has evolved for dealing with its climate hardships, and indications of useful intervention points in that system.

Because this is essentially a new approach, the chapter should be considered as an

R. S. Chen, E. Boulding, and S. H. Schneider (eds.), Social Science Research and Climate Change: An Interdisciplinary Appraisal, 205–207.

invitation to the reader to enter into the process of elucidating and applying the HOES system. Here, for starters, are some suggestions for further development of HOES which the reader might want to consider, offered briefly under the rubrics of (1) the concept of household, (2) household/organizational environment interface, (3) uniqueness of climatically induced stress compared to other stresses, (4) implications of interactive processes for central planning and (5) relevance of varying time frames.

The Household

It is not households that interact, but people living in households. Each household is in itself a microsystem. Responses to climate stress are forged from the special skills and knowledge stocks of individual family members, and these skills are strongly gender- and age-related, particularly in traditional societies. Intra-household systems analysis will be needed to supplement inter-household system analysis to uncover critical roles played by women, children and the elderly in climate stress situations. The rich literature on family adaptation to stress and crisis, ranging from depression-years studies of loss of employment and war-years studies of adjustments to wartime hardships to more recent studies of family response to natural diasters, provides a strong conceptual base for further research on household adaptation to climate-induced stress. The HOES framework can and should include that theme of analysis.

Household/Organizational Environmental Interfaces

Each household interfaces uniquely with the organizational environment, according to its unique internal composition, generating surpluses and deficits which enhance or deplete the stress-meeting resources of the aggregate. The interactions are complex and not necessarily beneficent. For example, Spitz (1981) points out a recurring historical phenomenon: incomes for better off farm households rise in drought while poorer farm households frequently lose everything. The interface processes that produce such differences within a single aggregate of households should be studied.

Uniqueness of Climatically Induced Stress

The household might almost be defined as a social invention to deal with the seasonal patterns of weather change that each social group experiences. Cushioning the alternations of hot and cold or wet and dry is what households are organized to do. Dealing with variability in the patterns is also what they are organized to do. When the variability overloads traditional coping mechanisms, the households may still be the "experts", unlike the situation in crises induced by disasters quite outside society's cultural experience. The implications of that fact for planners must be considered.

Interactive Processes and Central Planning

On the one hand the HOES focus on households may result in an exaggeration of the potential of households for autonomous action and a downplaying of the constraints placed on them by the larger system of which they are a part. On the other hand top-down development planning undertaken without knowledge of local skills and resources has done a lot of harm in both industrialized and nonindustrialized societies. One striking example is in the ignoring by development experts of the skills of nomadic families in using arid lands productively, particularly in the Sahel. HOES can be utilized to identify points of interaction between decision-makers and household aggregates, points at which the expertise of household aggregates can come into dialogue with the knowledge of decision-makers.

Relevance of Varying Time Frames

Every organizational environment is the product of the unique social rhythms of a given society, including the life span rhythms of individuals, households and household aggregates-cum-ancestors. Climatic stress is experienced locally in the context of the prevailing cultural time frames, which may be multigenerational, while the planner may be thinking only of this month, or this year, or this decade. The way time is experienced determines preferred solutions even in the short term. It should be possible for social rhythms and cultural time concepts to be treated as part of the HOES schema, making them a resource rather than a constraint in meeting crises.

It is my conviction that the HOES schema can encompass the processes and dimensions just discussed and more, and will contribute substantially to research and planning regarding response to climate change.

E. BOULDING

References

Spitz, Pierre: 1981, The Economic Consequences of Food and Climate Variability, Working paper. United Nations Research Institute for Social Development, Palais des Nations, Geneva.

Torry, William: 1979a 'Anthropological Studies in Hazardous Environments: Past Trends and New Horizons', *Current Anthropology* **20**, 517–540.

Torry, William: 1979b, 'Anthropology and Disaster Research', *Disasters* **3**, 43–52.

WILLIAM I. TORRY*

ANTHROPOLOGICAL PERSPECTIVES ON CLIMATE CHANGE

1. Introduction

Natural and physical scientists deserve credit for pointing out the prospects of global climate changes, but the properties of climate variation that fall within their scope of expertise are without any direct policy significance. To have any impact on planning and policy making cryospheric, atmospheric, and biospheric anomalies must be interpreted with a view to the question: How will this or that irregularity affect our social and economic security? Climate change issues in the realm of policy boil down therefore questions of societal adaptation. Social science comes into play here in some important ways. It redefines purely physical events in terms of societal factors. It associates these factors with specific security problems. And it identifies alternative courses of policy action compatible with the goals of averting and alleviating crisis or, in other words, protecting public security.

Climate stresses create public security problems when they alter the availability and the productivity of natural and man-made resources within and among geopolitical regions. The social scientist's brief is that of determining social/economic costs and benefits associated with such changes and gauging, at the very least, the fitness of social institutions to minimize costs and risks.

When we talk about impact prediction, two basic principles bear remembering. First, human exposure and response to climate stresses develop as products of social institutions. Second, specific institutions must cope with multiple stresses. That risk and loss (or, for that matter, gains) result from climate's interaction with many sources of stress, and that diverse stressors may evoke similar or identical adjustments, have important implications. What is called for is an extension of our terms of reference to situations besides those marked by climatic disturbances (e.g., droughts and floods) so that community level stresses observed in non-climate contexts can be treated as analogues for predicting likely impacts of climate. By ranging over a plurality of settings where adjustments between societies and their physical surroundings break down, we acquire a broader base of observations for evaluating the likely effects of climate stress.

We begin by setting forth a schema of major variables incorporated into resource management systems normally studied empirically by sociocultural anthropologists. The schema (1) delimits societal systems and sub-systems central to research in anthropology; (2) identifies resources essential for system security, and; (3) specifies modes by which principal resources managers in the system adjust to or instigate stresses from the environment. This exercise lays the foundation for the presentation of societal stress issues taken up in later sections. Even though the schema does not mesh fully with our analysis of issues, it stand as a fixed point of reference from which the anthropological dimensions of any given issue can be seen. Discussion centers on the developing

R. S. Chen, E. Boulding, and S. H. Schneider (eds.), Social Science Research and Climate Change: An Interdisciplinary Appraisal, 208–227.

world which is populated by a high proportion of persons acutely vulnerable to environmental perturbations.

2. Units and Targets of Analysis

2.1. Social Units

The household, in one form or other, represents society's most elemental unit of collective resource management. Anticipated and actual responses made by households to mass disturbances, climate or otherwise, dictate adjustments at higher organizational levels. Sociocultural anthropology has specialized in the study of the household, or more accurately, local associations of households, organized for maintenance within an environmental system. Hence, it is at this level, we believe, that anthropologists can make their greatest contribution to the study of climate impacts.

In the past, the anthropologist's terms of reference have tended to embrace household interactions within closely knit socio-spatial units, but more recent work on economic development and cultural change has had to adopt a larger perspective. Whatever might have been the case in the past, small communities no longer operate as adaptive isolates.

Markets and administrative agencies now assume many functions previously discharged by households and small communities. Virtually everywhere today therefore the household as consumer, producer and distributor of resources stands at the juncture of three interlocking sources of supply: market networks, state bureaus, and grass-roots organizations. Access to commodities and services supplied by these systems in turn mediates interactions between the household and its natural habitats. Household aggregates and their *organizational environments* (i.e., the networks of households, commercial institutions, and administrative agencies) in short provide the research setting within which most sociocultural anthropologists work. In rural sectors of developing countries, for instance, the organizational environment can include field offices and extension staff from central ministries, district and sub-district level administrators, party officials, members of parliament, entrepreneurs and petty traders of many stripes, cooperative societies and agricultural banks, indigenous councils and corporations, neighboring households, and the diaspora of kin and friends settled elsewhere in the cities, including those of Europe and America.

The expression "household organizational environment system" (HOES) is devised by us because we need something other than the conventional terms to signify this outreach. Constructs such as "community" or "people", as frequently applied by anthropologists and sociologists, connote: that the group under consideration is rural; that it is a unit of organization capable of taking concerted action; and that it exists somehow apart from the external agencies which impinge on it. To deal with the social environment within which households cope today, almost everywhere in the world, we need a more inclusive concept. Here we have a term that takes in a large if not complete assortment of units empirically studied by sociocultural anthropologists, and can be used in research settings representing a variety of cultural situations and climatic conditions.

2.2. *Resources and Resource Management and Security*

We mention "security" with some frequency. We associate security with the household's control over resources of two kinds, the first being products and services and the second being assets. Exogenous products and services originate in both the natural and organizational environments. Food, cash, employment, capital, and credit, for instance, are extracted from these environments by the household and converted into assets. Assets divide into productive and non-productive items. The former may include land, labor and working capital, and the latter jewelry, furnishings, leisure vehicles, insurance policies, and social alliances or other stores of value. Of course, non-tangibles, such as titles and credit, can fall into either bracket, depending on their use.

As a socio-economic entity, the household does what it can to maintain or to improve upon its current asset position. Its strategies depend upon the structure and stability inhering in the organizational environments, and the skills and opportunities by which domestic groups can influence the supply and distribution of resources. It would be well to emphasize that as anthropologists, our chief concern is not the household itself, nor its commercial and administrative environments *per se*. We regard our focal interest to be the behavior of households *as aggregates*. Our mission is finding principles governing how households as local assemblages maintain identifiable levels of assets, which translate into household security, *vis-a-vis* one another and the organizations with which they transact.

2.3. *Stress-Crisis Adjustment*

Asset regulation, as it relates to household units, provides an anthropological "handle" on the question of climate change. We assume that official concern about large-scale stresses (crises) will lead governments to seek responsive programs and policies. Informed planning for crisis intervention and prevention requires a diagnosis of symptoms by decision makers before they administer remedies. So, before promulgating controls or setting up agencies, the powers-that-be must acquire and accurately interpret data that specify patterns of risk and loss distribution throughout major sectors of society and ascertain how well these sectors can handle deprivation. Crisis, as we have conceptualized it, denotes the threat or occurrence of abnormally high levels of asset depletion experienced by significant numbers of locally aggregated households. In all likelihood we would expect many such congeries spread over large regions of the globe to be affected by radical changes of climate.

The strategy for anthropological research on the consequences of climatic change should address itself to the choices and consequences of adjustments made by households through their organizational environments to outside perturbations. Adjustments will be understood to signify the regulation of supply and quality of domestic assets within certain limits of admissibility.

The HOES stress/crisis adjustment process includes the following components:

2.3.1. *Household Asset Position*. Asset security relates to several asset classes (land, labor, income, savings, etc.) and three levels of availability, each a product of asset quality and volume: high, medium, and low. Actual and perceived availability may be at variance, but such distinctions would best be left until some future stage of project formulation.

A household's selection of crisis adjustment modes depends among other things on real or anticipated post-impact asset levels. Assets available before the impact and the balance of asset losses and increments existing during the crisis jointly set these levels. The modal pattern of response in the case of affluent households whose assets have not been or are not expected to be put on the line might be: do nothing, accumulate assets (sometimes at the expense of less fortunate neighbors); or perhaps redistribute property to other households gratis or at a charge. This latter option can lead to a transfer of resources to components within the system (e.g., friends, kin and charitable organizations).

Households with medium or small asset holdings may initiate several different responses ranging from comparatively low-cost, low-energy activities to drastic manuevers, terminating in the dissolution of the household. Finally, all adjustment steps, save the last, lead to calls upon external assistance. Responses elicited from public agencies to the household's stress/distress signals feed back to inform the household's choices of adjustment. These programs and agencies can supply products and services that are vital to the survival of the household.

2.3.2. *Administrative and Market Structures and Political Conditions*. The availability of resources furnished for household consumption by HOES agents comes under the control of influences originating outside the HOES domain. Some of the more prominent of these influences determining resource availability would be: politico-ideological precepts about public responsibility for guaranteeing all citizens certain minimal rights (e.g., to food and shelter); number and intensity of crises simultaneously draining the national treasury; fiscal solvency of regional and national governments; the scale, efficiency and structure of public administration at all levels in ordinary circumstances and during times of crisis; and levels of international aid consigned for immediate relief and long-term rehabilitation of victim communities.

2.3.3. *Buffers/Amplifiers*. These represent the raw materials, products and services households extract from their organizational environments.

The sections which follow describe prospective issues of socio-environmental stress that concern the household and administrative components of our HOES schema. While we do not attempt giving equal treatment to noteworthy interactions involving households and commercial institutions, we do move close enough to this nexus to adumbrate relevant problems worth delving into further.

3. Inter-Household Environment

The questions we raise in this section involve the social and economic factors that determine how households and aggregates react to conditions brought about by socio-environmental stresses and the pressures that promote actions that evoke stress. We begin by examining certain strengths and weaknesses typical of the way that local coping mechanisms operate in traditional and in modern problem solving contexts.

An assessment of the effectiveness of these coping mechanisms that takes into account their inner logic should constitute an initial step in laying plans for projects that combine rural development and stress control measures. We base our claims on several premises. First, indigenous institutions stand as tested solutions of organizational problems connected with the management of local resources. Even if old ways disintegrate, they may offer specific procedures or suggest important principles of organization on which agencies effecting reforms can build. Second, new programs are likely to be resisted or malabsorbed when planners/administrators fail to take cognizance of the reasoning that attaches target populations to existing ideas and practices even if these are ecologically unsound in view of current circumstances. Finally, crisis warning systems need to be able to detect where, in the progression of local adjustments to stress, conventional safeguards begin breaking down, and emergency relief programs need to be able to zero in on the coping problems faced by those in greatest need of assistance. Failures here jeopardize the long-term economic security of whole regions.

3.1. Stress-Adjustment Mechanisms: Past Strengths and Current Weaknesses

3.1.1. *Strengths.* The effectiveness of any adjustment mechanism is judged against the goals it has to fulfill. The farmer's mode of livelihood, for example, requires the physical survival of the household and the maintenance of its productive capacity at levels that allow farming to resume once the stress period ends. The process of coping with scarcity moves into gear when failure of the crops appears certain. Ceremonial life and food consumption constrict, sharing through mutual aid networks intensifies, extra effort is invested eeking out income from alternative forms of subsistence, and jewelry, household furnishings, and surplus livestock may be sold to purchase food. Mass migration and the liquidation of productive assets take effect when hardship reaches intolerable levels.

If the system of traditional buffers worked well during periods of stress, it was because: surpluses got redistributed from 'haves' to 'have nots:' people redistributed themselves in space, from blighted to more abundantly endowed tracts; and modes of livelihood were rearranged to compensate for the failure of principal crops (Torry, 1979; Jodha, 1975; Watts, 1980). The major strengths of these systems thus lay in the flexible ways they could reshuffle people and resources.

3.1.2. *Weaknesses.* Local societies with institutions that maintain their traditional stress buffering capacities are disappearing. The difficulties that beset traditional safeguards stem from two related factors: contradictions between former options and new constraints; institutional decay. Many conventions that once adapted human groups to localized pressures of population and environment now conflict directly with requirements tied to needs and interests identified with encompassing regional and national organizations. Infanticide and abortion, raiding and warfare, mass migration, hinterland colonization, and intensified use of unmanaged home-base resources clash sharply with norms of security imposed from the outside. Such practices consequently come under heavy attack and cease to function as they were intended. An even more pervasive pattern of decline involves the gradual crumbling away of *entire systems of social forms* owing to the aggregate's immersion in a cash economy and its subordination to legal and administrative agencies of the state.

What significance do these discontinuities have for the asset position of rural households and the economic viability of the aggregate? Where poverty sets in or gains a firmer foothold, what social structural and economic correlates reveal its causes and effects? Do patterns of risk and loss, for instance, develop through newly or more sharply drawn divisions of class and ethnicity? How do radical changes bear on the persistence of institutionalized exchanges of gifts, advances of loans, and the transfer of rights in productive resources? How is a worsening situation reflected in standards of housing, nutrition, and mental health? And what demographic and ecologic side-effects take place? Before setting their sights on stresses in store for rural populations of the future, policy makers would do well to probe the genesis of stresses that sap the economic vitality of large sectors of these populations today. To do so provides answers to questions such as those just raised.

3.2. Innovation

How successfully households work out ways for coping with socio-environmental stresses disruptive of established arrangements for living will depend ultimately upon their ingenuity for making constructive changes in set routines *voluntarily* and on their capacity for assimilating external innovations. Formation, transmission, adoption, and impacts of innovations are a vastly complex subject to which full justice cannot be done here. What we can do is identify selected problem areas of critical importance. These involve social structural and situational factors that influence the acceptance of innovations, and the social and economic ramifications of innovations that catch on. We focus in this section on the innovation process as it takes place at the level of the household and the aggregate, while programs of change imposed by the state come up for consideration when we discuss the household-administrative environment in a later section.

Through statistical analysis of cross-cultural data, Cancian (1979) has demonstrated a significant association between economic rank (wealth) and the inclination to adopt new farming practices. Higher ranking, or wealthier farmers are among the first and poorest farmers are among the last to adopt innovations while the "lower middle class" is likely to adopt innovations before the "upper middle class" does. These findings carry important implications for public policy. Cancian suggests that development programs earmarked for small farmers but actually reaching their upper-middle class neighbors, who have the leeway to capitalize on new inputs, may be doomed to failure. Others tie conservativeness/innovativeness to caste ranking, tenurial status, authority structures, levels of education, and so forth (cf., Barnett, 1953; Rogers and Shoemaker, 1971; Palmer 1975).

Social and ecological circumstances have a great deal to do with fostering attitudes of acceptance or rejection towards innovation. Scudder (1973) hypothesizes that households that move involuntarily from their homelands cope with the initial stresses of resettlement by avoiding both new and old activities which entail risks. During this recuperative phase, resettlement authorities might be well advised to go slow on introducing programs of development until the relocatees begin "feeling at home" (Scudder and Colson, 1980). Yet, displaced, or otherwise traumatized populations, are known to exhibit considerable initiative for experimenting with new forms of land use and social organization (Kiste, 1969; Lessa, 1964). A great deal of research remains to be done before social structural and situational interactions in diverse cultural settings can be analytically isolated and causally connected.

Technological advances once adopted may enhance the prosperity of some, but may depress the economic status of others and create long-term environmental hazards for all. Certain serious untoward consequences are as follows.

(1) An increase in existing concentrations of weath and land, and the subsequent widening of already large disparities in income and asset distribution (Frankel 1971; Wharton, 1969; Ahmad, 1972).

(2) Decreased demand for labor with a trend toward greater capital investment (references just cited).

(3) The breakdown of mutual aid between patrons and clients (Frankel, 1971; Scott, 1976).

(4) The despoilation of fragile soils resulting from tractorized farming (Bedoian, 1977; Jodha, 1972).

(5) Labor migration as a novel means for augmenting assets and income while simultaneously withdrawing manpower from village agricultural pursuits and thereby reducing yields and allowing agricultural infrastructure to decline (Birks, 1977).

There is enough documentation on innovation and rural change in LDCs now to make it possible to analyze both achievements and failures in specific terms that can inform the process of building climate stress scenarios.

3.3. Voluntary Migration

Migration is a strategy adopted commonly by rural households coping with resource shortages and other kinds of deprivation occasioned by socio-environmental stresses. Knowing how this process works is a step in the direction of making it amenable to regulation or even obviating it. Some issues that would have to be looked into are:

(1) The factors that weigh upon a household's plans to relocate in response to a premonition or to the actual occurence of an environmental stressor and the ways that neighbors, relatives, local leaders, and other sources of local influence shape the process of decision making.

(2) Should the household decide to migrate, the extent that choices of destination (e.g., hinterland villages, large towns or urban centers close to home, other districts or other countries) depend upon: class, religious, ethnic or other socio-economic factors; the intensity of distress experienced; and the relative attractions of alternate settlement sites.

(3) If displacement cannot be effected without a breakup of the household, the culturally defined rules that determine which household members move where and with whom.

(4) The impact of the diaspora of households upon the economic and political stability of the depopulated aggregate.

(5) The aggregates' capacity for absorbing large numbers of refugee households, or reasons for their inability to do so successfully.

3.4. Land Abuse as a Man-Made Stressor

Expanding needs tied to sharply rising rates of population growth have had a damaging effect on environments exploited by rural aggregates in many regions of the world. These destabilizing pressures on land forms create new environmental perils or make long-standing hazards, such as droughts and floods, more severe. What land-use practices, therefore, bring aggregates into ever more precarious associations with their habitats? And where and why have these tendencies been reversed? Let us have a closer look at these questions.

(1) Opportunity vacuums back home impel subsistence herders and farmers in many parts of Africa, Latin America and Asia toward sparsely settled but highly hazardous desert fringes, frequently innundated floodplains, levees, soil-thin mountain slopes or marshlands, where they perpetuate traditional land-use techniques ill-adapted to the harsh constraints imposed by the landscapes they invade. How do the yoked conditions of poverty and population pressure play a part in creating such situations?

(2) Where communities have awakened recently to the realization that their once bountiful habitats are degenerating because certain essential resources, such as water and woodland, have progressively diminished, what mix of factors will promote voluntary conservation and emergency preparedness within the aggregate (and which factors will hinder these initiatives), and along what lines will protective measures be organized?

(3) How do changing land ownership and management patterns, such as the progressive concentration of land parcels in the hands of absentee landlords disinclined toward costly but necessary capital investments in land improvements, or the overtaxing of soils instigated in part by inheritance rules that inexorably subdivide arable tracts into patchworks of miniscule plots, invite land abuses that trap households into vicious cycles of impoverishment *cum* environmental stress? What part does national agricultural policy and the development of multinational agribusiness play in this process?

3.5. Protest Movements

The socio-economic antecedents of revolutions and millinarian movements are very complex and take on patterns from case to case that vary greatly (Lloyd, 1971). In agrarian Southeast Asia, and other regions of the developing world, conditions that set subsistence crises into motion, Scott (1976) argues, account for revolutionary movements as well. Popkin (1979) challenges this proposition by marshalling evidence of peasant revolutions where protest issues bear no clear relationship to the subsistence threat. Even where the latter is a necessary cause, other factors, that include reliance on external markets and the presence of a revolutionary leadership capable of absorbing peasants and enlarging its base of power through recruitment, are equally decisive. The spawning grounds for millinarianism, according to Barkun's elaborately reasoned thesis, are to be found in disaster situations: "the collapsed social structure renders authority relationships less effective and traditional statuses less meaningful . . . [and] under these conditions, millinarian movements appear." (1975: 55)

While theorists' attributions of causality may differ, the data they draw on give the indication climate stresses enter as an enabling factor in the rise of many agrarian insurrections. A concern worth pointing out has to do with the policy maker's recognition of the social, political and economic ingredients that transform stable aggregates into insurrectionary tinder boxes, and their use of the planning apparatus for seeing to it that the inequities that arouse political unrest are minimized. Social scientific research can certainly reveal pressure points where divisive processes build up force. Or it can identify where discontent comes in for treatment through chiliastic retreats into fantasy. Even protest settings not ostensibly affected by climatic disturbances will be of interest, because a common core of social forces prefigures acts of protest of many kinds.

4. Household-Administrative Environment

Climate changes may alter the location and modify important features of agricultural regions, forest belts, pasturelands, and fisheries. It follows that established modes of using land and exploiting its products may become untenable where changes take place within a narrow span of time. New forms of cooperation that involve the social and spatial organization of settlements, the production and marketing of food, the tenure and protection of land, and the delivery of welfare benefits will be called for. By spearheading these cooperative ventures, governments will be building institutions at the level of the HOES which introduce special organizational and material demands and which invite new definitions of social identity. How do social and environmental feats of engineering auguring such far reaching consequences come about with minimal disruption to the economic security of the household and the aggregate?

For heuristic purposes, it is probably safe to postulate a significant relationship between HOES level intervention of this scope and experiments in community development taking place, at the instance of administrative bodies, today. We would expect that a common core of institutional predispositions and constraints influences the acceptance of innovations in most rural communities irrespective of the development programs introduced. Moreover, the fact of their probable enmeshment with all major aspects of aggregate activity would imply that climatic related programs of institutional modification, in purpose if not in format, would resemble many rural development program operating today. These programs therefore offer planners of the future instructive guidelines for judging the plasticity of societal forms in general and pinpointing conditions that might favor institutional reforms designed specifically for enhancing local system resilience to growing stresses of climate.

Rural cooperatives, resettlement schemes and land reform initiatives, and specialized technologies for augmenting agricultural production exemplify experiments that bring aggregates and their habitats into new patterns of association. The important sociological features characterizing these schemes deserve an examination. Floods, droughts, and frosts may occur in certain regions with greater frequency and result in scarcities that call for state assistance. Problem areas connected with the detection and management of such crises warrant mention, as they fit into the total scheme of issues that chart the course of planning for change. Natural resource conservation and population control programs may rank among the measures marshalled by states for reducing societal susceptibility to stress of climate. Accordingly, some critical questions pertaining to these two approaches by the state for intervening at the local level will be taken up.

4.1. Cooperatives

Cooperative movements exist in most if not all LDCs. The many types of agricultural cooperative settlements in operation differ widely with respect to size, degree of collectivization of property ownership, availability of technical, marketing and credit services, profit sharing arrangements and so forth. Common to all is the guiding aim of boosting the small farmer's productivity *cum* prosperity by a pooling of resources belonging to individual farmers and to the state (and possibly private agencies as well) and achieving through a regime of mutual aid economies of scale not normally attainable by small, autonomous operators. The more elaborate, multi-purpose schemes supply member households with land, credit, the use of seeds and operating equipment, technical training and counseling, administrative supervision, and welfare. Security of tenure, protection from moneylenders, market outlets for crops, some measure of insulation from wild price gyrations in the marketplace, and an assured source of income rank high among benefits member households expect to receive. The Israeli *kibbutz* and *moshav* (Baldwin, 1972; Weintraub *et al.*, 1969), the Philipine *compact farm* (Hunt, 1977), *ujamaa* villages of Tanzania (Cliffe, 1970), the *ejido* of Mexico (Wilkie, 1971), the *negdel* and *sum* of Mongolia (Humphrey, 1978), the *kolkhuz* and *soukhoz* of the Soviet Union (Stuart, 1972), and Chinese communes (Vermeer, 1977) would fall into this category of cooperatives.

Cooperatives also take on functions associated less directly with production and marketing of crops. They may be used by agencies of the state as organizing instruments in programs that bring added land into cultivation, that relocate households from densely to more lightly settled tracts, that mobilize public participation in water conservancy projects, and that thwart impending peasant revolutions through the conspicuous transfer of wealth to needy sectors. One particularly fascinating feature exhibited by Chinese collective organizations and found nowhere else is that they afford a regional infrastructure through which households are inducted into climate impact management projects of truly massive proportions (involving several millions of persons simultaneously) (Vermeer, 1977).

Cooperative movements suffer their share of setbacks and failures. Acts of coersion allow for some planning targets in the long-run, while forcing serious hardships on farmers in the short-run. Famine related directly to the impetuous Soviet collectivization drive of the 1930s claimed the lives of an estimated five million peasants and decimated livestock holdings to the extent that pre-collectivization levels did not reappear until the 1950s (Volin, 1970). Land in some instances, is apportioned in standardized plots irrespective of soil properties, slope, drainage, and proximity to permanent water (Hall, 1978; Hunt, 1977). Some schemes exclude the neediest of the rural poor, the landless laborers (Tadros, 1976; Elder *et al.*, 1976), and the emergence of an elite that builds its base of power in off-scheme enterprises promoted for influencing in-scheme committees, is not unheard of (Apthorpe, 1968; Sandberg, 1974). Credit may dwindle or become available so late in the progress of a crisis that borrowers find themselves once again at the mercy of landlords and moneylenders (UNRISD, 1972). Factionalism may result and conflicts erupt if aggregates with no history of friendly association suddenly find themselves settled together (Guillet, 1978; Chambers, 1970).

Collectivization may transform the traditional economy radically. The camel keeping

Somali, for example, began fishing and gardening (Lewis, 1978), the pastoral Swazi ventured into sugar cane production (Tuckett, 1977), and the Egyptian Nubians were directed to relinquish the custom of migrating to urban labor markets for a settled life on newly designed farms (Fernea and Kennedy, 1966). The collective experience often compels changes in knowledge involving physical characteristics of the land, in concepts of time, in the scheduling of economic tasks through detailed and strictly enforced timetables, in systems for transmitting property rights, in codes of sharing economic responsibilities with neighbors, in sexual division of labor, and in legal sanctions. Where these movements run into trouble, one often finds that the gestation period alloted for the assimilation of new skills, values and ideas is far too short, and the training and motivation component of program administration is weak (UNRISD, 1972; Worsley, 1971; Erasmus, 1977).

If climate changes mean an evolution of structures of community association in the rural sector, the outcome in some countries might be a more efficient use of surplus human resources and deficit natural resources attained through a regrouping of units of production. Should this happen, then some of the problems just identified ought to loom large in the minds of policy makers and planners. First is the overarching issue of "peasant mentality", a mentality which is biased in favor of private ownership and use of land. Bearing in mind that all countries except China and the U.S.S.R. produce "mixed systems" (Dorner and Kanel, 1977: 3) possessing elements of collective enterprise and the private ownership of productive assets and individual farming, then there will be conflicts that pit existing attitudes about land and the culture-bound concepts of individual and social identity which underpin them against new social and environmental realities. How do these conflicts get resolved? Since many environmental and economic risks would devolve upon government, would the economic security of the household become menaced if the state shirked its duty as a protector of local interests? Do larger collective units accelerate the adoption of new organizational forms and technical procedures? Do large cohesive units have an advantage over a number of smaller ones in building hazard protection works and in responding to mass emergencies? Can larger collective units more flexibly adopt "land use patterns to fit variations in soil and topography," (Dorner and Kanel, p. 6) where there is great diversity in land quality? Dorner and Kanel (*ibid.*, p. 6) suggest that this possibility "may become an increasingly important factor as population pressures on the land shorten the cultivation-fallow cycle in tropical areas, and as more farmers are forced to seek a subsistence living through cultivation of steep slopes subject to soil erosion." And does communal land tenure encourage the uncontrolled or weakly regulated exploitation of scarce pasturelands and forests, on the part of individual users (Sörbo, 1977; Thompson, 1977)?

4.2. Resettlement Schemes

Resettlement schemes redistribute rights in land through movements of population controlled by the state and evolve in many different ways (Chambers, 1969; United Nations, 1968; Palmer, 1979). Schemes tied in with land reform legislation may cede rights of ownership to tenants in land they cultivate in the aftermath of the breakup of colonial and royal estates or the large holdings of big landlords. Or nationalized land may divide into individual holdings or consolidate into development blocks managed by cooperative societies. For coming to terms with problems of a political, economic, and welfare-related nature, resettlement schemes assume some of the following functions.

(1) sedentarizing nomads

(2) settling and controlling the movements of refugees

(3) isolating dangerous or imperilled groups for reasons of security. The camps in which Japanese–Americans were interred during World War II and the "villigization" schemes introduced during the Mau Mau uprisings in Kenya served these functions

(4) settling victims of natural disasters

(5) relocating people displaced by dams, roads, and the construction of towns and agricultural settlements

(6) relieving pressure of population on land

(7) evacualing persons from areas infested with disease

(8) developing or reclaiming for cultivation land which was previously uncultivated or abandoned. This pattern might be illustrated by "back to the land" projects, exemplified by Vietnam's efforts at resettling more than half of all the residents of Ho Chi Minh City in rural communes or in their native villages, or Indonesia's new lands settlement schemes.

Distinctions in purpose, mode of inauguration, and structure of resettlement schemes are codified by numerous typologies (Palmer, 1979; Scudder and Colson, 1980; Apthorpe, 1968; Chambers, 1969, 1979; Yeld, 1968). If climates do deterioriate dramatically, there would be many different situations giving rise concurrently to the uprooting of human populations, and, accordingly, many varieties of resettlement would come up for consideration by public leaders. Bearing this possibility in mind, we advocate that typologies give way to general theories accounting for the adaptive mechanisms, evident in any array of resettlement settings, recasting the conventions of uprooted peoples to new varieties of social and economic challenge. Such formulations will give policy makers a coherent framework for thinking about parallels, weighing alternatives, and drawing conclusions from a much wider stock of observations than would be available to them were lessons sought from isolated experiences.

Theory might relate the successes or failures of specific schemes to: (1) the proximity of settlements to previous places of residence; (2) ethnic and class diversity of the settlement; (3) occupational opportunities in the new settlement and in neighboring communities; (4) affinities between customary work skills and values and those demanded of relocatees at their new place of residence; (5) literacy levels; (6) efficiency and sociological knowhow of resettlement administrators; (7) compensation received for possessions lost or damaged during relocation, in instances of involuntary resettlement; (8) host community attitudes towards relocatees; (9) receptivity to innovations – where resettlement is combined with development – as influenced by the degree of trauma experienced; (10) structural complexity of the pre-relocation social system. Getting at the thorny problems of rural settlement assumes a certain urgency in view of the rapid pace of colonization of unoccupied arable land being witnessed in almost every country of the developing world.

4.3. *Land-Use Legislation*

By being moved onto ecologically inferior tracts or denied full customary access to water, forestland or forage, many indigenous minorities see themselves increasingly hard pressed to wrest a suitable livelihood from the land. Politicians and powerful interest groups share the role of culprits in these crises. Although there is a tendency to associate such flagrant violations of human rights with LDCs, the U.S. government is not without blame. For instance, 55% of the nation's 255 reservations and almost 75% of its 370,000 reservation Indians occupy arid and semi-arid regions wanting in lakes and permanent rivers. Hundley (1977: 1) observes that:

> it is axiomatic that the conditions on reservations in the arid and semi-arid West cannot improve significantly unless the Indians are guaranteed an adequate supply of water and how does the Indian justify his rights to it?

Indians find themselves defending their rights because the resources in which these rights are vested become an object of competition among diverse interest groups each asserting claims deemed by it legitimate and incontrovertable. HOES level research might serve a useful function by affording administrators and policy makers a reasonably impartial and scientific basis for making informed judgements about the way each contesting party interprets its rights, the validity of the social and economic needs on which these rights are defended, and adverse consequences for claimants whose arguments get dismissed. Smaller minority groups, normally at a decisive disadvantage when challenged by larger adversaries, *may* have the most to gain from efforts of this sort. Collecting and weighing the evidence is especially important when the group that stands to lose occupies climatically hazardous environments and the resources in dispute are very scarce and critical to indigenous systems of livelihood. We might add that should the distribution and availability of productive resources shift geographically if and when a long phase of climate deterioration sets in, conflicts involving land and water rights will probably proliferate. Lessons learned from the research we endorse therefore may have significant implications in the decades ahead.

4.4. *National Birth Control Programs*

Planned programs of birth control legitimate intervention of the state in an activity embedded deeply in household and aggregate life, for the express purpose of achieving a balanced adjustment between population numbers and physical resources. There is something in general to be learned about the pliability of local institutions from the results of programs that vest agencies of the state with authority to reconstitute human environments. More significant from our point of view is the prospect that governments might actually enlist these experiments in situations spawned by stresses of climate.

Birth control programs in many developing countries encounter stubborn resistance, to some extent, because planners assume that their premise – that it is rational and in the vital interests of the household that it should adjust its numbers in accordance with the resources available – is shared by the peasant. Often, though, a very different brand of logic guides the poorer family's reproduction strategies. The household head(s) often reasons that large families mean more workers, and more workers enhance income earning options in the labor market (Mamdani, 1972; Gould, 1976). To understand the farmer's attitudes toward programs of birth control and family planning, therefore, is to specify the influence of material conditions on family size. These conditions include such factors as land tenure codes, land fragmentation patterns, employment opportunities, time and energy demands and economic returns associated with specific productive tasks, the division of labor by sex and age, and indebtedness levels and prevailing terms for securing loans (e.g., through debt service).

Families liberated from subsistence activities and concerned more with quality of life considerations rather than "do or die" survival compulsions usually look on the idea of family planning through contraception or abortion more favorably. It has been suggested that fertility management programs might better be concentrated on sectors of the population that want small families so that growth control can come about without changes in social structure (Freedman and Takeshita, 1969: 364). China, on the other hand, has made great strides toward achieving significant reductions in fertility, across all strata of the population, an unrivaled achievement that has come about through profound structural transformations in social institutions. Since 1971, the targets of programs of birth control have been, in the words of Chen and Miller (1975: 354):

(1) Late marriage: late twenties in urban areas (age 28 for men, age 25 for women); early to mid-twenties in rural areas (age 25 for men, age 23 for women).

(2) Childbirth spacing: four-to-five year intervals after the first child.

(3) Inculcation of small-family norms: two children in urban areas and three in rural areas, regardless of six.

The birth control program is integrated completely with other social and economic development programs through the provision of grass-roots educational services and propaganda work. Peer pressure is a major force arousing attitudes of allegiance to these programs. While recognizing that Chinese models cannot be transferred intact to countries that lack China's organizational resources and taking into account its total commitment to the successful outcome of programs of fertility control, Chen and Miller suggest certain features of China's policies that ought to merit careful consideration in most if not all LDCs. Those particularly relevant are as follows.

(1) Recognizing regional and local preferences in contraceptive methods.

(2) Integrating health and family planning services with the ordinary development functions of rural cooperatives.

(3) Rendering family planning services free or on a low cost basis and near the home or place of work.

(4) Recruitment of paramedics and non-professionals from local areas, and the use of married women, particularly mothers, in educational and motivational work and in the distribution of contraceptives.

(5) Building motivational appeals on reasoning relevant to cultural and environmental contexts (e.g., relating benefits of programs to land fragmentation, drought vulnerability, malnutrition and other hardships perceived as deriving from pressures of population).

(6) Using local leaders who adhere to late marriage and small-family norms as role models and deploying them when necessary for mobilizing public opinion.

(7) Evolving incentives and penalties that put thrust into program implementation.

(8) Providing childcare facilities that help keep women in the labor force.

(9) Legalizing all contraceptive operations and removing any stigma attached to them.

Cross-cultural/national research that explores the feasibility of measures such as these should have something of value to say in general about the receptivity of rural aggregates to outside innovations, and should indicate, in particular, the willingness by which these units might take an active part in relieving pressure on scarce resources.

4.5. Planning for Emergencies

Emergency operations form the terminal and one of the most crucial of links in the chain of measures necessary for preventing and mitigating crises of scarcity – such as those instigated by stresses of climate – and involve the distribution of supplies to victims in a manner that can benefit them significantly. If indigenous institutions barely provision poor rural households when supplies fluctuate around the "norm", they cannot do so strained beyond their limits under the daunting pressures of droughts or floods. This puts the welfare ball pretty much in the lap of the state. The state has essentially three ways that it can forge a response, one being feasible, and two not. Of the latter, it can do nothing, or else underreact. Or it can furnish no cost restitution for all losses suffered with a bonus added, and thereby overrespond. The Bengal famine of 1943 and dozens of catastrophes brought on by droughts in 19th century India and China illustrate the dangers regnant in a policy of inaction, or only half-hearted action. Many factors would veto a strategy of overresponse. First, no state can afford such largesse. Moreover, it would usurp the proper fiscal functions of business, and would encounter stiff resistance from donor groups. Besides, the affected community's resolve towards self-sufficiency in agriculture and related activities would be weakened. No wonder it is doubtful that this option has ever been tried.

Middle range, or feasible options, would incorporate *welfare, production*, or *development* components, or all three. Welfare keeps people reasonably well fed and healthy. Production means gearing up farm operations for the next sowing season. And development imparts resistance to the farmstead against future environmental and economic shocks. These three components, in combination with a provision that guarantees a return to the state on its investment, would constitute an "optimal feasible solution" to the problem of crisis.

As far as we know, case studies that offer documentary accounts of the system of operations agents of the state proceed through to render assistance to crisis stricken aggregates and that detail the aggregate's responses to these initiatives, have not been undertaken, or at least not published. Until such accounts become available, it will be difficult making up guidelines for improving upon the performance of these operations and evaluating the feasibility of adapting them in countries that lack a developed relief apparatus. The data that would be useful for these purposes should cover the following points.

(1) Stress indicators officially recognized by the state (e.g., grain prices, crop outturns, deaths, outbreaks of disease, land transfers, incidents of outmigration) and the speed and accuracy that go into the gathering and interpretation of relevant statistics.

(2) Types of assistance provided, proportion of the affected population that benefits, and cut-off dates on the delivery of aid. The programs of assistance might include low interest loans for production and consumption requirements; employment generated by public works; grants of seeds, fertilizers and fodder; grain sold at subsidized prices; and remission and suspension of revenues.

(3) The stability of cooperative credit societies under pressure.

(4) The degree of distress in evidence before the state fully mobilizes its resources.

(5) The extensiveness of long-term improvements in the productivity of local resources, through such emergency inputs as well drilling, road laying, building or extending irrigation facilities, introducing improved varieties of seeds and livestock, installing flood control works, and revegetating denuded tracts of land.

(6) The implementarion of schemes that integrate or replace emergency relief projects with perennial measures of protection, including guaranteed employment (Reynolds and Sundar, 1977), crop insurance (Dandekar, 1976), and food rationing for lower income groups (Agrawal, 1979; Gwatkin, 1979).

(7) The capacity of households to recover or to improve upon their preimpact asset positions.

4.6. Natural Resource Conservation

Where the overall supply of arable land is scarce, as it is today over most of the developing world, the stresses of climate exert added pressures of human and animal numbers on fragile resources. When drought sets in, for example, bark may be stripped from trees and eaten. Trees might be felled, as the indigent look to the forests for firewood and charcoal which fetch a small profit at the local marketplace. Livestock might graze down semi-barren pastures for want of supplemental feed. And quick return but destructive forms of land use, such as goat husbandry, may proliferate. Left unchecked, these pressures inexorably diminish the long-term productive capacity of the region. Policies chosen by agencies of the state that take the squeeze off the land and restore the productivity of fields, pastures and forests will succeed only as far as they find acceptance among the households for whom they are intended. If it is going to be forthcoming at all, assimilation by the household of an externally inspired ethic of conservation will hinge on certain planning tactics we turn to now.

Conservation regulations take effect generally when the process of resource depletion is well underway. This makes code enforcement especially difficult because of built-in attitudes of resistance fostered by the resource abusers. Short-term survival, for them, overshadows concerns about long-term disasters (Thompson, 1977). Discontent with earlier state policies perceived by the community as an infringement on its rights, further contributes to a posture of "obstinancy" or "unprogressiveness". Conservation programs therefore will have to be made comprehensible to the target population, involve these groups directly in important phases of project planning and implementation, and use their cooperation as an instrument for augmenting household purchasing power, for protecting household assets, and for preserving or resurrecting community self-respect. Each of these prescriptions should be heeded by framers of conservation policies and programs. The quasi-success of the famous Navajo reservation soil control program of the 1930s illustrates our point in an interesting way. The United States Bureau of Indian Affairs designed an extensive program of soil conservation. It checked the erosion and restored millions of acres of land and established a comprehensive program of range management. The project worked splendidly in the technical sense, but, planned without the prior consultation of Navajo leaders to determine local needs and customary rights, it created a long lingering resentment hindering government experiments in land development on other fronts (Spicer and Collier, 1952). More typically, the outcome of projects that fail is that land codes cannot be enforced rigorously (nor respected), and programs consequently fall short of their technical goals as well.

Encouragement can be taken from a growing number of modest success stories. The World Food Program has developed some innovative experiments in economic reform that combine soil conservation with food aid. Soil erosion caused mostly by a major drought followed by torrential rains in the early 1930s in Swaziland is now being combatted by a watershed management scheme and fish farms, stocked with carp. The latter fights erosion by retaining run-off water during heavy rains, furnishes an important source of protein for local consumption, and generates market profits (Craig, 1976). A WFP project in the once forested catchment areas of Hazara District, north Pakistan, is giving farmers maize as animal feed with the understanding that livestock graze outside the protected areas (Ellis, 1974). Livestock reduction is also being promoted through terracing of steep slopes, which now sustain thriving fruit orchards, and the building of market connected roads. Furthermore, WFP volunteers enlist themselves on reforestation schemes.

The Paper and Industrial Corporation of the Phillipines, a public corporation manufacturing wood and wood fiber products, has embarked on a program for halting the shifting cultivation practices that chip away valuable forestland surrounding its groves (Arnold, 1979). The local farmers are encouraged to grow trees as a cash crop on their land, which they sell the company for use as pulpwood at prevailing market prices. The company provides extension services for silvicultural training and for the improvement of farming practices. The *taungya* system of Java, which increases the peasants' production of food from forestlands through intercropping trees with food crops and yields fuel through the planting of fast growing species (Atmosoedarjo and Banyard, 1978), and the ambitious Drought Prone Area Program of India, which supplies selected villages in 74 of the country's 330 districts with an assortment of extension services, and creates jobs on soil conservation and reforrestation projects (Puttaswamiah, 1975; Government of Gujarat, 1976) further illustrate the strides being made in this direction.

State run conservation projects remain predominantly in the proving stage, and their scale tends to be small. It will be of interest to see how far the programs reported to date can be sustained and the proportion of the total population of "abusers" that eventually joins them. Indepth studies of socio-ecological impacts appear to be few. Probing social research, however, should take its place as a key component in any serious official attempt to arrive at an overall evaluation of the long-term costs and benefits accruing from conservation programs, as they bear on the process of planning for environmental *and* community development.

5. Conclusion

Environment and society interface through feedback networks so vast in their complexity and variability that no one can realistically expect achieving a capability anytime soon of predicting with fine precision the institutional effects upon or reactions to climate changes taking place at community and regional levels. Those in positions of having to worry about the human dimensions of climate change prospects focus the burden of their concern on a recognition that some affected populations will be worse off, economically and politically, than before. Further, they recognize the need for taking decisive precautionary steps toward forestalling risks and destruction by introducing changes in social conventions existing in these vulnerable societies. *In the final analysis, planning for climate change means planning for social change*. Planning problems in Third World countries, where anthropologists often chose to do fieldwork, revolve chiefly around programs of economic development.

Social change rarely if ever accompanies atmospheric stresses alone, since all environmental influences filter through layers of societal institutions. Therefore we do not confine our analysis of social change to any one source of stress. Our survey addresses several crisis situations calling for a restructuring of existing local institutional arrangements to conform to changing material realities, and it outlines some achievements and failures resulting from the imposition of government initiated development programs. We argue that more systematic cross-cultural exercises of this kind will show what degree of resilience certain kinds of societies equip themselves with for facing stresses, including stresses related to the natural environment. Cross-cultural investigation can also reveal the capacity of societies to assimilate programs that strengthen this resilience or restore it after it breaks down.

Department of Sociology/Anthropology
West Virginia University

Note

* I wish to thank Professor Elizabeth Colson for generously donating her time to read this and earlier drafts of this paper, and, in both instances, for providing many helpful editorial comments.

References

Agrawal, N. S.: 1979, 'Instrument for Consumer Protection: the Indian Experience', *Ceres* **72**, 33–37.

Ahmad, Z. M.: 1972, 'Social and Economic Implications of the Green Revolution in Asia', *International Labor Review* **105**, 9–34.

Apthorpe, R.: 1968, 'Planned Social Change and Land Resettlement', in R. Apthorpe and J. Kuper (eds.), *Land Resettlement and Rural Development in Eastern Africa* NKAGA Editions No. 3. Kampala: Transition Books, pp. 5–13.

Arnold, M.: 1979, 'A Habitat for More than Trees'. *Ceres* **12** (5), 32–37.

Atmosoedarjo, S. and Banyard, S. G.: 1978, 'The Prosperity Approach to Forest Community Development in Java'. *The Commonwealth Forest Review* **57** (2) No. 172, 89–96.

Baldwin, E.: 1972, *Differentiation and Cooperation in an Israeli Veteran Moshav*, Mancester: Manchester University Press.

Barkun, M.: 1974, *Disaster and the Millennium*, New Haven: Yale University Press.

Barnett, H. G.: 1953, *Innovation: the Basis of Cultural Change*, N.Y.: McGraw-Hill Book Co. Inc.

Bedoian, W. H.: 1977, 'Human Use of the Pre-Saharan Ecosystem and Its Impact on Desertization', Paper (revised) for the 143rd Annual Meeting of the American Association for the Advancement of Science, Denver, Colorado.

Birks, J. S.: 1977, 'The Reaction of Rural Populations to Drought: A Case Study from South East Arabia'. *Erdkunde* **31** (4), 299–305.

Cancian, F.: 1979, *The Innovator's Situation: Upper Middle-Class Conservatism in Agricultural Communities*, Stanford: Stanford University Press.

Cahambers, R.: 1969, *Settlement Schemes in Tropical Africa*, London: Routledge and Kegan Paul.

Chambers, R.: 1970, *The Volta Resettlement Experience*, London: Pall Mall Press.

Chambers, R.: 1979, 'Rural Refugees in Africa: What the Eye Does Not See'. *Disasters* 3 (14), 381–92.
Chen, Pi-Chao and Miller, A. L.: 1975, 'Lessons from the Chinese Experience: China's Planned Birth Program and Its Transferability'. *Studies in Family Planning* 6 (10): 354–66.
Cliffe, L.: 1970, 'Traditional Ujamaa and Modern Producer Co-operatives in Tanzania', in C. G. Widstrand (ed.), *Co-operatives and Rural Development in East Africa*. N.Y.: Africana Publishing Corp, pp. 38–60.
Craig, D.: 1976, 'Bonus: Erosion Control Helps Provide Fresh Fish'. *World Food Program News* (March); 8–9.
Dandekar, V. M.: 1976, 'Crop Insurance in India'. *Economic and Political Weekly* 11 (26): A61–A80.
Dorner, P. and Kanel, D.: 1977, 'Some Economic and Administrative Issues in Group Farming', in P. Dorner (ed.), *Cooperatives and Communes: Group Farming in the Economic Development of Agriculture*. Madison: University of Wisconsin Press, pp. 3–14.
Elder, J. W. *et al.*: 1976, *Planned Resettlement in Nepal's Terai: A Social Analysis of the Khajura/Bardia Punarvas Projects*, Tribhuvan University.
Ellis, M.: 1974, 'Mountain Dwellers of North Pakistan Fight Erosion with Help of Food Aid'. *World Food Program News* (July): 7–9.
Erasmus, C.: 1977, *In Search of the Common Good: Utopian Experiments Past and Future*, N.Y.: The Free Press.
Fernea, R. A. and Kennedy, J. G.: 1966, 'Initial Adaptations to Resettlement: a New Life for Egyptian Nubians'. *Current Anthropology* 7, 349–54.
Friedman, R. and Takeshita, J. Y. (eds.): 1969, *Family Planning in Taiwan: an Experiment in Social Change*, Princeton: Princeton University Press.
Gould, K. H.: 1976, 'The Twain Never Met: Sherpur, India, and the Family Planning Program', in J. F. Marshall and S. Polgar (eds.), *Culture, Nature, and Family Planning*. Monograph 21, Carolina Population Center. Chapel Hill: University of North Carolina Press, pp. 184–203.
Government of Gujarat: 1976, 'General Administration Department (Planning) Development Program 1976–77', Budget Publication 14.
Guillet, D.: 1978, 'Peasant Participation in a Peruvian Agrarian Reform Cooperative'. *J. of Rural Cooperation* 6 (1), 21–34.
Gwatkin, D. R.: 1979, 'Food Policy, Nutrition Planning and Survival'. *Food Policy* 4 (4): 245–58.
Hall, A. L.: 1978, *Drought and Irrigation in North-East Brazil*, Cambridge: Cambridge University Press.
Humphrey, C.: 1978, 'Pastoral Nomadism in Mongolia: The Role of Herdsmen's Cooperatives in the National Economy'. *Development and Change* 9 (1): 133–60.
Hundley, N.: 1977, *Special Water Rights for Indians: Principle Versus Conclusion*, Navajo Nation Energy Conference, Tsaile, Arizona, May 2. Los Alamos Scientific Laboratory, Los Alamos, N.M.
Hunt, C. L.: 1977, 'The Phillipine Compact Farm: Right Answer or Wrong Question?' *J. of Rural Cooperation* 5 (2): 121–40.
Jodha, N. S.: 1972, 'Agricultural Development and Problems of Nomadic Tribals in Rajasthan', in M. L. Patel (ed.), *Agro-Economic Problems in Tribal India*. Bhopal: Progress Pub. Co., pp. 23–26.
Jodha, N. S.: 1975, 'Famine and Famine Policies: Some Empirical Evidence'. *Economic and Political Weekly* 10 (41): 1609–23.
Kiste, R.: 1968, *Kili Island: a Study of the Relocation of the Bikini Marshallese*, Eugene: University of Oregon, Department of Anthropology.
Lessa, A.: 1964, 'The Social Effects of Typhoon Ophelia (1960) on the Ulithi'. *Micronesia* 1: 1–47.
Lewis, I. M.: 1978, 'The Somali Democratic Republic: an Anthropological Overview', MS. USAID.
Lloyd, P. C.: 1971, *Classes, Crises and Coups: Themes in the Sociology of Developing Countries*, London: MacGibbon and Kee.
Mamdani, M.: 1972, *The Myth of Population Control: Family, Caste, and Class in an Indian Village*, NY: Monthly Review Press.
Palmer, G. B.: 1979, 'The Agricultural Resettlement Scheme: A Review of Cases and Theories', in B. Berdichewsky (ed.), *Anthropology and Social Change in Rural Areas*. Mouton: The Hague, pp. 149–86.

Palmer, I.: 1975, *The New Rice in the Phillipines, Studies on the "Green Revolution"* – No. 10. United Nations Research Institute for Social Development. Geneva.

Popkin, S. L.: 1979, *The Rational Peasant*, Berkeley: UC Press.

Puttaswamaiah, K.: 1975, *Towards Drought Proofing in Karnataka*, Bangalore, Government of Karnataka.

Reynolds, N. and Sundar, P.: 1977, 'Maharashtra's Employment Guarantee Scheme: A Programme to Emulate?' *Economic and Political Weekly* **12** (29): 1149–58.

Rogers, E. and Shoemaker, F. F.: 1971, *Communication of Innovations: A Cross-Cultural Approach*, N.Y.: The Free Press.

Sandberg, A.: 1974, 'Socio-Economic Survey of the Lower Rufiji Flood Plain: Rufiji Delta Agricultural System', Bureau of Resource Assessment and Land Use Planning. Research Paper No. 34. Dar es Salaam.

Scott, J.: 1976, *The Moral Economy of the Peasant*. New Haven: Yale University Press.

Scudder, T.: 1973, 'The Human Ecology of Big Projects: River Basin Development and Resettlement', *Annual Review of Anthropology*, Palo Alto: Annual Review, Inc. pp. 45–55.

Scudder, T. and Colson, E.: 1980, 'Conclusion', MS.

Sörbo, G.: 1977, 'Nomads on the Scheme – a Study of Irrigation Agriculture in Eastern Sudan', in P. O'Keefe and B. Wisner (eds.), *Land Use and Development*. African Environment Special Report 5. International African Institute, London, pp. 132–50.

Spicer, E. H. and Cocher, J.: 1952, 'Sheepmen and Technicians: a Program of Soil Conservation on the Navajo Indian Reservation', in E. H. Spicer (ed.), *Human Problems in Technological Change*, N.Y. Russell Sage Foundation, pp. 185–207.

Stuart, R. C.: 1972, *The Collective Farm in Soviet Agriculture*, Toronto: Lexington Books.

Tadros, H. R.: 1976, 'Problems Involved in the Human Aspects of Rural Resettlement Schemes in Egypt', in A. Rappoport (ed.), *The Mutual Interaction of People and Their Built Environments: A Cross-Cultural Perspective*, Mouton Pubs.: The Hague, pp. 453–84.

Thompson, J. T.: 1977, 'Ecological Deterioration: Local-Level Rule Making and Enforcement Problems in Niger', in M. H. Glantz (ed.), *Desertification: Environmental Degradation in and Around Arid Lands*. Boulder: Westview Press, pp. 57–76.

Torry, W. I.: 1979, 'Anthropological Studies in Hazardous Environments: Past Trends and New Horizons', *Current Anthropology* **20** (3): 517–40.

Tuckett, J. R.: 1977, 'Vuvulane Irrigated Farms, Swaziland: a Report on the First Ten Years', *Agricultural Administration* 4 (1): 80–96.

United Nations: 1968, 'Report of the World Land Reform Conference', Rome June 20-July 2, 1966.

United Nations Research Institute for Social Development: 1972, *Cooperatives and Development in Asia*, Geneva.

Vermeer, E. B. 1977, *Water Conservancy and Irrigation in China*, Leiden: Leiden University Press.

Volin, L.: 1970, *A Century of Russian Agriculture: From Alexander II to Krushchev*, Cambridage: Harvard University Press.

Watts, M. J.: 1980, 'The Political Economy of Climatic Hazards: A Village Perspective on Drought and Peasant Economy in a Semi-Arid Region of West Africa', Paper presented to the Congress of the International Geographical Union, Tokyo, Japan, September 3 rd, 1980.

Weintraub, D., Lissak, M., and Azman, Y.: 1969, *Moshava, Kibbutz, and Moshav: Patterns of Jewish Rural Settlement and Development in Palestine*, Ithaca: Cornell University Press.

Wharton, C. R., Jr.: 1969, 'The Green Revolution: Cornucopia or Pandora's Box', *Foreign Affairs* 47: 464–76.

Wilkie, R.: 1971, *San Miguel: A Mexican Collective Ejido*, Stanford: Stanford University Press.

Worsley, P.: 1971, 'Introduction', in *Two Blades of Grass: Rural Cooperatives in Agricultural Modernization*. Peter Worsley (ed.), Machester: Manchester University, pp. 1–43.

Yeld, R.: 1968, 'The Resettlement of Refugees', in R. Apthorpe (ed.), *Land Settlement and Rural Development in Eastern Africa*, NKANGA Editions No. 3. Kampala: Transition Books, pp. 33–37.

INTRODUCTION TO: INTERDISCIPLINARY RESEARCH AND INTEGRATION: THE CASE OF CO_2 AND CLIMATE

In the fields of societal and environmental impact assessment nearly everybody talks about "interdisciplinary research", but very few have ever written about it. In 1977, for example, I organized a session at the American Association for the Advancement of Science Annual Meeting in Denver, Colorado on the issue of interdisciplinary research process. Fortunately for both the participants and posterity, Margaret Mead served as a discussant to the papers, which had been given by physical scientists who had worked in interdisciplinary projects. Although no proceedings were envisioned for that gathering, those responsible for tape recording a few pre-selected AAAS seesions ignored their original roster and simply assumed that any meeting with Margaret Mead on the program should be recorded. And it was a happy accident, for when Chen began his search for writings on the *process* of interdisciplinary research, he found that the formal literature was exceedingly thin; but the general literature contained several informal sources, including these unscheduled recordings that proved to be one of his principal references. Indeed, although Dr. Mead chided the young interdisciplinarians for largely rediscovering methods that had worked or failed in numerous projects long before, she admitted that there were few written accounts of the success stories which could have been used to help structure interdisciplinary research groups. It is largely because of this pressing need to document some of the record of interdisciplinary research efforts that Chen's timely review is so welcome in this volume. Not only does he overview and distill a good part of the frequently informal "literature" on interdisciplinary research, but he goes beyond this to discuss how individual research components from interdisciplinary workers can be integrated in the context of CO_2/climate impact assessment. For all of these reasons, I believe that this contribution is one of the most significant in our volume.

S. H. SCHNEIDER

R. S. Chen, E. Boulding, and S. H. Schneider (eds.), Social Science Research and Climate Change: An Interdisciplinary Appraisal, 229.

ROBERT S. CHEN*

INTERDISCIPLINARY RESEARCH AND INTEGRATION: THE CASE OF CO_2 AND CLIMATE

Abstract. The possibility that increasing atmospheric carbon dioxide concentrations may lead to significant climate changes poses a problem of unusual breadth and complexity to society. Research on this problem, and on ways society can respond to it, needs to be carefully organized and managed in an interdisciplinary and flexible manner. New means of integrating research results and ensuring their usefulness for policy decisions must be explored. Research on the CO_2 problem should also be closely 'tied-in' with research on other social and environmental issues.

1. Introduction

The possibility of significant climate changes due to increasing carbon dioxide (CO_2) in the atmosphere presents an unprecedented challenge to modern society. The challenge is to respond effectively to the 'CO_2 problem' in advance, despite many remaining uncertainties. Possible responses range from a 'wait-and-see' policy to immediate actions to curb CO_2 emissions. To put decisionmaking on a firmer basis will require substantial natural- and social-science input. One key task for social scientists is to examine: (1) how and with what success people and their societies have responded to environmental stress in the past and present; (2) how they might respond to the prospect or advent of CO_2-induced changes in the future; and (3) what policy options are available or can be developed to prevent or mitigate potentially adverse impacts of CO_2 increases. Given the wide variety of possible environmental stresses, societal responses, and policy options and in view of the complexity of interactions among these, research to deal with CO_2 increases must involve extensive collaboration among scientists from many different academic disciplines. Considerable effort will also be needed to combine and translate individual research results into realistic, useful policy information. How *interdisciplinary* research can be fostered and *integration* of research results achieved is the focus of this paper.

2. Critical Research Considerations in the CO_2 Context

Several unusual aspects of the CO_2 problem and society's possible responses to it raise important issues that need to be considered in planning and conducting research.

First, the problem is *unique* in both scale and complexity. Increasing carbon dioxide in the atmosphere is generally believed to result from over a century of past fossil-fuel

* The views expressed in this paper are those of the author and do not necessarily represent the views of the Climate Board, the National Academy of Sciences, or the University of North Carolina.

R. S. Chen, E. Boulding, and S. H. Schneider (eds.), Social Science Research and Climate Change: An Interdisciplinary Appraisal, 230–248.

use, and possibly some land uses, on which modern industrial and agricultural development depends (Geophysics Study Committee, 1977). Present-day populations rely heavily on fossil-fuel energies for their survival and, in light of continuing population growth, are likely to continue to do so for at least several decades (OECD, 1979a: 39; WAES, 1977: 242–255). CO_2 increases may lead to potentially major global changes to the environment, with attendant effects – both beneficial and adverse – on human activities and welfare. Responses to the advent or prospect of CO_2-induced changes also could be global, and, in turn, could have widespread feedbacks to the environment and society (Climate Board, 1980).

Second, the CO_2 problem is inescapably *normative.* Climate changes will affect different people in different ways at different times. Even the interpretation of some specific effect as either beneficial or adverse is likely to vary greatly among individuals, depending on their own perceptions, values, and priorities (Chen, 1980). A drought-induced rise in food prices, for example, may help farmers but hurt consumers. Moreover, many value-laden issues arise quickly in responses to the CO_2 problem, as, for example, in making explicit tradeoffs between reducing CO_2 emissions and providing needed energy. Such issues include equity in the distribution of impacts and in the distribution of benefits from CO_2-producing activities, responsibility for adverse changes, the rights of future versus present generations, and the 'proper' relationship between humanity and the global environment. These normative aspects extend the bounds of the CO_2 problem well beyond the 'factual', technical realms of the natural sciences into consideration of social and ethical factors.

Finally, the challenge to society posed by the CO_2 problem is, as has been mentioned, *unprecedented.* Unlike many past situations, society has the opportunity to deal with the problem explicitly and in advance – not by default (e.g., see Brown, 1980; WMO, 1979: 61). Many different policy options exist, ranging from 'doing nothing different' or 'more study' to active measures to 'build resilience' and even 'reduce the CO_2 insult' by altering energy, agricultural, or land-use policies (Schneider and Chen, 1980). Choosing among these options will be difficult, in part because of the unique and normative facets of the CO_2 problem and the great uncertainties that exist at present in the likelihood, magnitude, and timing of CO_2-induced climate changes, and in part because of the many different parties that are likely to be involved in any globally effective responses.

These unique, normative, and unprecedented aspects of the CO_2 problem have important implications for research. Obviously, no single academic discipline, nor several working by themselves, can provide all the answers needed. Many different disciplines must be called upon, and their expertise blended into an interdisciplinary effort. Some barriers to and requirements for effective interdisciplinary research are discussed in the next several sections. A second implication is that research must specifically address the decisions society may have to or may be able to make and the choices or policy options available to it. The role of various decision makers and their perceptions, evaluations, and choices thus becomes important. Moreover, it is notable that research is itself part of our response to the prospect of CO_2-induced climate change. We therefore need to understand what the benefits and limitations of research are, in terms of how research could materially change our understanding and perception of the problem and our ability to develop and implement timely, effective responses to it. These issues are discussed in more detail in Sections 7 and 8.

Another implication is that research on the CO_2 problem must be 'tied-in' with research on other environmental and social issues. Given limited resources of time, money, and scientific talent, we must take advantage of past and present research on similar problems such as the societal impact of present-day climate fluctuations or non-climatic, slowly evolving environmental stresses on society, e.g., soil erosion. Indeed, an important by-product of research on the CO_2 problem may be the new knowledge gained that can be applied to these other problems – regardless of how critical the CO_2 problem itself becomes. In particular, we must look for new knowledge concerning societal adaptability to environmental stress. This subject is treated more extensively in Section 9.

3. Societal Response to Increasing CO_2: An Interdisciplinary Research Problem

Many different social-science disciplines can make important contributions to the understanding of potential responses to CO_2-induced climate changes or policies implemented to prevent or mitigate them. As is true of many other important human problems, the possibility of CO_2-induced changes is primarily of concern because of its potential stresses on society with consequent adverse impacts on human welfare. Human experience in responding to such stress and impacts is obviously immense and much of it could be directly applied to the CO_2 problem. In the fields of history and geography, the effects of climatic variability and change on various aspects of society have been extensively, if not consistently, examined through detailed case studies and comparisons among cases from different regions, cultures, and time periods (e.g., Rabb, 1983; Warrick and Riebsame, 1981). Economics provides a number of tools capable of analyzing some of the stresses on local, regional, national, and international economies that might arise; economics methods also help to quantify the potential monetary costs and benefits for different groups of people (e.g., Smith, 1980). Political science and law can help us to predict the ability of local, national, and international institutions to assess and respond to the CO_2 problem based on past experience with other environmental and social problems (e.g., Mann, 1983; Weiss, 1983). Anthropology and sociology can relate economic and institutional changes to their likely inplication for people and various kinds of social organization (e.g., Torry, 1983; Panel IV, 1980). Psychology is important because of the key role of perception, evaluation, and choice in individual and group responses to actual or perceived stress (e.g., Fischoff and Furby, 1983). The humanities (e.g., philosophy) are needed to help clarify some of the difficult normative issues raised with regard to such fundamental questions as humanity's relationship to nature and the rights of present versus future generations (Schneider and Morton, 1981).

Research undertaken solely within each isolated discipline would be of relatively little value to understanding the CO_2 problem as a whole. The 'net' effect of climate changes, or of actions taken to adjust to them, will emerge from the interaction of many different processes. Changes in climate or in energy and agricultural policies will have diverse consequences for, and result in diverse responses by, different groups in society.

These responses will in turn cause their own impacts and might lead to responses, which will in turn produce further repercussions, and so on. Influencing these responses will be the various perceptions of CO_2 issues held by individuals within a hierarchy of different decision units, including households, corporate bodies, governments, and international and non-governmental organizations. Furthermore, trends or changes in social and environmental factors, such as growing populations, increasing worldwide industrialization, altered economic interdependence, and resource depletion, are certain to affect the patterns of societal responses.

A research task is thus to explore the extremely complex and dynamic *system of interactions* among climate, other aspects of the environment, and society (e.g., see Butzer, 1980; SCOPE, 1978). Research efforts should not be restricted entirely to present divisions of scientific methods and knowledge represented by traditional academic disciplines. Rather, research should draw from diverse disciplinary expertise and combine and integrate available capabilities in ways that directly address the interactions among environmental and societal factors. Such an *interdisciplinary* approach will help match our evolving analytic abilities with the complexity of various aspects of the CO_2 problem.

Interdisciplinary research on the CO_2 problem could take different forms. Fischhoff (1980), for example, identifies four modes of interdisciplinary work:

(1) clarification of assumptions and paradigms used in traditional disciplinary research to generalize results;
(2) inclusion of factors other than those normally considered in disciplinary research;
(3) 'borrowing' of methods and approaches developed in other disciplines; and
(4) collaboration among scientists from different disciplines.

Interdisciplinary research may thus involve one scientist or many, one-way or two-way communication between disciplines, and different degrees of interaction among scientists. A higher degree of interaction is often encouraged by forming a research 'team' of scientists from different disciplines in one location to enable them to learn to work together effectively over an extended period (a year or more) (Fischhoff, 1980; Mar *et al.*, 1976). Alternatively, a 'pool' model is sometimes employed in which a center or institute serves as a locus and source of talent for a variety of interdisciplinary (or multi-disciplinary) activities (Mar *et al.*, 1976). With regard to the CO_2 problem, although the first three of Fischhoff's modes of interdisciplinary work will be important in some instances, extensive collaboration and interaction among people of diverse backgrounds will be indispensable to effective research on the CO_2 problem (Panel IV, 1980; DOE, 1980).

4. Obstacles to Interdisciplinary Research

Interdisciplinary research on the CO_2 problem is not routine. Past experience with interdisciplinary research in other areas has shown that many barriers and problems exist. Although the formal literature on this subject is somewhat limited, a summary of some key issues that have been identified in both formal and informal literature seems useful at this point.

Perhaps the major barrier to interdisciplinary research is that so little of it has been done – and even less has been well documented. No 'traditional' models or guidelines exist on how to approach a problem in an interdisciplinary fashion (Fischhoff, 1980). No well-accepted, general criteria exist for evaluating results (Schneider, 1977a, b). Few refereed interdisciplinary or even multi-disciplinary journals are published. Interdisciplinary positions in research institutions are rare, and few scientists have extensive experience in interdisciplinary work (Schipper, 1977). Indeed, few scientists are trained in more than one discipline, perhaps because of the massive effort required to learn the techniques and details of just one field of knowledge thoroughly.

Furthermore, scientists are not generally encouraged to work with those from other fields. Instead, they are frequently discouraged, whether intentionally or not, by the existing peer-review system, hiring practices, promotional structures, publication outlets, time restrictions, inflexible accounting systems, and other institutional idiosyncracies that tend to favor disciplinary research to the exclusion of other things (Corwin and Arnstein, 1977). For example, an interdisciplinary proposal, if reviewed only from the traditional perspective of a single discipline, is not as likely to be rated as highly as a disciplinary proposal, since the former will usually have less promise of providing disciplinary advances and also will likely cover topics beyond the familiarity of the reviewer (Schneider, 1977a). An interdisciplinary scientist or research team may thus find it difficult to obtain financial support for interdisciplinary projects, especially in times of budget limitations when such projects must compete with under-funded – at least in reviewers' perceptions – disciplinary research (Mar *et al.*, 1976). Similarly, existing hiring and promotional structures typically reserve major rewards for individually authored publications in respected disciplinary journals; and disciplinary originality is the principal criteria of excellence. Thus, interdisciplinary research, particularly team research, is not now as likely to offer in an academic setting the same professional and financial benefits as might more traditional research (Schneider, 1977b). In addition, inflexible accounting systems can inhibit funding of interdisciplinary projects by multiple sponsors and limit independence and originality of scientific work (MacLane, 1980).

The division of scientific research along disciplinary lines is reinforced by – and in turn reinforces – 'communications gaps' between scientific fields. An isolated discipline develops its own sets of jargon, implicit assumptions, analytic methods, specialized literature, and meetings that a scientist from another field may find difficult to access, interpret, or understand. Scientists from different disciplines often 'ask different questions; use different concepts; use different terms for the same concept and the same term with different meanings; explicitly or implicitly make different assumptions; and perceive different opportunities for empirical verification' (Koopmans, 1979: 1). In collaborative efforts, scientists must therefore generally trust the quality of their colleagues' work within their respective disciplines (Fischhoff, 1980). This situation may lead to difficult quality-control problems such as errors arising from a superficial appreciation of jargon (White, 1979) or even plagiarism (Broad, 1980a, b).

Lack of communication may also breed other problems. Mar *et al.* (1976: 651) cite the "disdain of scientists for engineers, or mathematicians for physicists, or pure scientists for applied scientists." Schipper (1977) tells of the physical and social separation between laboratory scientists, engineers, and social scientists at the University of California at Berkeley. An arbitrary distinction between 'basic' and 'applied' research is often made in research institutions and funding agencies (Churchmann, 1977). Some natural scientists apparently feel that 'the social sciences are not really 'Science' at all' (OECD, 1979b: 7; see also, Nagel, 1961: 460–466). Among social scientists, a mutual bias sometimes emerges between those who advocate quantitative analysis and those who emphasize qualitative approaches (Bohrnstedt, 1980).

Similar problems exist relative to interdisciplinary research. Schneider (1977a: 1), for example, points out the 'widespread perception in the scientific community that a lot of interdisciplinary work is shallow'. He believes that this is in part due to the lack of established evaluative criteria and rigorous quality reviews for interdisciplinary work. Weingart (1977) notes that since many interdisciplinary research problems are also 'real-world' problems – involving, for example, energy, pollution, resource, and population issues – they often are inseparable from policy considerations. This causes conflicts between those who believe scientists should avoid such considerations and those who do not. Indeed, scientists may arbitrarily consider the 'scientific' part of an interdisciplinary problem – namely, their own fields of interest – to be more intrinsically important than other aspects of the problem such as its policy implications.

Another problem is that not everyone may be suitable for interdisciplinary research. Mead (1977) asserted that 'digital' thinkers may be too narrow in focus to deal with broad, cross-cutting issues. 'Analogic' thinkers, she argued, can better perform the kinds of integration of knowledge necessary. Personal traits such as tolerance for criticism, articulateness, and genuine interest in the problem at hand can be very important (Hahn, 1977; White, 1979). Mar *et al.* (1976: 651) observe that 'it is not unusual for 50 or more individuals to flow through a group before a handful of individuals 'settle out' and begin the integration process'.

Research in some interdisciplinary areas may be hampered by disciplinary paradigms and traditions. For instance, Dunlap (1980: 6) points out that 'thoughout the social sciences the adaptation of modern societies to the biophysical environment is typically seen as nonproblematic (indeed, the issue is seldom even considered)'. He attributes this avoidance of a major interdisciplinary area in part to the strongly anthropocentric worldview of western culture which, after several centuries of rapid economic and technological development, has led to the tendency 'to treat human societies as if they were exempt from ecological constraints' (Catton and Dunlap, 1980: 25). Additional impetus may have been provided by the rejection by most social scientists of the 'naive and racialistic theories' (Biswas, 1979: 241) put forward by 'environmental determinists' such as E. Huntington (1915). Unfortunately, this rejection may have inhibited consideration of important environmental factors in much social-science research (Biswas, 1979; Dunlap, 1980).

Interdisciplinary research involving both the natural and social sciences may be hampered by disparities between their analytic capabilities. Many social sciences lack data, models, and theories that can be easily combined with the data, models, and theories of other disciplines. Measurements of social phenomena are often crude, subjective, and limited, inhibiting detection of causal relationships between social and other parameters. Thorny problems sometimes arise because of different standards for information quality, especially tradeoffs between data quality and its timeliness (Abt, 1979). Undue emphasis may be given to factors that are easier to quantify than others (Sheldon and Parke, 1975; see also, Daly, 1980). Unlike many natural sciences, the social sciences have relatively limited opportunities for 'controlled investigations'; that is, laboratory or field experiments for testing hypotheses or building models (Nagel, 1961: 452–457). Observational studies tend to be hampered by uncontrolled variables, measurement problems, long time scales, resource constraints, ethical questions, and difficulties in replicating situations and minimizing the effects of the observer. The consistency and generality of hypotheses are difficult to test due to the complexity of social-science problems and their specificity in regard to particular cultural contexts, time periods, and locations (Nagel, 1961: 459–466; Beck, 1949). Indeed, the social sciences lack any sort of generally recognized theory or paradigm of human behavior and social change on which to base 'objective' interpretations and reliable predictions (Dunlap, 1980). These disparities between the natural and social sciences are reflected in controversies over the nature of science (e.g., Kuhn, 1970; Toulmin, 1977; Feyerabend, 1978).

As evidenced by the many issues raised in this brief review, a wide range of obstacles, both institutional and substantive, surround interdisciplinary research. Resolution of these difficulties will involve considerable cooperation and innovation within the scientific community.

5. Management of Interdisciplinary Research on the CO_2 Problem

How can interdisciplinary research on the CO_2 problem be encouraged and managed effectively? Several interrelated needs are evident from the previous discussion of difficulties. Although specifically cast in the context of CO_2 research, these needs also apply to interdisciplinary research in general.

First, new or modified institutional arrangements are clearly needed to deal with the special problems posed by the interdisciplinary nature of the CO_2 problem. For example, interdisciplinary experience and publications should be given explicit credit in hiring and promotional policies. Considerable attention should be devoted to finding interdisciplinary thinkers and workers, and to developing disciplinary balance within compatible teams of researchers. New internal and external quality-control mechanisms should be established that are capable of generating constructive criticism and ensuring the high quality of the interdisciplinary, as well as disciplinary, aspects of research. This should include the development of clearcut standards for the peer review of interdisciplinary research, such as those purposed by Schneider (1977: 41) and used as a basis for peer reviews in the interdisciplinary journal, *Climatic Change*: '(a) disciplinary accuracy, (b) clarity of cross-disciplinary communications, and (c) utilization and combination of existing knowledge from many fields to help solve a problem or to raise or advance knowledge about a new issue'.

Second, continuing high-level managerial commitment to interdisciplinary research on the CO_2 problem is needed. Management should help overcome problems caused by traditional, discipline-oriented attitudes and institutions, and try to increase support for interdisciplinary CO_2 research among researchers, their scientific colleagues, and funding organizations (Horvitz and Evans, 1977). In particular, management of interdisciplinary CO_2 research should itself be interdisciplinary, at least in the sense of having an appreciation for and a working knowledge of the breadth and complexity of the problem, including its implications for policy and the approaches necessary to analyze it. Often, effective management depends on the vigorous leadership of an interdisciplinary individual. Another way to ensure interdisciplinary management is to establish advisory or steering groups consisting of individuals experienced in interdisciplinary research, familiar with the broad aspects of the CO_2 problem, and willing to commit adequate time to keeping up with – and criticizing – research developments.

Third, stable financial support for interdisciplinary CO_2 research is a prerequisite. It must be recognized that the development of effective interdisciplinary research efforts may take large investments of time and resources – for a small group of three researchers, at least one year and many tens of thousands of dollars each, according to Mar *et al.* (1976). Individuals and groups may undergo an 'incubation period' of at least a year, and perhaps several, during which they may produce little in the way of interdisciplinary products. In some instances, money alone may not suffice. Many universities, for example, limit the time that their faculty members can spend on outside projects. Resources and services such as office space, laboratory facilities, travel funds, support staff, and financial assistance for students (e.g., assistantships, scholarships, and grants)* may be important incentives (Mar *et al.*, 1976). Support should also extend to continuing quality control, interaction with users, and dissemination of results both inside and outside of the scientific community.

Fourth, the 'infrastructure' of interdisciplinary research on the CO_2 problem needs increasing support. Included in this infrastructure are networks of scientists engaged in interdisciplinary efforts such as environmental and social impact assessment and technology assessment; interdisciplinary and multi-disciplinary journals, newsletters, and other publications; and assorted workshops, symposia, courses, grants, and other educational activities supported by various public and private organizations. This infrastructure will be important to fostering communication and reducing institutional and methodological barriers among scientists from different disciplines. Such infrastructure could form the basis for careful, conscientious peer review of interdisciplinary work on the CO_2 problem, and help to improve overall scientific credibility and acceptance. Moreover, it could help attract the talent needed for interdisciplinary research on the CO_2 problem and begin the process of building a solid community of scientists involved in and supportive of such research.

Finally, improvements in the 'science' of interdisciplinary research as applied to the CO_2 problem are needed (UNEP, 1980; Kates, 1980). Past examples of related research,

* The author's own involvement in interdisciplinary research on the CO_2 problem was made possible in part by an unrestrictive graduate fellowship.

such as that for the stratospheric ozone issue, should be closely examined and publicized (e.g., Glantz *et al.*, 1981). The advantages and disadvantages of various possible modes of interdisciplinary research such as those identified by Fischhoff (1980) and various ways of organizing interdisciplinary collaborative efforts should be assessed critically (Mar *et al.*, 1976; Scribner and Chalk, 1977; see also, Jones, 1979). Basic assumptions and methods within disciplines should be continually reviewed and reassessed (e.g., Dunlap, 1980; Bohrnstedt, 1980; Torry, 1979). Techniques such as scenario analysis that can help integrate diverse knowledge should be reviewed and documented (e.g., Ericksen, 1975; Lave, 1981). A variety of workshops, conferences, commisioned studies, and other activities should contribute greatly to such efforts (e.g., DOE, 1980).

Satisfaction of the five needs listed above cannot of course guarantee the success of interdisciplinary research on the CO_2 problem. However, unless steps to encourage such research are taken, the collaboration and interaction critical to useful results will be difficult to achieve. Indeed, for the CO_2 question and perhaps other issues, a major problem may be to gain sufficient scientific interest, especially among social scientists, to attract competent workers to the problem. Clearly, a long-term commitment to the funding of interdisciplinary research will be a principal stimulus of such interest. This issue is discussed further in Section IX.

6. Integration of Research Results in the CO_2 Context

An interdisciplinary approach to the CO_2 problem will help to ensure that research on important questions is analyzed from a broad perspective using a variety of techniques when necessary. Some means is also needed to formulate research questions and integrate the "answers" that are obtained into useful policy information.

A useful starting point is the analytic framework suggested by geographers Warrick and Riebsame (1981). Their conceptualization attempts to isolate explicitly many interactions among climate, other aspects of the environment, and society. It gives explicit recognition to the role of information, evaluation, and choice at many points in the model. Associated with this framework is a corresponding set of key research questions which divide the broad problem into (hopefully) more researchable parts. The framework and associated questions are designed to permit a consistent and systematic integration of knowledge about the CO_2 problem in several ways. Even though contemporary society as a whole has never before experienced the sort of massive climate change projected as possible from increasing carbon dioxide, past cultures have been subjected to major changes in climate. Additionally, some parts of modern society have experienced other kinds of environmental stress and shorter-term, more severe climatic variations which could provide useful analogues. With such experience as a 'laboratory', it may be possible to develop and test various physical, environmental, economic, and social models of society/environment/climate interactions. One approach, for example, would be to study particular past or present situations in which significant climate/society linkages are more easily discernible, for instance, marginal agricultural systems in arid lands or habitation in Arctic regions (Panel IV, 1980). If reasonable hypotheses of human/climate interactions can be developed from these case studies and applied to the more general research questions, they can then be incorporated into the overall framework. The proposed framework and associated questions thus provide one possible mechanism for helping individual researchers to keep their efforts focused on the most important issues and to combine diverse research findings into an integrated understanding of possible societal responses and their likely consequences. Other mechanisms should also be explored.

It is also important to note that the study of climate-society interaction can itself be approached at different levels of abstraction and complexity. Kates (1980), for example, postulates a hierarchy of models of the interaction between climate and society (see Figure 1): the input-output model, the interactive model, the interactive model with feedback, and the interactive model with feedback and underlying process. As in the natural sciences (e.g., climatology), there are clearly tradeoffs among such models between their tractability, verifiability, and realism. Ideally, one would want a highly realistic model which could generate highly accurate predictions of societal/environmental interactions efficiently. In practice, this is, of course, difficult in all but a few simple examples. While it may be possible to quantify partially the direct effects of some change in a climatic parameter on a particular economic unit or sector, indirect impacts arising from intersectoral interactions or societal changes and responses are much more difficult to analyze quantitatively – or even qualitatively. Yet these indirect impacts could outweigh or mitigate the direct impacts. Models that fail to include important pathways or mechanisms for such impacts could yield extremely misleading results.

The key may be to identify the strengths and limitations of models at each level of organization or complexity by performing "sensitivity experiments" with a hierarchy of such models. In other words, we would learn from the differences among models. Even though a more complex model may fail to give fully satisfactory answers, it may indicate the relative importance of processes neglected in simplified models. We can then estimate whether, and for what purpose, various feedbacks and interactions have been taken sufficiently into account. Conversely, a hierarchy of well-understood simpler models can increase confidence in predictions of more complicated ones that might be more realistic but more difficult to interpret and verify (cf., Schneider and Dickinson, 1974).

A hierarchy of analytic models and a carefully designed organizing framework and accompanying set of research questions are some promising approaches toward the difficult, but important task of integrating research on the CO_2 problem. Clearly, other approaches to integrating research should also be pursued. For example, "scenario-analysis" techniques have been used to assist decisions about responses to natural hazards (Ericksen, 1975; White and Haas, 1975) and have been linked with models of human judgment (Hammond *et al.*, 1977). Scenario-building exercises focusing on the CO_2 problem have recently been proposed (Ausubel *et al.*, 1980; Ausubel, 1980; DOE, 1980; WMO, 1981: 25–27; Robinson and Ausubel, 1981; Lave, 1981).

(a) Input-Output Model

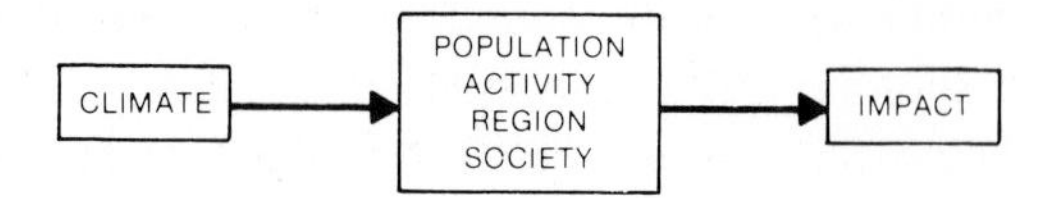

(b) Interactive Model

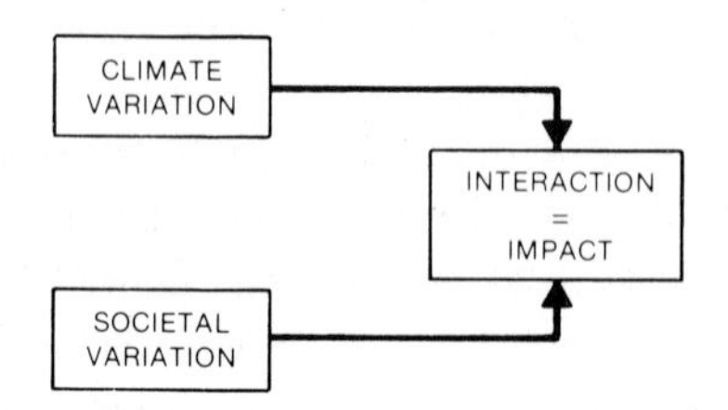

(c) Interactive Model with Feedback

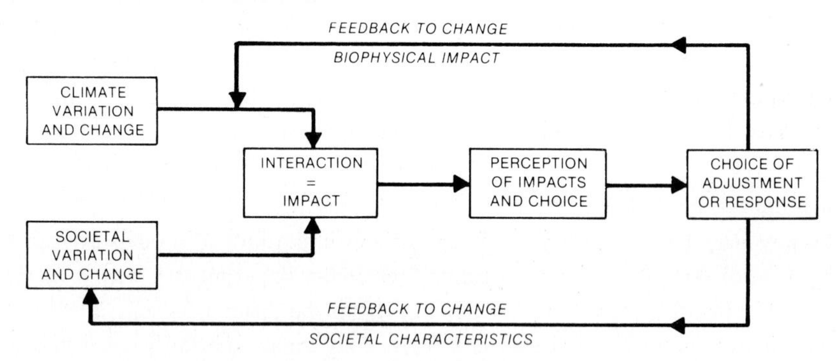

(d) Interactive Model with Feedback and Underlying Process

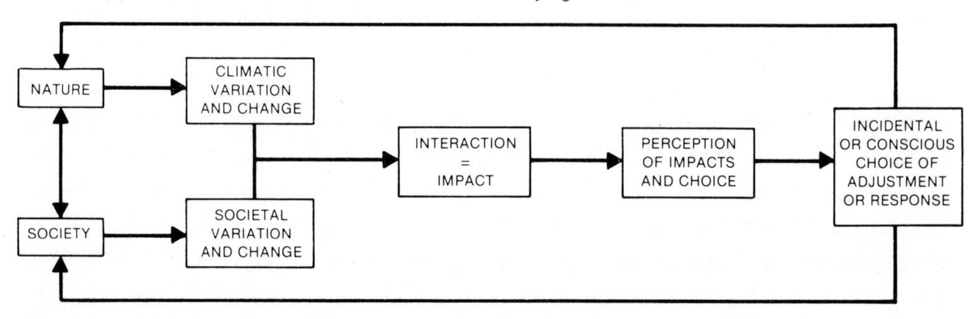

Fig. 1. A hierarchy of models of Climate-Society relationships (from Kates, 1980).

7. Perception, Evaluation, and Choice in Responses to the CO_2 Problem

The identification of the roles of perception, evaluation, and choice is a critical part of an analysis of policy responses to the CO_2 problem. Even if all costs and benefits of CO_2-

induced climate changes, or alternative policies in response to its prospect, could be determined by analytic and empirical methods, it might still be difficult to alter the course of societal decision making and effectively implement desirable actions. Every day, billions of people make billions of individual decisions, both implicit and explicit, to burn fossil fuels, cut down and burn trees, or consume irreplaceable resources. Changing their behavior, even if they were informed fully about the long-term consequences of CO_2 emissions, could be difficult and slow. Moreover, since climate changes are likely to affect people and groups with different values and priorities in different ways at different times, the desirability of particular actions may also be hard to measure.

Thus, if we are to accomplish the 'unprecedented' task of addressing the CO_2 problem explicitly and of providing timely inputs into the decision-making process, we must learn how decisions *at all levels* are made, who participates and what perceptions, values, and constraints are involved. We must examine the level and origins of our knowledge, the problems of communication and interpretation, the interplay of science, ethics, and judgment, and the primary modes of societal response. We also need to analyze the past record with regard to effecting massive societal change or adapting to massive (or potentially massive) environmental change, in order to look for successful or innovative ways of doing so in the future.

For example, carbon dioxide can in many respects be considered a pollutant, for which a substantial body of national and international law and practice already exists (Weiss, 1983). This suggests that current international efforts to regulate usage of fluorocarbons, arising in part from their potential effects on the earth's ozone layer, may provide a valuable legal precedent and international framework for possible actions in response to the CO_2 problem (UNEP, 1980). Similarly, climate can itself be viewed as a natural resource (Landsberg, 1946; Climate Board, 1981), which suggests that certain of present institutions for allocating resources may be useful reservoirs of experience relevant to dealing with a scarcity of 'good' climate (Weiss, 1983). Other approaches to various aspects of the CO_2 problem are provided by such social-science disciplines as psychology, anthropology, sociology, political science, history, and geography (e.g., Fischhoff and Furby, 1983; Torry, 1983; Mann, 1983; Rabb, 1983; Warrick and Riebsame, 1981).

However, because of the unique elements of the CO_2 problem, it is important to recognize that past experience often cannot be applied directly. In the fluorocarbon case, although many similarities to the CO_2 problem exist, there are crucial differences between fluorocarbon use and fossil-fuel consumption, particularly in the magnitude and distribution of benefits and the availability of alternatives. For instance, since fossil fuels are generally perceived to be a necessary ingredient for economic development, and since most energy alternatives are still highly uncertain in costs and risks, the CO_2 question is much more deeply embedded in controversies over worldwide development goals and strategies. Moreover, it is not likely that existing institutions, most of which operate on short-term horizons, can effectively deal with a slow, cumulative, and potentially irreversible change of anthropogenic origin (Glantz, 1979; Mann, 1983). Such institutions might not be able to deal with issues such as decade-long lags between CO_2 emissions and noticeable impacts, counter-intuitive short-term phenomena (e.g., a local short-term cooling trend embedded in a long-term global warming), gradual changes in the frequency of weather events or episodes, and difficulties in detecting or projecting changes and in attributing them to specific causes with a scientific consensus. Considerable care in the use and interpretation of past experiences will thus be needed to draw applicable analogies – or to avoid drawing inapplicable analogies – between past situations and future CO_2 responses.

8. Research as a Response to the Prospect of CO_2 Impacts

The present disparity between disciplinary analytic abilities and the unprecedented number of interconnected dimensions of the CO_2 problem raise several important questions regarding the role of research. Research on the CO_2 problem is itself a response to the *perception* that CO_2-induced climate changes could adversely affect people and society in the future. It is essentially one way society can find out more about the implications of the CO_2 problem. We must therefore ask whether research by itself is an adequate response – or perhaps even a harmful one. Will sufficient information of sufficient certainty be available in time for adequate actions to be taken to prevent, mitigate, or adapt to CO_2-induced changes? Is it likely that 'knowledge can probably be made to grow faster than the problem', as one report asserts (Climate Board, 1980: 9)? To delay actions stronger than research (i.e., actions that could mitigate CO_2 increases or their impacts) merely commits posterity to a larger 'dose' of CO_2 than if those mitigating actions were taken now. Are the risks imposed by the delay of such more active responses worth taking in light of the present benefits from CO_2-emitting activities? Will the new knowledge that may be gained from research be likely to negate or increase the need for action or improve the effectiveness of future actions? Finally, are there actions which could be taken now that might allow society more options and choices in the future? These might include vigorous development of alternative energy systems or active efforts to make societal adaptation to environmental change easier (Schneider and Chen, 1980; Kellogg and Schware, 1981).

Obviously, inasmuch as words such as 'sufficient', 'adequate', or 'worth' are normative, these issues are value-laden policy questions. They cannot be answered by research alone. Nevertheless, research can certainly help to assess the options now available, their possible consequences, and the present areas of uncertainty. Research should be able to give a better idea of present and likely future levels of certainty and the usefulness of more knowledge and increased certainty in guiding future policy decisions. It could, by examining past cases, also identify the potential dangers and benefits of making decisions under uncertainty versus delaying them. Finally, and perhaps most importantly, research could help clarify the opportunities and constraints which condition society's decisions and could help begin the process of expanding the options available to it. Despite the cautions raised about unavoidable normative issues, research should nevertheless be able to help strengthen the scientific basis for present and future decisions about the CO_2 problem – provided the research results are disseminated in such a way as to allow non-specialists to easily separate issues of 'fact' from those of 'values'.

9. Research 'Tie-ins'

Not only must we be aware of the role of research in societal responses to the CO_2 problem, we must also recognize the links between research on the CO_2 problem and research on other social issues. Limitations in time, money, and scientific talent may restrict the amount of effort that can be expended on the CO_2 problem alone. However, the CO_2 problem is itself inextricably connected with such important social issues as future economic development in the developing and developed nations and population pressures on the environment and society. On the one hand, economic development and population growth, if made possible by increasing fossil-fuel use, wood burning, and massive clearing of land, are likely to be key factors in how fast CO_2-induced climate changes develop and how serious they become – if they occur at all. Developments such as breakthroughs in renewable energy technologies, resource depletion, or restrictions on fossil-fuel trade could also be important. On the other hand, the most serious impacts of CO_2-induced climate changes may well be the intensification of existing social problems such as hunger, poverty, poor health and sanitation, mass migration, environmental degradation, social unrest, and political tension. For example, greater climate variability, including more severe weather and seasonal extremes such as droughts, heat waves, cold spells, and persistent rains, could severely reduce food production, increase energy requirements, and aggravate water-related problems (e.g., shortages, floods, and contamination of drinking water). Such impacts are now visible at present levels of climate variability (e.g., Center for Environmental Assessment Services, 1980). Non-climatic events or episodes such as volcanic eruptions, earthquakes, trade embargoes, and wars also cause stresses on society, and engender responses, similar to those that might result from climate changes.

These two-way connections between CO_2 and other problems suggest that research efforts should be closely linked, or 'tied-in', with each other (Panel IV, 1980). Research on the CO_2 problem should be able to proceed more quickly and effectively by taking advantage of past and present research on these other problems and the policies used to deal with them. Conversely, society may well benefit from the new knowledge obtained from research on the CO_2 problem, and policies to prevent or ameliorate it, that can be applied to other problems. We might, for example, learn how to modify existing social institutions to make society more "resilient" to environmental stress or learn what social indicators are good precursors of social change. Research on policies such as the development of non-fossil fuel alternatives or food reserves would be applicable to other important social issues as well as to the CO_2 problem. Thus, even if the CO_2 problem proves moot, whether because of some new information or some breakthrough in energy-supply systems that reverses the trend of increasing fossil-fuel use, research on the CO_2 problem should itself have wide-reaching uses outside of the CO_2 context. This possibility may be one of the most important factors in generating interest in the CO_2 problem among scientists, who are likely to have other demands competing for their time, and particularly among those in control of research resources, who may feel that other social problems like poverty and nuclear proliferation are more important and deserving of research resources. Indeed, a major payoff of research on the CO_2 problem may in fact be the interdisciplinary and international linkages among scientists and others that it generates, which could provide a basis – and hopefully a successful precedent – for international cooperation on global, short- or long-term environmental and social problems.

10. Conclusion

The great variety of potential societal responses to the prospect or advent of CO_2-induced climate changes poses an unprecedented research problem of extraordinary breadth and complexity in the physical, biological, and social sciences and also in the humanities. Experience in integrating inputs from the many disciplines of relevance is limited, and the analytic techniques available are often crude. A comprehensive, interdisciplinary research approach will therefore be crucial to obtaining realistic, credible research results that can help society respond to the problem effectively. Development of such a research approach will require innovations, especially in the organization, nurture, and management of interdisciplinary research efforts. Many of these innovations may fail. Thus, interdisciplinary research should be viewed as a flexible, 'trial-and-error' type of evolutionary process. Indeed, enough of such endeavors may, at best, lead eventually to the development of a comprehensive theory of climate and society interactions; at the least, they should help put CO_2-related decisionmaking on a firmer scientific basis.

As a starting point, some key needs can be identified which any coordinated program of research on the CO_2 problem should attempt to meet. These are:

- the need to define, and continually refine, conceptual frameworks of climate/society interactions;
- the need to split the overall problem into a set of focused, disciplinary and interdisciplinary research questions which divide the problem into more tractable parts and explicitly address the role of information, evaluation, and choice at various levels;
- the need to develop a variety of approaches and techniques that may transcend traditional disciplinary boundaries and are able to address the climate/society system at different levels of organization and complexity; and
- the need to develop flexible, innovative approaches to research management, with special emphasis on quality control, stable funding, professional opportunities, and interdisciplinary supervision.

Ways of meeting these needs, though likely to be based initially on disciplinary methods and perspectives, will require ongoing refinements, largely and perhaps unavoidable through 'learning by doing'. Efforts must especially be made to ensure that research questions are relevant to the overall climate/society system and that they are posed so as to be intelligible to a variety of disciplines. Research should be flexible enough to direct a wide range of human experience toward the unique, unprecedented, and normative aspects of the CO_2 problem. It must stress the identification of mechanisms that can help us to respond to the prospect or advent of CO_2-induced changes, including consideration of the potential impact of the research itself. Such research should produce results that are directly applicable to many other problems involving interactions between society and the environment. Similarly, it is important that experience gained from research on such other problems be utilized in CO_2 research. These 'tie-ins' call for research efforts that are particularly sensitive to advances and retreats in the progress of other research across the physical, biological, and social sciences. Such efforts will require the disciplinary and interdisciplinary skills of scientists of extraordinary tolerance and creativity and the managerial talents of research sponsors of extraordinary breadth and judgment.

Acknowledgements

The author gratefully acknowledges the many comments and suggestions provided by the members of the Social and Institutional Responses Working Group, AAAS Climate Project, and the prompt and efficient typing of Doris Bouadjemi.

Department of Geography
University of North Carolina
Chapel Hill, N.C. 27514, U.S.A.

References

Abt, C. C.: 1979, 'Social Science Research and the Modern State', *Daedalus, Proc. Am. Acad. Arts Sci.* **108**, 89.

Ausubel, J.: 1980, 'CO_2: An Introduction and Possible Board Game', Working Paper WP-80-153, International Institute for Applied Systems Analysis, Laxenburg, Austria, 31 pp.

Ausubel, J., Lathrop, J., Stahl, I., and Robinson, J.: 1980, 'Carbon and Climate Gaming', Working Paper WP-80-152, International Institute for Applied Systems Analysis, Laxenburg, Austria, 16 pp.

Beck, L. W.: 1949, 'The "Natural Science Ideal" in the Social Sciences', *Sci. Mon.* **68**, 386.

Biswas, A. K.: 1979, 'Climate, Agriculture and Economic Development', in Biswas, M. R. and Biswas, A. K. (eds.), *Food, Climate and Man*, Wiley, New York, pp. 237–259.

Bohrnstedt, G. W.: 1980, 'Social Science Methodology: The Past Twenty-Five Years', *Am. Behav. Sci.* **23**, 781.

Broad, W. J.: 1980a, 'Imbroglio at Yale (I): Emergence of a Fraud', *Science* **210**, 38.

Broad, W. J.: 1980b, 'Imbroglio at Yale (II): A Top Job Lost', *Science* **210**, 171.

Brown, Jr., Hon. G. E.: 1980, 'The CO_2 Problem: Unprecedented Challenges and Opportunities', in *Workshop on Environmental and Societal Consequences of a Possible CO_2-Induced Climate Change*, CONF-7904143, Carbon Dioxide Effects Research and Assessment Program, Rept. No. 009, U.S. Department of Energy, Washington, D.C., pp. 122–5.

Butzer, K. W.: 1980, 'Civilizations: Organisms or Systems?', *Am. Sci.* **68**, 517

Catton, Jr., W. R. and Dunlap, R. E.: 1980, 'A New Ecological Paradigm for Post-Exuberant Sociology', *Am. Behav. Sci.* **23**, 15.

Center for Environmental Assessment Services: 1980, 'Climate Impact Assessment: United States', Annual Summary 1980, National Oceanic and Atmospheric Administration, Washington, D.C. 66 pp.

Chen, R. S.: 1980, 'Impacts of Carbon Dioxide Induced Climate Change', in *Proceedings, Bio-Energy '80: World Congress and Exposition*, Bio-Energy Council, Washington, D.C. pp. 544–7.

Churchman, C. W.: 1977, 'Toward a Holistic Approach', in Scribner, R. A. and Chalk, R. A. (eds.), *Adapting Science to Social Needs: Conference Proceedings,* American Association for the Advancement of Science, Rept. No. 76-R-8, pp. 11–24.

Climate Board: 1980, Letter Report of the *Ad Hoc* Study Panel on Economic and Social Aspects of Carbon Dioxide Increase, National Academy of Sciences, Washington, D.C., 11 pp.

Climate Board: 1981, *Managing Climatic Resources and Risks,* Report of the Panel on the Effective Use of Climate Information in Decision Making, National Academy Press, Washington, D.C., 151 pp.

Corwin, R. and Arnstein, S.: 1977, 'Motivation and Reward Structures: What are the Incentives and Risks in Doing Problem-Oriented Research?', in Scribner, R. A. and Chalk, R. A. (eds.), *Adapting Science to Social Needs: Conference Proceedings*, American Association for the Advancement of Science, Rept. No. 76-R-8, pp. 107–13.

Daly, H. E.: 1980, 'Growth Economics and the Fallacy of Misplaced Concreteness: Some Embarassing Anomalies and an Emerging Steady-State Paradigm', *Am. Behav. Sci.* **24**, 79.

DOE: 1980, *Environmental and Societal Consequences of a CO_2-Induced Climate Change: A Research Agenda, Vol. 1*, DOE/EV/10019-01, Carbon Dioxide Effects Research and Assessment Program, Rept. No. 013, U.S. Department of Energy, Washington, D.C., 125 pp.

Dunlap, R. E.: 1980, 'Paradigmatic Change in Social Science: From Human Exemptions to an Ecological Paradigm', *Am. Behav. Sci.* **24**, 5.

Ericksen, N. J.: 1975, *Scenario Methodology in Natural Hazards Research*, University of Colorado Institute of Behavioral Science, Boulder, CO.

Feyerabend, P.: 1978, *Against Method*, Schocken, New York.

Fischhoff, B.: 1980, 'No Man is a Discipline', in Harvey, J. H. (ed.), *Cognition, Social Behavior, and the Environment*, Eribaum, Hillsdale, N.J., pp. 579–583.

Fischhoff, B. and Furby, L.: 1983, 'Psychological Dimensions of Climatic Change', in Chen, R. S., Boulding, E. M., and Schneider, S. H. (eds.), *Social Science Research and Climate Change: An Interdisciplinary Appraisal*, D. Reidel, Dordrecht, Holland, 177–204 (this volume).

Geophysics Study Committee: 1977, *Energy and Climate*, National Academy of Sciences, Washington, D.C., 158 pp.

Glantz, M.: 1979, 'A Political View of CO_2', *Nature* **280**, 189.

Glantz, M. H., Robinson, J., and Krenz, M. E.: 1981, 'Report of a Workshop on Improving the Science of Climate Impact Study: An Assessment of Five Major Climate Impact Studies of the 1970's', 30 June – 2 July 1981, Institute for Energy Analysis, Oak Ridge, TN.

Hahn, W. A.: 1977, 'Observations on Interdisciplinarity: Its Need, Management and Utilization', in Scribner, R. A. and Chalk, R. A. (eds.), *Adapting Science to Social Needs: Conference Proceedings*, American Association for the Advancement of Science, Rept. No. 76-R-8, pp. 253–63.

Hammond, K. R., Mumpower, J. L., and Smith, T. H.: 1977, 'Linking Environmental Models with Models of Human Judgment: a Symmetrical Decision Aid', *IEEE Trans. on Systems, Man, and Cybernetics* SMC-7, 358–67.

Horvitz, D. and Evans, N.: 1977, 'Recommendations for Creating Effective Management Styles for Interdisciplinary Research', in Scribner, R. A. and Chalk, R. A. (eds.), *Adapting Science to Social Needs: Conference Proceedings*, American Association for the Advancement of Science, Rept. No. 78-R-6, pp. 239–42.

Huntington, E.: 1915, *Climate and Civilization*, Yale U., New Haven, CT.

Jones, C. O.: 1979, 'If I Knew Then. . . (A Personal Essay on Committees and Public Policy)', *Policy Analysis* **5**, 473.

Kates, R. W.: 1980, 'Improving the Science of Impact Study', Project Summary for the Scientific Committee on Problems of the Environment, International Council of Scientific Unions, Paris, France, 19 pp.

Kellogg, W. W. and Schware, R.: 1981, *Climate Change and Society: Consequences of Increasing Atmospheric Carbon Dioxide*, Aspen Institute for Humanistic Studies and Westview Press, Boulder, CO, 178 pp.

Koopmans, T. C.: 1979, 'Economics Among the Sciences', *Am. Econ. Rev.* **69**, 1.

Kuhn, T. S.: 1970, *The Structure of Scientific Revolutions*, 2nd ed., University of Chicago Press, Chicago.

Landsberg, H.: 1946, 'Climate as a Natural Resource', *Sci. Mon.* **63**, 293, (reprinted in Climate Board, 1981: 17.).

Lave, L. B.: 1981, 'Mitigating Strategies for CO_2 Problems', Collaborative Paper CP-81-14, International Institute for Applied Systems Analysis, Laxenburg, Austria, 10 pp.

MacLane, S.: 1980, 'Total Reporting for Scientific Work', *Science* **210**, 158.

Mann, D.: 1983, 'Research on Political Institutions and Their Response to the Problem of Increasing CO_2 in the Atmosphere', in Chen, R. S., Boulding, E. M., and Schneider, S. H. (eds.), *Social Science Research and Climate Change: An Interdisciplinary Appraisal*, D. Reidel, Dordrecht, Holland, 116–146 (this volume).

Mar, B. W., Newell, W. T., and Saxberg, B. O.: 1976, 'Interdisciplinary Research in the University Setting', *Env. Sci. Tech.* **10**, 650.

Mead, M.: 1977, Discussant, 'Can Research Institutions Accommodate Interdisciplinary Researchers?', Symposium at 143rd Annual Meeting of the American Association for the Advancement of Science, Denver, CO, 20–25 February 1977 (taped transcript available from AAAS).

Nagel, E.: 1961, *The Structure of Science: Problems in the Logic of Scientific Explanations*, Harcourt, Brace, and Jovanovich, New York.

OECD: 1979a, *Interfutures: Facing the Future*, Organization for Economic Co–Operation and Development, Paris, France, 425 pp.

OECD: 1979b, *Social Sciences in Policy Making*, Organization for Economic Co-Operation and Development, Paris, France, 56 pp.

Panel IV: 1980, 'Social and Institutional Responses', in *Workshop on Environmental and Societal Consequences of a Possible CO_2-Induced Climate Change*, CONF-7904143, Carbon Dioxide Effects Research and Assessment Program, Rept. No. 009, U.S. Department of Energy, Washington, D.C., pp. 79–103.

Rabb, T. K.: 1983, 'Climate and Society in History: A Research Agenda', in Chen, R. S., Boulding, E. M., and Schneider, S. H. (eds.), *Social Science Research and Climate Change: An Interdisciplinary Appraisal*, D. Reidel, Dordrecht, Holland, 61–70 (this volume).

Robinson, J. and Ausubel, J.: 1981, 'A Framework for Scenario Generation for CO_2 Gaming', Working Paper WP-81-34, International Institute for Applied Systems Analysis, Laxenburg, Austria, 46 pp.

Schipper, L.: 1977, Discussant, 'Can Research Institutions Accomodate Interdisciplinary Researchers?', Symposium at 143rd Annual Meeting of the American Association for the Advancement of Science, Denver, CO, 20–25 February 1977 (taped transcript available from AAAS).

Schneider, S. H.: 1977a, 'Quality Review Standards for Interdisciplinary Research', paper presented at 'Can Research Institutions Accommodate Interdisciplinary Researchers?', Symposium at 143rd Annual Meeting of the American Association for the Advancement of Science, Denver, CO, 20–25 February 1977 (taped transcript available from AAAS).

Schneider, S. H.: 1977b, 'Climate Change and the World Predicament: A Case Study for Interdisciplinary Research', *Climatic Change* **1**, 21.

Schneider, S. H. and Chen, R. S.: 1981, 'Carbon Dioxide Warming and Coastline Flooding: Physical Factors and Climatic Impact', *Ann. Rev. Energy* **5**, 107.

Schneider, S. H. and Dickinson, R. E.: 1974, 'Climate Modeling', *Rev. Geophys. Space Phys.* **12**, 447.

Schneider, S. H. and Morton, L.: 1981, *The Primordial Bond: Exploring Connections Between Man and Nature through the Humanities and Sciences*, Plenum, New York, 324 pp.

SCOPE: 1978, 'SCOPE Workshop on Climate/Society Interface', December 10–14, 1978, Scientific Committee on Problems of the Environment, International Council of Scientific Unions, Paris, France, 37 pp.

Scribner, R. A. and Chalk, R. A. (eds.): 1977, *Adapting Science to Social Needs: Conference Proceedings*, American Association for the Advancement of Science, Rept. No. 78-R-8, Washington, D.C., 312 pp.

Sheldon, E. B. and Parke, R.: 1975, 'Social Indicators', *Science* **188**, 693.

Smith, V. K.: 1980, 'Economic Impact Analysis and Climate Change: An Overview and Proposed Research Agenda', Final Report to the National Climate Program Office, National Oceanic and Atmospheric Administration, Contract NA-79-SAC-00754, U.S. Dept. of Commerce, Washington, D.C., 55 pp.

Torry, W. I.: 1979, 'Anthropological Studies in Hazardous Environments: Past Trends and New Horizons', *Current Anthropology* **20**, 517.

Torry, W. I.: 1983, 'Anthropological Perspectives on Climate Change', in Chen, R. S., Boulding, E. M., and Schneider, S. H. (eds.), *Social Science Research and Climate Change: An Interdisciplinary Appraisal*, D. Reidel, Dordrecht Holland, 205–228 (this volume).

Toulmin, S.: 1977, *The Philosophy of Science*, Harper Row, New York.

UNEP: 1980, 'Report of the UNEP Expert Group Meeting on Climate Impact Studies', United Nations Environment Programme, UNEP/WG.38/4, Nairobi, Kenya, 27 pp.

WAES: 1977, *Energy: Global Prospects 1985–2000*, Report of the Workshop on Alternative Energy Strategies, McGraw-Hill, New York, 291 pp.

Warrick, R. A. and Riebsame, W. E.: 1981, 'Societal Responses to CO_2-Induced Climate Change: Opportunities for Research', *Climatic Change* **3**, 387.

Weingart, J.: 1977, 'Transdisciplinary Research: Some Recent Experiences with Solar Energy Conversion', paper presented at 'Can Research Institutions Accomodate Interdisciplinary Researchers?,' Symposium at 143rd Annual Meeting of the American Association for the Advancement of Science, Denver, CO, 20–25 February 1977 (taped transcript available from AAAS).

Weiss, E. B.: 1983, 'International Legal and Institutional Implications of an Increase in Carbon Dioxide: A Proposed Research Strategy', in Chen, R. S., Boulding, E. M., and Schneider, S. H. (eds.), *Social Science Research and Climate Change: An Interdisciplinary Appraisal*, D. Reidel, Dordrecht, Holland, 147–176.

White, G. F. and Haas, J. E.: 1975, *Assessment of Research on Natural Hazards*, M.I.T. Press, Cambridge, MA, 487 pp.

White, I. L.: 1979, 'Interdisciplinarity', *Environmental Professional* **1**, 51.

WMO: 1980, *Outline Plan and Basis for the World Climate Programme 1980–1983*, Rept. No. 540, World Meteorological Organization, Geneva, Switzerland, 64 pp.

WMO: 1981, 'Report of the Joint WMO/ICSU/UNEP Meeting of Experts on the Assessment of the Role of CO_2 on Climate Variations and Their Impact', World Meteorological Organization, Geneva, Switzerland, 35 pp.

INDEX